CAMBRIDGE LIBRARY COLLECTION

Books of enduring scholarly value

Darwin

Two hundred years after his birth and 150 years after the publication of 'On the Origin of Species', Charles Darwin and his theories are still the focus of worldwide attention. This series offers not only works by Darwin, but also the writings of his mentors in Cambridge and elsewhere, and a survey of the impassioned scientific, philosophical and theological debates sparked by his 'dangerous idea'.

The Philosophy of Zoology

John Fleming (1785–1857) was a minister of the Church of Scotland, but in his time at the University of Edinburgh he had also studied geology and zoology. In the tradition of the country parson who was also a talented and knowledgeable naturalist, he published his first works on the geology of the Shetland Islands while serving there as a minister. His subsequent works led to his being offered the chair of natural philosophy at the University of Aberdeen, and subsequently at the newly created chair of natural history at the Free Church College in Edinburgh. The two-volume Philosophy of Zoology was published in 1822, and the young Charles Darwin is recorded as borrowing it from the library of Edinburgh University in 1825/6. His intention in the book was to 'collect the truths of Zoology within a small compass, and to render them more intelligible, by a systematical arrangement'.

Cambridge University Press has long been a pioneer in the reissuing of out-of-print titles from its own backlist, producing digital reprints of books that are still sought after by scholars and students but could not be reprinted economically using traditional technology. The Cambridge Library Collection extends this activity to a wider range of books which are still of importance to researchers and professionals, either for the source material they contain, or as landmarks in the history of their academic discipline.

Drawing from the world-renowned collections in the Cambridge University Library, and guided by the advice of experts in each subject area, Cambridge University Press is using state-of-the-art scanning machines in its own Printing House to capture the content of each book selected for inclusion. The files are processed to give a consistently clear, crisp image, and the books finished to the high quality standard for which the Press is recognised around the world. The latest print-on-demand technology ensures that the books will remain available indefinitely, and that orders for single or multiple copies can quickly be supplied.

The Cambridge Library Collection will bring back to life books of enduring scholarly value (including out-of-copyright works originally issued by other publishers) across a wide range of disciplines in the humanities and social sciences and in science and technology.

The Philosophy of Zoology

Or a General View of the Structure, Functions, and Classification of Animals

VOLUME 2

JOHN FLEMING

CAMBRIDGE UNIVERSITY PRESS

Cambridge, New York, Melbourne, Madrid, Cape Town, Singapore,
São Paolo, Delhi, Dubai, Tokyo

Published in the United States of America by Cambridge University Press, New York

www.cambridge.org
Information on this title: www.cambridge.org/9781108001663

© in this compilation Cambridge University Press 2009

This edition first published 1822
This digitally printed version 2009

ISBN 978-1-108-00166-3 Paperback

This book reproduces the text of the original edition. The content and language reflect
the beliefs, practices and terminology of their time, and have not been updated.

Cambridge University Press wishes to make clear that the book, unless originally published
by Cambridge, is not being republished by, in association or collaboration with, or
with the endorsement or approval of, the original publisher or its successors in title.

THE

PHILOSOPHY

OF

ZOOLOGY.

THE
PHILOSOPHY
OF
ZOOLOGY;
OR
A GENERAL VIEW OF THE
STRUCTURE, FUNCTIONS, AND CLASSIFICATION OF ANIMALS.

By JOHN FLEMING, D. D.
MINISTER OF FLISK, FIFESHIRE,
FELLOW OF THE ROYAL SOCIETY OF EDINBURGH, OF THE WERNERIAN NATURAL HISTORY SOCIETY, &c.

IN TWO VOLUMES.

WITH ENGRAVINGS.

VOL. II.

EDINBURGH:
PRINTED FOR ARCHIBALD CONSTABLE & CO. EDINBURGH:
AND HURST, ROBINSON & CO. LONDON.

1822.

THE
PHILOSOPHY
OF
ZOOLOGY.

THE observations which we have ventured to offer in the former volume, relate to what may be termed the Motive, the Sentient, the Nutritive, and Reproductive Functions of Animals. The various Organs of the animal frame have been described, their actions investigated, and the important purposes of life, to which they are subservient, have at the same time been pointed out. An equally extensive field of Zoological Science remains to be investigated.

Animals are related to one another, and to the objects which surround them, in such a manner, as to be dependent on a variety of circumstances for the preservation of their existence, their dispersion over the globe, and their power of accommodation to the changes of the seasons. They are likewise to be viewed as admitting of division into classes and subordinate groups, according to the external or internal characters which they exhibit. In the investigation of these characters, a variety of methods are employed, and many rules have been prescribed, to regulate the principles of zoological nomenclature.

In order to enter more fully into these important subjects, we shall distribute the present volume into Four Parts. In the first, we shall consider the Condition of Animals in reference to their Duration, Distribution, and Economical Uses. In the second, we shall treat of the Methods of Investigation employed to ascertain their structure and actions. In the third, we shall examine the Rules of Nomenclature; while the fourth will embrace a General View of the Classification of the Objects of the Animal Kingdom.

PART FIRST.

ON THE CONDITION OF ANIMALS.

CHAP. I.

ON THE DURATION OF ANIMALS.

EACH species of Animal is destined, in the absence of disease and accidents, to enjoy existence during a particular period. In no species, however, is this term absolutely limited, as we find some individuals outliving others, by a considerable fraction of their whole life. In order to find the ordinary duration of life of any species, therefore, we must take the average of the lives of a number of individuals, and rest satisfied with the approximation to truth which can thus be obtained.

There is but little resemblance, in respect of longevity, between the different classes, or even species of animals. There is no peculiar structure, by which long-lived species may be distinguished from those which are short-lived. Many species, whose structure is complicated, live but for a few years, as the rabbit; while some of the testaceous mollusca, with more simple organization, have a more extended existence. If longevity is not influenced by structure, neither is it modified by the size of the species. While the horse, greatly larger than the dog, lives to twice its age, man enjoys an existence three times longer than the former.

The circumstances which regulate the term of existence in different species, exhibit so many peculiarities, corresponding to each, that it is difficult to offer any general observations on the subject. Health is precarious, and the origin of diseases generally involved in obscurity. The condition of the organs of respiration and digestion, however, appears so intimately connected with the comfortable continuance of life, and the attainment of old age, that existence may be said to depend on the due exercise of the functions which they perform.

Whether animals have their blood aerated by means of lungs or gills, they require a regular supply of oxygen gas. But as this gas is extensively consumed in the processes of combustion, putrefaction, vegetation and respiration, there is occasionally a deficiency in particular places for the supply of animal life. But, in general, where there is a deficiency of oxygen, there is also a quantity of carbonic acid or carburetted hydrogen present. These gases not only injure the system by occupying the place of the oxygen which is required, but exercise on many species a deleterious influence. To these circumstances may be referred the difficulty of preserving many fishes and aquatic mollusca in glass jars or small ponds; as a great deal of the oxygen in the air contained in the water, is necessarily consumed by the germination and growth of the aquatic cryptogamia, and the respiration of the infusory animalcula. In all cases, when the air of the atmosphere, or that which the water contains, is impregnated with noxious particles, many individuals of a particular species, living in the same district, suffer at the same time. The disease which is thus at first endemic or local, may, by being contagious, extend its ravages to other districts.

The endemical and epidemical diseases which attack horses, sheep and cows, obtain in this country the name of *murrain*, sometimes also *the distemper*. The general term, however, for the pestilential diseases with which these and other animals are infected is Epizooty, (from επι upon, and ζωον an animal.

The ravages which have been committed among the domesticated animals, at various times, in Europe, by epizooties, have been detailed by a variety of authors. Horses, sheep, cows, swine, poultry, fish, have all been subject to such attacks; and it has frequently happened, that the circumstances which have produced the disease in one species have likewise exercised a similar influence over others.

That these diseases arise from the deranged functions of the respiratory organs, is rendered probable by the circumstance, that numerous individuals, and even species, are affected at the same time; and this opinion is strengthened, when the rapidity with which they spread is taken into consideration.

Many diseases, which greatly contribute to shorten life,

take their rise from circumstances connected with the organs of digestion. Noxious food is frequently consumed by mistake, particularly by domesticated animals. When cows, which have been confined to the house, during the winter season, and fed with straw, are turned out to the pastures in the spring, they eat indiscriminately every green plant presented to them, and frequently fall victims to their imprudence. It is otherwise with animals in a wild state, whose instincts guard them from the common noxious substances of their ordinary situation.

The shortening of life, in consequence of the derangement of the digestive organs, is chiefly produced by a scarcity of food. When the supply is not sufficient to nourish the body, it becomes lean, the fat being absorbed to supply the deficiency,—feebleness is speedily exhibited,—the cutaneous and intestinal animals rapidly multiply, and, in conjunction, accelerate the downfal of the system.

The power of fasting, or of surviving without food, possessed by some animals, is astonishingly great. An eagle has been known to live without food five weeks,—a badger a month,—a dog thirty-six days,—a toad fourteen months, and a beetle three years. This power of outliving scarcity for a time, is of signal use to many animals, whose food cannot be readily obtained; as is the case with beasts of prey, and rapacious birds. But this faculty does not belong to such exclusively. Wild pigeons have survived twelve days, an antelope twenty days, and a land tortoise eighteen months. Such fasting, however, is detrimental to the system, and can only be considered as one of those singular resources which may be employed in cases where, without it, life would speedily be extinguished.

In situations where animals are deprived of their accustomed food, they frequently avoid the effects of starvation, by devouring substances to which their digestive organs are not adapted. Pigeons can be brought to feed on flesh, and hawks on bread. Sheep, when covered with snow, have been known to eat the wool off each other's backs.

The various diseases to which animals are subject, tend greatly to shorten the period of their existence. With the methods of cure employed by different species, we are but little acquainted. Few accurate observations appear to have been made on the subject. Dogs frequently effect a cure of their sores, by licking them. They eat grass to excite vomiting; and probably to cleanse their intestines from obstructions, or worms, by its mechanical effects. Many land animals promote their health by bathing, others by rolling themselves in the dust. By the last operation, they probably get rid of the parasitical insects with which they are infected.

But independent of scarcity, or disease, comparatively few animals live to the ordinary term of natural death. There is a wasteful war every where raging in the animal kingdom. Tribe is divided against tribe, and species against species, and neutrality is nowhere respected. Those which are preyed upon, have certain means which they employ to avoid the foe; but the rapacious are likewise qualified for the pursuit. The exercise of the feelings of benevolence may induce us to confine our attention to the former, and adore that goodness which gives shelter to the defenceless, and protection to the weak, while we may be disposed to turn, precipitately, from viewing the latter; lest we discover marks of cruelty, where we wished to contemplate nothing but kindness. These feelings are usually the companions of circumscribed and partial observation, and fall far short of the object at which they aim.

It would be impious in us to inquire why the waster has been created to destroy. It is enough if we know that rapacious animals occupy a station in the scale of being. And, while we eagerly explore the various methods employed by the defenceless, to secure themselves from danger, and evade the threatened death; it is suitable for us likewise to contemplate the various means employed by carnivorous animals to gain the means of their subsistence. When we see a hawk in pursuit of a lark, we are apt to admire exclusively, the dexterity of the latter in avoiding destruction, and to triumph when it has obtained the requisite protection in a thicket. We seem to forget that the digestive organs of the hawk are fitted only for carrion; and we lose sight of the benevolence and wisdom exhibited, in giving to its wings a power of inflicting a deadly blow, and rendering the claws suited for grasping, and the bill for tearing in pieces the quarry. We are not therefore to take confined views of the animal kingdom, if we wish to read the lessons concerning the Providence of God which it teaches. He that causeth the grass to grow for the cattle, and herb for the service of man; likewise giveth meat in due season to the young lions which roar after their prey; and feedeth the ravens, though they neither sow nor reap. We see rapacious and defenceless animals existing, yet we do not observe the former successful in extirpating the latter. Limits are assigned to the ravages of this universal war. The excess only of the population is cut off,—and this excess, on whose production so many animals depend for subsistence, is as uniform as the means used to restrain its limits.

These various circumstances which we have now enumerated as limiting the duration of animals, preserve the balance of life, restrain within suitable bounds the numbers of the individuals of a species, and give stability to that system, the wise arrangements of which can only be discovered by a close examination of the whole.

CHAP. II.

ON THE DISTRIBUTION OF ANIMALS.

In examining the zoological productions of different countries, we observe, that the species which are commonly met with in one district, are rare, or not to be detected, in the others. If we confine our attention to any one species, we shall observe, that there is some particular country where the individuals are most numerous, and where the energies of life are exerted with the greatest activity. As we recede from this district, the individuals become less numerous, their increase goes on at a slower rate, and those which are produced are rather of dwarfish stature: at length, we reach the limits beyond which they do not extend. The geographical distribution of each species, therefore, may be represented by a circle, towards the centre of which, existence can be comfortably maintained; but as we approach the circumference, restraints multiply, and life at last becomes impracticable. Each species has a range peculiar to itself, so that the circles of different species intersect one another in every possible relation.

The extent of the earth's surface over which the individuals of a species are dispersed, can only be ascertained after a long series of observations, conducted by naturalists in different countries. Hitherto the geographical limits of but few species have been satisfactorily determined. These chiefly belong to the larger species of quadrupeds, as the African and Asiatic elephants, the ass and the quagga, the lion, hippopotamus, and polar bear. In the tribes of the less perfect animals, the species of which have been investigated by few, the extent of their GEOGRAPHICAL DISTRIBUTION has been very imperfectly determined.

Before proceeding to the examination of the laws which regulate the geographical distribution of any one species, it is expedient that we previously make ourselves acquainted with the range of country it inhabits, the situations in which it has been observed, and the peculiar characters it exhibits in these different situations. But while this minute and varied information is requisite for the purpose of investigating fully the physical history of any one species, it is enough, for ordinary investigations, that we ascertain those districts and situations where the individuals are most healthy and most prolific, and those where they do not exist. By comparing the physical circumstances of the former with those of the latter, it will be no difficult matter to discover those conditions which promote the vigour of life in the one, and restrain or destroy its energies in the other. What, then, are those conditions which limit the geographical distribution of species? They appear to be limited to circumstances connected with Temperature, Food, Situation, and Foes.

I. Temperature.

We have already stated, that the degree of heat at the equatorial regions appears to be most favourable for the increase of living beings, and that they diminish in numbers as we approach the poles. There is no latitude, however, which the perseverance of man has yet reached, where living beings have not been observed. The icy shores of the arctic regions are peopled as well as the arid plains or shaded forests of tropical climates. When, however, an inhabitant of the colder regions is transported to a warmer district, the increased temperature is painful, the functions become deranged, and disease and death ensue. The inhabitants of the warmer regions, when transported to the colder districts, experience inconvenience from the change of temperature, equally hurtful to the system, and fatal to its continuance. The polar bear appears to be accommodated to live in a region, whose mean annual temperature is below the freezing point. In the summer temperature of Edinburgh, however well supplied with food, he appears to languish in misery. Cold spring-water poured upon him seems to revive him for a little; but all relief is temporary, the climate is too hot for the enjoyment of life. Destined to live in a climate where the system is required to secrete heat chiefly, it seems incapable of generating the cold requisite to counteract the effects of even a temperate climate. The inhabitants of the torrid regions, on the other hand, seem capable of generating cold chiefly, all their organs being adapted for resisting high temperatures; and hence, when brought to cold districts, they are incapable of generating the requisite degree of heat.

In those districts where the individuals of a species are most vigorous and prolific, the temperature most suitable for existence prevails. The native country of the horse is probably Arabia. There he exists in a wild state in the greatest numbers. In the Zetland Islands, where he is nearly in a state of nature, he is approaching the polar limits of his distribution. He has become a dwarf. He does not reach maturity until his fourth year, seldom continues in vigour beyond his twelfth, and the female is never pregnant above once in two years. At the line where the energies of the horse terminate, however, the rein-deer becomes a useful substitute. Its equatorial limits do not reach the shores of the Baltic.

The variations of the seasons, which bring along with them corresponding changes of heat and cold, exercise a powerful influence on the distribution of animals, in reference to temperature. Some species appear to possess a considerable range of temperature, within which life can be easily preserved, and all its functions regularly performed. We do not mean to intimate, that there is any animal which can live in our climate, for example, and remain uninfluenced by a difference of temperature of upwards of twenty degrees between summer and winter. The constitutional arrangement suited to the one season, would be prejudicial during the continuance of the other. But there are many animals which live in the same district both in summer and winter, and even in districts differing considerably in their mean annual temperature. What, then, are the means employed by these species to preserve life in the midst of such vicissitudes? The power of producing heat or cold, is a property obviously possessed by the warm-blooded animals, and probably in an inferior degree by those which are termed cold-blooded. But in all the efforts made by the system to secrete extraordinary degrees of either heat or cold, there is so great a portion of vital energy expended, that exhaustion and death follow its long continuance. In all cases where the influence of the seasons are to be resisted by efforts of this kind, it would be requisite to continue them uninterrupted for many months. These efforts, however, are diminished in extent and duration by a variety of the most wonderful arrangements, exhibiting the infinite resources of that Wisdom which planned the constitution and continuance of the animal kingdom. To the chief of these compensating or counteracting circumstances we shall now briefly advert.

1. *Changes take place in the Quantity of the Clothing.*— The same circumstances which enable the Negro to go about in a state approaching to nakedness, and impel the inhabitants of the arctic regions to cover themselves with woollen cloth or skins, operate in regulating the clothing of quadrupeds and birds. In the warmer regions, it is requisite to suffer the temperature of the body to be diminished, while,

in the colder regions, the very opposite object is aimed at. In the former case, the hair or feathers are thinly spread over the body, while, in the latter, they form a close and continuous covering. In the dogs of Guinea, and in the African and Indian sheep, the fur is so very thin that they may almost be denominated naked. In the Siberian dog and Iceland sheep, on the other hand, the body is protected by a thicker and longer covering.

The clothing of animals, living in cold countries, is not only different from that of the animals of warm regions in its quantity, but in its arrangement. If we examine the covering of swine of warm countries, we find it consisting of bristles or hair of the same form and texture; while the same animals which live in colder districts, possess not only common bristles or strong hair, but a fine frizzled wool next the skin, over which the long hairs project. Between the swine of the south of England, and the Scottish Highlands, such differences may be observed. Similar appearances present themselves among the sheep of warm and cold countries. The fleece of those of England consists entirely of wool; while the sheep of Zetland and Iceland possess a fleece, containing, besides the wool, a number of long hairs, which give to it, when on the back of the animal, the appearance of being very coarse. The living races of Rhinoceros and Elephant, inhabitants of the warm regions, have scarcely any hair upon their bodies; while those which formerly lived in the northern plains of Europe, the entire carcases of which have been preserved in the ice in Siberia, were covered with fur similar to the Icelandic Sheep, consisting of a thick covering of short frizzled wool, protected by long coarse hairs. These species, now extinct, possessed clothing suiting them for the climate where they lived, and where they became at last enveloped in ice. Had they been transported by any accident from a warmer region, they would have exhibited in the thinness of their covering, unequivocal marks of the climate in which they were reared.

By means of this arrangement, in reference to the quantity of clothing, individuals of the same species can maintain life, comfortably, in climates which differ considerably in their average annual temperature. By the same arrangements, the individuals residing in a particular district, are able to provide against the varying temperature of the seasons. The covering is diminished during summer and increased in winter, as may be witnessed in many of our domestic quadrupeds.

Previous to winter, the hair is increased in quantity and length. This increase bears a constant ratio to the temperature; so that when the temperature decreases with the elevation, we find the cattle and horses, living on farms near the level of the sea, covered with a shorter and thinner fur than those which inhabit districts of a higher level. Cattle and horses, housed during the winter, have shorter and thinner hair than those which live constantly in the open air. The hair is likewise shorter and thinner in a mild, than during a severe winter.

This winter covering, if continued during the summer, would prove inconveniently warm. It is, therefore, thrown off by degrees as the summer advances; so that the animals which were shaggy during the cold months become sleek in the hot season.

This process of *casting the hair* takes place at different seasons, according to the constitution of the animal with respect to heat. The mole has, in general, finished this operation before the end of May. The fleece of the sheep, when suffered to fall, is seldom cast before the end of June. In the northern islands of Scotland, where the *shears* are never used, the inhabitants watch the time when the fleece is ready to fall, and pull it off with their fingers. The long hairs, which likewise form a part of the covering, remain for several weeks, as they are not ripe for casting at the same time with the fine wool. This operation of pulling off the wool, provincially called *rooing*, is represented by some writers, more humane than well-informed, as a painful process to the animal. That it is not even disagreeable, is evident from the quiet manner in which the sheep lie during the pulling, and from the ease with which the fleece separates from the skin.

We are in general inattentive with respect to the annual changes in the clothing of our domestic animals; but when in search of those beasts which yield us our most valuable *furs*, we are compelled to watch these operations of the seasons. During the summer months the fur is thin and short, and is scarcely ever an object of pursuit; while during the winter, it possesses in perfection all its valuable qualities. When the beginning of winter is remarkable for its mildness, the fur is longer in *ripening*, as the animal stands in no need of the additional quantity for a covering; but as soon as the rigours of the season commence, the fleece speedily increases in the quantity and length of hair. This increase is sometimes very rapid in the hare and the rabbit, the skins of which are seldom ripe in the fur until there is a fall of snow, or a few days of frosty weather; the growth of hair in such instances being dependent on the temperature of the atmosphere.

The *moulting* of birds is another preparation for winter, which is analogous to the casting of the hair in quadrupeds. During summer, the feathers of birds are exposed to many accidents. Not a few spontaneously fall; some of them are torn off during their amorous quarrels; others are broken or damaged; while in many species they are pulled from their bodies to line their nests. Hence their summer dress become thin and suitable. Previous to winter, however, and immediately after the process of incubation and rearing of the young is finished, the old feathers are pushed off in succession by the new ones, and in this manner the greater part of the plumage of the bird is renewed. During this process of moulting, the bird seems much enfeebled, and, if previously in a weak state, is in danger of dying during the process. In consequence of this renewal of the feathers, the winter covering is rendered perfect, and the birds prepared for withstanding all the rigours of the season. In those birds whose plumage changes colour with the seasons, the moulting takes place in subserviency to the purposes of these variations, as we shall shortly have occasion to notice.

By this addition to the non-conducting appendices of the skin, quadrupeds and birds are enabled to preserve the heat generated in their bodies, from being readily transmitted to the surrounding air, and carried off by its motions and diminished temperature. But along with a change of quantity, there is frequently also a change of colour.

2. *Changes take place in the Colour of the Clothing.*— The distribution of colour in the animal kingdom, appears to be connected with latitude as correlative with temperature. In the warmer districts of the earth, the colours of man, quadrupeds and birds, exhibit greater variety, and are deeper and brighter, than in the natives of colder countries.

Among the inhabitants of the temperate and cold regions there are many species which, in reference to the colour of their dress, do not appear to be influenced by the vicissitudes of the seasons. In others, a very marked difference prevails between the colour of their summer and winter garb. A few of the more obvious instances of these changes, in British species, may be here produced.

Among quadrupeds, the Alpine hare (*Lepus variabilis*) is a very remarkable example. It is found, in this country, on the high mountains of the Grampian range. Its summer dress is of a tawny grey colour; but, about the month of September, its fur gradually changes to a snowy whiteness. It continues in this state during the winter, and resumes its plainer covering again in the month of April or May, according to the season. The *ermine* is another of our native quadrupeds which exhibits in its dress similar changes of colour according to the season. It frequents the outskirts of woods and thickets. During the summer months, its hair is of a pale reddish brown colour; in harvest it becomes clouded with pale yellow; and, in the month of November, with us, it is of a snow white colour. Its winter dress furnishes the valuable fur called ermine. Early in spring, the white becomes freckled with brown, and in the month of May it completely resumes its summer garb.

Among the feathered tribes such instances of change of colour in the plumage during winter are numerous. They greatly perplex the ornithologist, and have been the means of introducing into the system several spurious species. The white grous or ptarmigan (*Tetrao lagopus*) may be produced as a familiar example of this kind of hybernation. This bird, like the Alpine hare, inhabits the higher Grampians, and is never found at a great distance from the limits of the snow. In summer its plumage is of an ash colour, mottled with small dusky spots and bars. At the approach of winter the dark colours disappear, and its feathers are then found to be pure white. In remarkably mild winters the change is sometimes incomplete, a few dusky spots of the summer dress remaining. In spring its winter garb becomes again mottled, and the bird loses much of its beauty. Even the young birds in their autumnal dress resemble their parents in the mottled plumage, which likewise becomes white at the approach of winter.

Among the aquatic birds, similar changes in the colour of the plumage have been observed. The black guillemot (*Uria grylle*), so common on our coasts, is of a sooty black colour during the summer, with a white patch on the wings. During winter, however, the black colour disappears, and its plumage is then clouded with ash-coloured spots on a white ground. In the winter dress it has been described by some as a distinct species, under the name of the *spotted guillemot*. In the more northern regions, as in Greenland, for example, this bird, in winter, becomes of a pure white colour.

These changes of colour, which we have already mentioned, extend throughout the whole plumage of the bird; but, in some instances, the change takes place on a small part only of the plumage. Thus the little auk (*Alca alle*), during summer, has its cheeks and throat of a black colour, but in winter these parts become of a dirty white. In this its winter garb, it is often shot on our coasts. Its summer dress induced PENNANT to consider it as a variety, and as such to give a figure of it in his British Zoology. The black-headed gull (*Larus ridibundus*) has a black head during summer, as its English name intimates. During the winter, however, the black colour on the head disappears; and, when in this dress, it has been regarded by many as a distinct species, under the name of the *Red-legged Gull*.

In many other birds there is a remarkable difference, in point of colour, between the summer and the winter plumage, although not so striking as in those which we have already noticed. The colours of the summer feathers are rich and vivid; those of the winter obscure and dull. This is well illustrated in the dunlin (*Tringa alpina*), whose summer plumage is much intermixed with black and rufous colour, but whose winter plumage is dull and cinereous. In its winter dress it has been described as a distinct species, under the name of *T. cinclus*, or Purre. Similar instances might be produced in the wagtails, linnets, and plovers, and a great many other birds.

The circumstances under which these changes are observed to take place, indicate their dependence on temperature, as connected with the season. The deep colours of the summer dress are exchanged for the lighter or whiter colours of the winter, with a rapidity and extent proportional to the changes of the seasons. During a mild autumn, the shifting of the dark for the light coloured dress proceeds at a very slow pace; and when the winter also continues mild, the white dress is never fully assumed. In some species, as the black guillemot, the white winter dress is never acquired in this climate, although its ash-coloured plumage intimates a tendency to the change. In the climate of Greenland, on the other hand, the change is complete, and the plumage is of a snowy whiteness; as we had an opportunity of observing in the collection of the Dublin Society in 1816, in a specimen in its winter dress, brought from Greenland by an intelligent and enterprising naturalist, Sir Charles Giesecke'.

Having thus seen that the colour of the clothing of many animals changes with the season, and that, however diversified the summer dress may be, the colour during winter approaches to white, it may now be asked, What benefit is derived from this arrangement *?

---

* Some species of gulls exhibit in their winter plumage very striking deviations from this general rule. Montagu, in his Supplement to the Orni-

The rate at which bodies cool is greatly influenced by their colour. The surface which reflects heat most readily, suffers it to escape but slowly by radiation. Reflection takes place most readily in objects of a white colour, and from such, consequently, heat will radiate with difficulty. If we suppose two animals, the one of a black colour, and the other white, placed in a higher temperature than that of their own body, the heat will enter the one that is black with the greatest rapidity, and elevate its temperature considerably above the other. These differences are observable in wearing black and light coloured clothes during a hot day. When, on the other hand, these animals are placed in a situation, the temperature of which is consider-

---

thological Dictionary, article Common Gull, says,—" We have had this species alive for some years, and observed, that when it had attained its full mature plumage, in the second year, the head and neck is pure white during the summer; but, like the herring gull, these parts become streaked, and spotted with brown, in autumn, which is continued all the winter; and in the spring become again pure white." When speaking of the herring gull, he says,—" This gull is now living, and in high health, being thirteen years old. It begins moulting about the middle of August, when it annually assumes the mottled head and neck ; and about the middle of February, the partial spring moulting commences, the mottled feathers are discharged, and succeeded by pure white." A herring gull, at present six years old, in the garden at Canonmills, of my esteemed friend Mr P. NEILL, has, for the last three years, regularly acquired the mottled plumage of the head and neck, in the month of August. It did not acquire the pure white head and neck in spring and summer, before the third year. Captain SABINE, in his valuable Memoir on the Birds of Greenland, Linn. Trans. vol. xii. p. 544, when describing the changes of plumage which the Larus *glaucus* exhibits, adds,— " In winter, the mature bird has the head and neck mottled with brown, as is usual with all the white-headed gulls." In a specimen of L. *marinus*, shot in winter, I observed on the head, and chiefly in front of the eyes, a few black hairs, which were formed from the produced ends of the shafts.

ably lower than their own, the black animal will give out its heat by radiation to every surrounding object colder than itself, and speedily have its temperature reduced; while the white animal will part with its heat by radiation at a much slower rate. The change of colour in the dress of animals is therefore suited to regulate their temperature by the radiation or absorption of caloric.

While it is requisite that the temperature of some species should be preserved as equably as possible, the cooling effects of winter are likewise resisted by an additional quantity of heat being generated by the system. An increase in the quantity of clothing takes place to prevent that heat being dissipated by communication with the cold objects around, and the dress changes to a white colour, to prevent its loss by radiation. In summer, the pernicious increase of temperature is prevented by a diminished secretion of heat or the secretion of cold, increased perspiration, the *casting* of a portion of the winter covering, and by a superior intensity of colour in the remainder giving it a greater radiating power. This last character would, in the sunshine, by absorbing heat, prove a source of great inconvenience, were its effects not counterbalanced by other arrangements, and by the opportunity of frequenting the refreshing shade, or bathing in the stream.

In those cases, where particular parts only of the clothing change their colour, there are probably local circumstances connected with the secretions, or the sexual system, which render such arrangements necessary. Hair growing from a part which has been wounded, is always paler coloured than that which is produced on the sounder parts, intimating the operation of local causes on the colouring secretions, or local purposes to be served by the change.

It is probably for the purpose of preventing a wasteful dissipation of the heat of the system, that the dress of many animals becomes lighter coloured in old age, and that the human hair turns grey. Young animals seldom present the same dress and vivid colours, &c. which they assume upon arriving at maturity *.

The change of colour which takes place in the dress of some animals during winter, is supposed to serve other purposes than the regulation of their temperature. The white garb which they assume, assimilates them to the colour of the snow, and in this way they are considered as better able to escape the observation of their foes.

All our conclusions concerning final causes, ought to be the result of very extended observations, lest we delineate arrangements which would be productive of pain and ruin to many species, where we intended to unfold the marks of wisdom and benevolence. If the white dress of the alpine hare and ptarmigan concealed them from their enemies, the eagle, the cat, and the fox, these last, by being deprived of their ordinary food, would be in danger of starvation and death. But this variation of colour is not confined to weak or defenceless animals. Beasts and birds of prey are likewise subject to the change. Hence, if it yielded protection to some, it would enable others to prey with greater certainty of success on their defenceless neighbours. Many of these rapacious animals, (as the ermine for example, which is at all times well qualified to provide for its wants by its determined boldness, extreme agility, and exquisite smell), do not stand in need of such assistance. If this change extends to the rapacious as well

---

* Young gulls and solan geese, however, present very obvious exceptions. Their darkest colours are those of immaturity. Yellow seems a predominant tint of infancy. The hoofs of quadrupeds, and the beaks of birds, are usually more or less tinged with this colour at birth.

as the defenceless, it may likewise be observed in aquatic as well as in terrestrial animals. In reference to acquatic animals, we would ask, What protection is afforded to the black guillemot, during the winter, by its mottled plumage, or to the little auk, by its white chin, since the whiter their dress, so much the more unlike the dark coloured water of the clouded season in which it is exhibited? The popular opinion on the subject must be relinquished as untenable; especially as the change of colour from dark to white does not vary, however different the habits or even stations of the animals may be in which it takes place.

An interesting inquiry yet remains to be made regarding the manner in which this change in the colour of the dress is effected. The attention of naturalists has, of late years, been directed to this subject, and several important observations have been made, equally interesting to the physiological and systematical zoologist.

From the belief which is generally entertained, that in hair and feathers there is no circulation, neither secretion nor absorption, a conviction arose in the minds of many naturalists, that the change of colour which takes place in the dress of some animals according to the season, was not the effect of any organical change in the hair or feathers, but accompanied a renewal of the whole. The late GEORGE MONTAGU, Esquire, who had long attended to the characters and habits of the feathered tribes, delivers his opinion on this subject in the following terms: " Some species of birds seem to change their winter and summer feathers, or at least in part; in some, this is performed by moulting twice a-year, as in the ptarmigan, in others, only additional feathers are thrown out. But we have no conception of the feathers changing colour, although we have been informed of such happening in the course of one night *." Staggered with the statements of such a frequent renewal of the dress of animals, accompanied by such a wasteful expenditure of vital energy, and guided by multiplied observations, we ventured to offer the following remarks on the subject in the Edinburgh Encyclopedia, under the article " Hybernation," vol. xi. p. 387, published in 1817.

" It has been supposed by some, that those quadrupeds which, like the alpine hare and ermine, become white in winter, cast their hair twice in the course of the year; at harvest when they part with their summer dress, and in spring when they throw off their winter fur. This opinion does not appear to be supported by any direct observations, nor is it countenanced by any analogical reasonings. If we attend to the mode in which the hair on the human head becomes grey as we advance in years, it will not be difficult to perceive that the change is not produced by the growth of new hair of a white colour, but by a change in the colour of the old hair. Hence there will be found some hairs pale towards the middle, and white towards the extremity, while the base is of a dark colour. Now, in ordinary cases, the hair of the human head, unlike that of several of the inferior animals, is always dark at the base, and still continues so during the change to grey; hence we are

---

* " Ornithological Dictionary," Introduction, p. 25. London, 1802. The same intelligent observer continued, even after the publication of the " Supplement" to the above work, in 1813, where the subject frequently came under his notice, to entertain the same opinion. He was disposed to admit two, and in some cases three successive moultings in the course of a year. In a letter which I received from him on the subject, dated " Knowle, 7th December 1814," he adds, " But I have no conception of a change of colour in the same feather or hair, (because the colouring matter has been disposed in embryo), except by length of time, as our hair is changed by age."

disposed to conclude from analogy, that the change of colour, in those animals which become white in winter, is effected, not by a renewal of the hair, but by a change in the colour of the secretions of the rete mucosum, by which the hair is nourished, or perhaps by that secretion of the colouring matter being diminished, or totally suspended.

" But as analogy is a dangerous instrument of investigation in those departments of knowledge which ultimately rest on experiment or observation, so we are not disposed to lay much stress on the preceding argument which it has furnished. The appearances exhibited by a specimen of the ermine now before us are more satisfactory and convincing. It was shot on the 9th May (1814), in a garb intermediate between its winter and summer dress. In the belly, and all the under parts, the white colour had nearly disappeared, in exchange for the primrose-yellow, the ordinary tinge of these parts in summer. The upper parts had not fully acquired their ordinary summer colour, which is a deep yellowish-brown. There were still several white spots, and not a few with a tinge of yellow. Upon examining those white and yellow spots, not a trace of interspersed new short brown hair could be decerned. This would certainly not have been the case if the change of colour is effected by a change of fur. Besides, while some parts of the fur on the back had acquired their proper colour, even in those parts numerous hairs could be observed of a wax-yellow, and in all the intermediate stages from yellowish-brown, through yellow, to white.

" These observations leave little room to doubt, that the change of colour takes place in the old hair, and that the change from white to brown passes through yellow. If this conclusion is not admitted, then we must suppose that this animal casts its hair at least seven times in the year. In spring, it must produce primrose-yellow hair ; then hair of a wax-yellow; and, lastly, of a yellowish-brown. The same process must be gone through in autumn, only reversed, and with the addition of a suit of white. The absurdity of this supposition is too apparent to be farther exposed.

" With respect to the opinion which we have advanced, it seems to be attended with few difficulties. We urge not in support of it, the accounts which have been published of the human hair changing its colour during the course of a single night; but we think the particular observations on the ermine warrant us in believing, that the change of colour in the alpine hair is effected by a similar process. But how is the change accomplished in birds?

" The young ptarmigans are mottled in their first plumage similar to their parents. They become white in winter, and again mottled in spring. These young birds, provided the change of colour is effected by moulting, must produce three different coverings of feathers in the course of ten months. This is a waste of vital energy, which we do not suppose any bird in its wild state capable of sustaining; as moulting is the most debilitating process which they undergo. In other birds of full age, two moultings must be necessary. In these changes, the range of colour is from blackish grey through grey to white, an arrangement so nearly resembling that which prevails in the ermine, that we are disposed to consider the change of colour to take place in the old feathers, and not by the growth of new plumage; this change of colour being independent of the ordinary annual moultings of the birds.

" Independent of the support from analogy which the ermine furnishes, we may observe, that the colours of other parts of a bird vary according to the season. This is frequently observable in the feet, legs, and bill. Now, since a change takes place in the colouring secretions of these organs, What prevents us from supposing that similar changes take place in the feathers? But even in the case of birds, we have before us an example as convincing as the ermine already mentioned. It is a specimen of the little auk (*Alca alle*), which was shot in Zetland in the end of February 1810. The chin is still in its winter dress of white, but the feathers on the lower part of the throat have assumed a dusky hue. Both the shafts and webs have become of a blackish grey colour at the base and in the centre, while the extremities of both still continue white. The change from black to white is here effected by passing through grey. If we suppose that in this bird the changes of the colour of the plumage are accomplished by moulting, or a change of feathers, we must admit the existence of three such moultings in the course of the year—one by which the white winter dress is produced, another for the dusky spring dress, and a third for the black garb of summer. It is surely unnecessary to point out any other examples in support of our opinion on this subject. We have followed nature, and our conclusions appear to be justified by the appearances which we have described*."

Since the preceding observations were communicated to the public, Professor JAMESON has obligingly pointed out to me the following passage in CARTWRIGHT's Journal of Transactions on the Coast of Labrador (3 vols. 4to. Newark 1792), vol. i. p. 278, as containing an expression of the same opinion which I had formed upon the subject. " 28th September 1773.—This morning I took a walk upon the

---

* The Reverend Mr WHITEAR, in his " Remarks on the Changes of the Plumage of Birds," in Linn. Trans. vol. xii. p. 524, read April 6. 1819, has recorded several striking facts observed in the plumage of the mallard, chaffinch, and some other birds, which corroborate the opinion advanced above.

hills to the westward, and killed seven brace of grouse. These birds are exactly the same with those of the same name in Europe, save only in the colour of their feathers, which are speckled with white in summer, and perfectly white in winter, fourteen black ones in the tail excepted, which always remain the same. When I was in England, Mr BANKS (now Sir JOSEPH BANKS), Doctor SOLANDER, and several other naturalists, having inquired of me respecting the manner of these birds changing colour, I took particular notice of those I killed, and can aver for a fact, that they get at this time of the year a very large addition of feathers, all of which are white; and that the coloured feathers at the same time change to white. In spring, most of the white feathers drop off, and are succeeded by coloured ones; or, I rather believe, all the white ones drop off, and that they get an entire new set. At the two seasons they change very differently; in the spring, beginning at the neck, and spreading from thence; now, they begin on the belly, and end at the neck. There are also ptarmigans in this country, which are in all respects the same as those I have killed on some high mountains in Scotland."

The total absence of every thing like demonstration of the truth of the assertion regarding the autumnal change of the colour of the plumage of the grouse, and the language of hesitation which he employed when speaking of the changes observed to take place in spring, probably induced those naturalists who had perused CARTWRIGHT's work, to reject statements so opposite to the opinions generally entertained on the subject. The conjecture that all the white feathers drop off in spring, even those which were produced in the previous autumn, to be succeeded by coloured ones, has not been verified by any observations which we have been able to make on the subject.

The moulting of birds takes place in all cases gradually, and in those species whose plumage changes colour with the season, the different moultings take place at corresponding periods. In the autumn, we find that the black feathers on the head of the *Larus ridibundus* change to a white colour. But besides the altered feathers, others spring up, of a white colour, to increase the quantity of clothing. This gull has, therefore, during the winter, some of the feathers of the head old, and others young. Again, in spring the white feathers of the winter become black, and a few new feathers make their appearance, likewise of a black colour, to supply the place of the older ones which drop off in succession. Some of the feathers on the head of this gull are half a year older than others; and consequently, we may infer, will fall off sooner than those of more recent growth. From these, and similar facts, furnished by several species of British birds, we are disposed to conclude, that the feathers which are produced in autumn and the beginning of winter, and which correspond with the conditions of the season, change their colour in spring, and continue in this state until they are shed in autumn. The feathers which are produced in spring, continue of the same hue during the summer, change their colour in winter, and fall off again upon the approach of spring. In this manner, the quantity of the plumage fit for the different seasons of the year is easily regulated, and it is only necessary that the change of colour in each feather should take place but once in the course of its connection with the bird. By these arrangements, the welfare of the individual is promoted by the simplest means.

Many animals, which are unable to provide against the vicissitudes of the seasons, by varying the quantity and the colour of their dress, are, nevertheless, protected by being able to shift their quarters, so as to live throughout the whole year in a temperature congenial to their constitution.

3. *Periodical Migration.*—Quadrupeds in general, from their limited power of locomotion, cannot migrate from one country to another, with ease and safety, in order to avoid the variations of temperature which accompany the changes of the seasons. In the same country partial migrations may take place, as we witness in the stag and the roe, which leave the alpine regions at the approach of winter, and seek protection in the more sheltered plains. In America some species of the genus *Dipus* perform still more extensive migrations. Those quadrupeds, however, which have the faculty of flying, as the bats, or of swimming, as the seals and whales, may overcome the obstacles which oppose a change of place, and execute safely periodical migrations. Individuals of these tribes are accordingly observed to avoid the extremes of heat and cold, by shifting their situations according to the seasons. The great bat (*Vespertilio noctula*), which inhabits England during the summer, is known to spend its winters in a torpid state in Italy. The Greenland seal forsakes the icy shores which it has frequented during summer, and migrates southwards, at the approach of winter, to Iceland.

The facts which have been ascertained in reference to the periodical migration of quadrupeds, are too few for enabling us to point out the laws by which they are regulated. The movements of birds, however, furnish more interesting results.

The migrations of the feathered tribes have been the object of popular observation, since the days of the prophet Jeremiah : " For the stork in the heaven knoweth her appointed times; and the turtle, and the crane, and the swallow, observe the time of their coming." (ch. viii. ver. 7).

The systematical naturalists have likewise collected many scattered observations, and the subject appears now to be capable of receiving a satisfactory illustration. It is not our intention to enter into any minuteness of detail, regarding the migrations of particular species, but to ascertain the *laws* of migration, and the circumstances under which it takes place.

Before entering farther upon the subject, it may be proper to state, that the same species which is migratory in one country, is in some cases stationary in another; as the linnet, which is migratory in Greenland, but stationary in Britain. In Britain, both male and female chaffinches are stationary, while in Sweden, the latter are migratory *. Some species of the same natural genus are migratory, while others are stationary. Thus the fieldfare is migratory, while the blackbird is permanently resident.

Migrating birds may be divided into two classes, from the different seasons of the year in which they arrive or depart. To the first class will belong those birds which arrive in this country in the spring, and depart from it in autumn, and are termed *Summer Birds of Passage*. The second will include those which arrive in autumn, and depart in spring, and are called *Winter Birds of Passage*.

THE SUMMER BIRDS OF PASSAGE are not confined to any particular order or tribe; nor are they distinguished

---

* ECKMARK, when speaking of the migrations of this bird, informs us, " *Mares* inter primas sunt aviculas, quæ sonum suum hieme usitatum in cantum vertunt jucundissimum : vere primo, sub initium *mensis regelationis*, arboribus ad pagos insidentes garruli, *fœminis* adhuc *absentibus*, ver indicant adstans. Redeuntibus denique turmis maximis, quæ cœlum fere abscondunt, *fœminis*, omnes conjuges requirunt, quibus conjuncti sylvas petunt, ibi ut nidulos construant et multiplicentur. Initio *mensis defoliationis* mares suos, apud nos remanentes, *fœminæ* deserunt mutabiles, solæ regiones petentes peregrinas." Amæn. Acad. iv: 595.

by similarity of habits. Some of them belong to the division of *Water Fowls*, as the terns and gulls; while others are *Land Birds*, as the swallow and corn-crake. They differ also remarkably with regard to their food. Thus, the hobby is carnivorous; the gulls and terns, piscivorous; the swallow, insectivorous; and the turtle dove and the quail, granivorous. They, however, present one point of resemblance. All of them, during their residence in this country, perform the important offices of pairing, incubation, and rearing their young; and hence may, with propriety, be termed the natives of the country. We hail their arrival as the harbingers of spring, and feel the blank which they leave on their departure, although it is in some measure supplied by another colony of the feathered race, who come to spend with us the dreary months of winter.

The Winter Birds of Passage have more points of resemblance among themselves than those of the former division. They chiefly belong to the tribe of water-fowls. None of them are insectivorous, and very few are granivorous. They chiefly frequent the creeks and sheltered bays of the sea, and the inland lakes, or they obtain their food in marshy grounds, or at the margins of springs. When the rigours of the season are over, and when other birds which are stationary are preparing for incubation, these take their departure, to be again succeeded by our summer visitants.

We have stated generally, that our summer and winter birds of passage visit us at stated seasons of the year; that the summer visitants arrive in spring and depart in autumn; and that the winter visitants arrive in autumn and depart in spring. But the different species do not all observe the same periods of arrival and departure. Thus, among the summer birds of passage, the wheat-ear always precedes the swallow, while the swallow arrives before the martin, and the martin before the corncrake. Among the winter birds of passage, similar differences in the time of arrival are observable. Thus the woodcock precedes the fieldfare, and the fieldfare the redwing. The periods of departure have not been observed with such attention, as the subjects have then lost their novelty, so that we do not readily perceive their absence. It is probable, however, that in their departure, as well as their arrival, each species has its particular period.

The periods of arrival and departure, even in the same species, do not always take place at exactly the same day, or even month of the year. In different years these vary several weeks or even months, and evidently depend on very obvious circumstances. The meanest rustic, in regard to the summer birds of passage, is aware, that cold weather prevents the arrival of these messengers of spring; and that the early arrival of our winter birds of passage indicates a proportionally early winter. The same circumstances of temperature which retard our summer visitants also check the progress of vegetation. Hence, in all probability, we might be able to prognosticate the arrival of these birds, by attending to the time of the leafing or flowering of particular trees or plants. As the state of vegetation depends on the temperature of the season, and the life of insects on the state of vegetation, we may safely conclude, that the movements of the phytivorous and insectivorous birds must be dependent on the condition of plants.

LINNÆUS bestowed some attention on these connected circumstances, in his Calendar of Flora for Sweden; and STILLINGFLEET in that of England. LINNÆUS observed, that the swallow returned to Sweden when the bird-cherry came into leaf, and when the wood-anemone flowered. He also found the arrival of the nightingale accompanied with the leafing of the elm. STILLINGFLEET says, that the swallow re-

MIGRATION.

turns to Norfolk with the leafing of the hazel, and the nightingale with the leafing of the sycamore. It has also been observed, that the cuckoo sings when the marsh-marigold blows. It would tend greatly to increase our knowledge of this subject, were observations of this sort multiplied. We earnestly recommend the subject to the attention of the practical naturalist.

Having thus offered a few observations on the periods of arrival and departure of migrating birds, let us now enquire after *the places from whence they come, and to which they return.* In doing this, it will be proper to bestow some attention on the migrations of those birds which merely shift from one part of the island to another. The movements of such birds, though confined within narrow bounds, are probably regulated by the same laws, which, in the other species, produce more extensive migrations, and have the obvious advantage of being easily investigated.

In the inland districts of Scotland, the lapwings make their appearance about the end of February or the beginning of March, and, after fulfilling the purposes of incubation, hasten to the sea-shore, there to spend the winter, and to support themselves by picking up the small crustacea from among the rejectamenta of the sea. These birds seldom, however, remain all winter on the Scottish shores, though they are always to be found at that season on the southern English shores. In that part of the island they do not perform such extensive migrations, but may with propriety be considered as resident birds. The curlew arrives at the inland districts along with the lapwing, and departs in company about the beginning of August. The curlew, however, remains on the Scottish shores during the winter. The oyster-catcher, though it breeds in Scotland, retires to the English shores during the winter, and joins those which have remained there during the breeding season. The black headed gull breeds both in England and Scotland; but it retires from the last mentioned country, while it continues resident in the former.

From the examples quoted, it appears that some birds, which are stationary in one district, are migratory in another. But that which chiefly merits our consideration is the circumstance of those birds, whose annual migrations are confined to our own shores, forsaking the high grounds on the decline of summer, and seeking for protection at a lower level, and in a warmer situation. When these migrations become more extensive, the bleak moors and shores of Scotland are exchanged for the warmer and more genial climate of England. Hence it happens, that some of our Scottish summer visitants come from England, while some of the English winter visitants come from Scotland; the summer birds of passage coming from the south, and the winter passengers from the north. Do those birds that perform more extensive migrations obey the same laws?

As the summer birds of passage are more interesting to us, since they perform the great work of incubation in our country, than the winter birds of passage, which are the harbingers of storms and cold, and only wait the return of spring to take their leave of us, we shall endeavour to find out the winter residence of the former, before we attempt to discover the summer haunts of the latter. Natural history, it is true, is still in too imperfect a state, to enable us to point out with certainty the retreats of all those birds which visit us during summer. But enough appears to be known to enable us to ascertain the laws by which these migrations are regulated in a number of birds; and as the points of resemblance in the movements of the whole are numerous, we can reason from analogy on safer grounds with regard to the remainder.

The swallow, about whose migrations so many idle stories have been propagated and believed, departs from Scotland about the end of September, and from England about the middle of October. In the latter month M. ADANSON observed them on the shores of Africa after their migrations from Europe. He informs us, however, that they do not build their nests in that country, but only come to spend the winter. M. PRELONG has not only confirmed the observations of ADANSON, in reference to swallows, but has stated, at the same time, that the yellow and grey wagtails visit Senegal at the beginning of winter. The former (*Motacilla flava*) is well known as one of our summer visitants *. The nightingale departs from England about the beginning of October, and from the other parts of Europe about the same period. During the winter season it is found in abundance in Lower Egypt, among the thickest coverts, in different parts of the Delta. These birds do not breed in that country, and to the inhabitants are merely winter birds of passage. They arrive in autumn and depart in spring, and at the time of migration are plentiful in the islands of the Archipelago. The quail is another of our summer guests, which has been traced to

---

* " J'ai observé, comme ADANSON, que nos hirondelles et nos bergeronnettes arrivoient dans la zone torride huit ou dix jours après l'epoque où elles quittent nos climats. En 1788, j'ai vu les bergeronnettes du printems et les bergeronnettes grises arriver à Gorée le 14. Septembre. ADANSON dit qu'il a vu arriver les hirondelles au Senegal 9. Octobre ; or je me rapelle qu'elles quittent le department des Hautes Alpes vers la fin de Septembre, ce qui s'accorde parfaitement."—" Memoir sur les Iles de Gorée et du Senegal, par le cit. PRELONG." *Annales de Chemie*, t. xviii. p. 272.

Africa. A few, indeed, brave the winters of England, and in Portugal they appear to be stationary. But in general they leave this country in autumn, and return in spring. They migrate about the same time from the eastern parts of the Continent of Europe, and visit and revisit in their migrations the shores of the Mediterranean, Sicily, and the islands of the Archipelago.

While these birds perform those extensive migrations which we have here mentioned, others are contented with shorter journeys. Thus, the razor-billed auk (*Alca torda*), and the puffin (*Alca arctica*), frequent the coast of Andalusia during the winter season, and return to us in the spring.

These facts, and many others of a similar nature, which might have been stated, enable us to draw the conclusion, that our summer birds of passage come to us from southern countries, and, after remaining during the warm season, return again to milder regions. A few of our summer visitants may winter in Spain or Portugal; but it appears that in general they migrate to Africa, that unexplored country possessing every variety of surface, and consequently great diversity of climate. It is true that we are unacquainted with the winter retreats of many of our summer birds of passage, particularly of small birds; but as these arrive and depart under similar circumstances with those whose migrations are ascertained, and as the operations which they perform during their residence with us are also similar, we have a right to conclude that they are subject to the same laws, and execute the same movements. What gives weight to this opinion, is the absence of all proof of a summer bird of passage retiring to the north during the winter season.

In proof of the accuracy of the preceding conclusion, we may observe, that it is a fact generally acknowledged, that the summer birds of passage visit the southern parts of the country a few days, or even weeks, before they make their appearance in the northern districts. Thus, the common swallow (*Hirundo rustica*) appears in Sussex about the beginning of the third week of April; while in the neighbourhood of Edinburgh it is seldom seen before the first of May. The cuckoo appears in the same district about the last week of April; in Edinburgh seldom before the second week of May. The reverse of this holds true with these summer visitants at their departure. Thus, dotterels (*Charadrius morinellus*) forsake the Grampians about the beginning of August, and Scotland by the end of that month; while they return to England in September, and remain there even until November. A difference of nearly a month takes place between the departure of the goatsucker (*Caprimulgus Europæus*) from Scotland and from the south of England.

Having thus ascertained the winter haunts of our summer birds of passage, let us now endeavour to find out the summer retreat of our winter visitants. The conclusions which we have already established dispose us to look for these birds in countries situated to the northward. And as we are much better acquainted with the ornithology of those countries than of Africa, it will be in our power to prosecute our researches with greater certainty of success.

The snow-bunting (*Emberiza nivalis*), which is among the smallest of our winter guests, retires to the hoary mountains of Spitzbergen, Greenland, and Lapland, and there executes the purposes of incubation, making its nest in the fissures of the rocks. In these countries it is therefore a summer visitant, as it retires southward in autumn, to spend the winter in more temperate regions. To the sea-coasts of the same countries, the little auk (*Alca alle*), and the black-billed auk (*Alca pica*), repair for similar purposes as the snow-flake. The *woodcock* winters with us, but retires in the spring to Sweden, Norway, and Lapland *.

The fieldfare and the redwing resemble the woodcock in their migrations, depart at the same season, and retire for similar purposes to the same countries †.

These instances may suffice to support the conclusion, that all our winter birds of passage come from northern countries, and that the winter visitants of the south of Europe become the summer visitants of its northern regions. This is evidently an arrangement depending on the same law by which the African winter visitants become the summer birds of passage in Europe.

In support of this conclusion it may be mentioned, that, in their progress southward, the winter visitants appear first in the northern and eastern parts of the island, and gradually proceed to the southward and westward. Thus the snow-bunting arrives in the Orkney islands about the end of August, and often proves destructive to the corn fields. It then passes into the mainland of Scotland, and is seldom seen in the Lothians, even in the high grounds, before November. In like manner, the woodcock, which crosses the German Ocean, is first observed on the eastern side of the island, and then by degrees disperses towards the west and south.

---

* ECKMARK says of this bird, as a Swedish summer bird of passage, " Pullis in sylvis nostris exclusis, mare transmigrans, in Angliam avolat; ut ex Austria in Italiam. Vere autem novo, dum blatire incipit Tetrao tetrix, illinc descedunt, matrimonio junctæ ad nos revertentes." — *Amœnitates Academicæ,* iv. 591.

† Mr BULLOCK, however, has informed me, that he found the redwing breeding in the Island of Harris, one of the Hebrides, in 1818. It probably frequents other islands, but has hitherto been confounded with the thrush.

That these periodical movements take place, in order to guard against the vicissitudes of the seasons, must appear obvious to all, from the consideration of the facts which have been stated. An early winter brings the migrating birds from the north to this country before their usual time, and an early spring hastens the arrival of our summer visitants. In the beginning of winter the snow-bunting is found only in the high grounds, and it descends to a lower level with the increasing severity of the season *.

During the autumn, we thus observe a latitudinal movement of many species of birds towards the equator, in search of the temperature congenial to their constitutions, and which the winter of the district of their summer residence could not afford. The autumnal shifting of the feathered tribes, may therefore, with propriety, be termed the *Equatorial Migration;* all those species in which it is observed, returning from the pole towards the equator, each according to limits peculiar to itself.

The vernal shifting takes place with the increasing temperature of the high latitudes, and may be termed the *Polar Migration,* as all the species in which it is observed recede from the equator and approach towards the pole.

---

* Attempts have been made to preserve these birds during the summer season in this country, but, although liberally supplied with food, they have not survived. The experiment has succeeded, however, in America, with General DAVIES, who informs us, (Linn. Trans. vol. iv. p. 157.) that the snow-bird of that country always expires in a few days, (after being caught, although it feeds perfectly well), if exposed to the heat of a room with a fire or stove; but being nourished with snow, and kept in a cold room or passage, will live to the middle of summer: a temperature much lower than our summer heat proving destructive to these birds. The swallow, on the other hand, seems to delight in the temperature of our summer, and at that heat to be able to perform the higher operations of nature. When attempted to be kept during our winter, besides a regular supply of food, care must be taken to prevent it from being benumbed with cold.

The extent of degrees of latitude traversed in these migrations, differs, as we have seen, according to the species, and even in the same species in different parallels of longitude. Thus, the nightingale, in its polar migrations, does not reach the 55° of north latitude in Britain, while in Sweden it reaches to the 60°. Anomalies of this kind cease to excite our surprise, when it is considered, that the *isothermal lines* (or the latitudinal lines under which the mean annual temperature is the same), are not parallel with the sun's course, or do not observe a regular increase or diminution with the difference of latitude. Even the *isotherial lines*, (under which the mean heat of summer is the same), and the *isotheimal lines*, (under which the mean heat of winter is the same), are neither parallel to one another, nor to the isothermal lines. These differences, which HUMBOLDT has investigated with so much success, exercise a powerful influence on the distribution of plants and animals, and regulate the limits of those periodical migrations we have now been considering.

The preceding remarks relate to the equatorial and north polar migrations. Movements depending on the same circumstances, in all probability, take place on the other side of the equator towards the south pole. The Cape swallow (*Hirundo Capensis*), according to the observations of Captain CARMICHAEL, arrives at the southern extremity of Africa in the month of September, the commencement of the summer of that district, and departs again in March or April, on the approach of winter *. Reasoning from the analogies of the north polar migrations, we may conclude, that this species of swallow resides the remaining part of the year near the equator, and that its south polar migrations extend to the Cape of Good Hope.

---

* Edinburgh Phil. Journ. vol. i. p. 421.

MIGRATION.

It appears from these movements of birds, that, in the cold season, the polar regions are deserted by some species, and that there is an accumulation of life towards the equator. At another season, the equatorial regions are, in some measure, deserted by their temporary inhabitants, and the polar districts become peopled by the change.

Having now ascertained the period and the direction of these migrations, let us next attend to the act of migration itself, and the circumstances attending the flight.

Migrating birds, before they take their departure, in general collect together in flocks. This is very obviously the case with the swallow, and is even still better known with woodcocks, terns, puffins, and shearwaters. Woodcocks arrive in this country in great flocks about the same time; and should adverse winds occur at the period of their departure, they accumulate in such numbers on the eastern shores, as to furnish the fowler with excellent sport. Geese too, dotterels, and many others, during their migratory flights, always keep in company.

But there are many migrating birds which have never been observed to congregate previous to their departure. Thus the cuckoo, seldom seen in company with his mate even during the breeding season, is, to all appearance, equally solitary at the period of migration. These birds are supposed by naturalists to go off in succession.

It is certainly a very curious, and perhaps unexpected occurrence, that the males of many species of migrating birds appear to perform their migrations a few days before the females. This is remarkably the case with the nightingale. The bird-catchers in the neighbourhood of London, procure males only on the first arrival of this bird. The females do not make their appearance for a week or ten days after. Similar observations have been made with respect to the wheat-ear.

Those birds which feed during the night may be expected to perform their migrations during the same interval, it being the season of their activity; while those birds which feed during the day, may be expected to migrate with the help of light. The migrations of the woodcock and quail confirm this conjecture. The woodcocks arrive in this country during the night, and hence they are sometimes found in the morning after their arrival, in a neighbouring ditch, in too weak a state to enable them to proceed. Poachers are aware that they migrate during the night, and sometimes kindle fires on the coast, to which the woodcocks, attracted by the light, bend their course, and in this manner great numbers are annually destroyed. Quails, on the other hand, perform their migrations during the day, so that the sportsman in the islands of the Mediterranean can use his dog and gun.

It has often excited surprise in the minds of some, how migrating birds could support themselves so long on wing, as to accomplish their journeys, and at the same time live without food during their voyage. These circumstances have induced many to deny the existence of migration, and have excited others to form the most extravagant theories on the subject, to account for the preservation of these birds during the winter months. But the difficulties which have been stated, are only in appearance, and vanish altogether if we attend to the rapidity of the flight of birds.

The rapidity with which a hawk and many other birds occasionally fly, is probably not less than at the rate of 150 miles in an hour. Major CARTWRIGHT, on the coast of Labrador, found, by repeated observations, that the flight of an eider duck *(Anas molissima)* was at the rate of 90 miles an hour. Sir GEORGE CAYLEY computes the rate of flight, even of the common crow, at nearly 25 miles an hour; and SPALLANZANI found that of the swallow completed about 92 miles, while he conjectures that the rapidity of the swift is nearly three times greater. A falcon which belonged to HENRY the Fourth of France, escaped from Fountainbleau, and in twenty-four hours afterwards was found at Malta, a distance computed to be no less than 1350 miles; a velocity nearly equal to 57 miles an hour, supposing the falcon to have been on wing the whole time. But as such birds never fly by night, and allowing the day to be at the longest, his flight was perhaps equal to 75 miles an hour. It is probable, however, says MONTAGU, that he neither had so many hours of light in the twenty-four, to perform his journey, nor that he was retaken the moment of his arrival. But if we even restrict the migratory flight of birds to the rate of 50 miles an hour, how easily can they perform their most extensive migrations! And we know, in the case of woodcocks, and perhaps all other migrating birds, that they in general take advantage of a fair wind with which to perform their flights. This breeze perhaps aids them at the rate of 30 or 40 miles an hour; nay, with three times greater rapidity, even in a moderate breeze, if we are to give credit to the statement of aërial navigators, who seem to consider the rate of the motion of winds as in general stated too low.

It has been already observed, that many species do not perform their migrations at once, but reach the end of their journey by short and easy stages. There is little exertion required from such; while those who execute their movements at one flight, (if there be any that do so), may in a very short time, perhaps a day, by the help of a favourable breeze, reach the utmost limits of their journey. Many birds, we know, can subsist a long time without food; but there appears to be no necessity for supposing any such abstinence, since, as CATESBY remarked, every day affords an increase of warmth, and a supply of food. Hence we need not perplex ourselves in accounting for the continuance of their flight, or their sustenance in the course of it. Such journeys would be long indeed for any quadruped, while they are soon performed by the feathered tribes.

It is often stated as a matter of surprise, how these birds know the precise time of the year at which to execute their movements, or the direction in which to migrate. But this is merely expressing a surprise, that a kind and watchful Providence should bestow on the feathered creation powers and instincts suited to their wants, and calculated to supply them. How, we ask, does the curlew, when perched upon a neighbouring muir during the flowing of the tide, know to return at the first of the ebb, to pick up the accidental bounty of the waves? How are the sea-fowl, in hazy weather, guided to the sea-girt isles they inhabit, with food to their young, which they have procured at the distance of many miles? " The inhabitants of St Kilda," says MARTIN, " take their measures from the flight of these fowls, when the heavens are not clear, as from a sure compass; experience shewing, that every tribe of fowls bend their course to their respective quarters, though out of sight of the isle. This appeared clearly in our gradual advances; and their motion being compared, did exactly quadrate with our compass."

In the course of these annual migrations, birds are sometimes overtaken by storms of contrary wind, and carried far from their usual course. In such cases, they stray to unknown countries, or sometimes are found at sea in a very exhausted state, clinging to the rigging of ships. Such accidents, however, seldom happen, as these birds, year after year, arrive in the same country, and even return to the same spot. The summer birds of passage return not, it is true, in such numbers as when they left us; but, amidst all the dangers of their voyage, the race is preserved.

We thus see, that animals possess various resources, to enable them to accommodate themselves to the variations of temperature corresponding with the seasons. But these appear in some species to be inadequate for their protection, and another is provided for their safety.

4. *Torpidity.*—This is one of the most curious subjects in zoology, and has long occupied the attention of the natural historian and the physiologist. All animals we know require stated intervals of repose to recruit exhausted nature, and prepare for farther exertion,—a condition which is termed Sleep. But there are a few animals, which, besides this daily repose, appear to require annually some months of continued inactivity, to enable them to undergo the common fatigues of life during the remaining part of the year. These animals exhibit, therefore, two kinds of sleep,—that which they enjoy daily during the season of their activity, and that which they experience during their brumal retirement. This last kind of sleep is generally denominated *torpidity*, and is also known by the term *hybernation*, as it is evidently designed to afford protection against the cold of winter.

As the phenomena which torpid animals exhibit are somewhat different, according to the classes to which they belong, it will be more convenient for us to treat of the animals of each class separately, beginning with QUADRUPEDS.

The quadrupeds which are known to become torpid, belong exclusively to the unguiculated division. Some species are found among the *feræ*, as the different kinds of bats; the hedgehog and the tanric; while among the *glires* the torpid species are numerous, and their habits have been studied with the greatest attention, as the marmot, the hamster, and the dormouse.

The *food* of these animals is very different, according to the orders or genera to which they belong. The bats support themselves by catching insects, and those chiefly of the lepidopterous kinds; the hedgehog lives on worms and snails; while others, as the marmot and hamster, feed on roots, seeds, and herbs. They are nearly all nocturnal, or crepuscular feeders.

It is usually supposed that torpid animals are confined to the cold regions of the earth. That they abound in such regions must be admitted; but their range of latitude does not appear to be so limited as to prevent their occurrence in warm countries. Thus the *Dipus sagitta*, is equally torpid during the winter months in Egypt as in Siberia. In the former country it is more easily revived by a very slight increase of temperature, its lethargy not being so profound. The tanric (*Centenes caudatus*), which is an inhabitant of India and Madagascar, becomes torpid even in those countries, and continues so during nearly six months of the year.

The precise period of the year in which these animals retire to their winter quarters and become torpid, has not been ascertained with any degree of precision. The jumping mouse of Canada (*Gerbillus Canadensis*) is said to enter its torpid state in September, and to be again restored to activity in the month of May. The torpid animals of this country usually retire in October, and reappear in April. It appears probable, however, that the different species do not all retire at the same time, but, like the migrating birds, perform their movements at separate periods. It is also probable, that the time of retirement of each species varies according to the mildness or severity of the season. In general, however, they retire from active life when their food has become difficult to obtain, when the insects have fled to their hiding places, and the cold has frozen in the ground the roots and the seeds on which they subsist. At the period of their reviviscence, the insects are again sporting in the air, and the powers of vegetable life are exerted in the various processes of germination and vegetation.

Previous to their entrance into this state of lethargy, these animals select a proper place, in general assume a particular position, and even in some cases provide a small stock of food.

All the torpid animals retire to a *place of safety*, where, at a distance from their enemies, and protected as much as possible from the vicissitudes of temperature, they may sleep out, undisturbed, the destined period of their slumbers. The bat retires to the roof of gloomy caves, or to the old chimneys of uninhabited castles. The hedge-hog wraps itself up in those leaves of which it composes its nest, and remains at the bottom of the hedge, or under the covert of the furze, which screened it, during summer, from the scorching sun or the passing storm. The marmot and the hamster retire to their subterranean retreats, and when they feel the first approach of the torpid state, shut the passages to their habitations in such a manner, that it is more easy to dig up the earth any where else, than in such parts which they have thus fortified. The jumping mouse of Canada seems to prepare itself for its winter torpidity in a very curious manner, according to the communications of Major-General Davies, on the authority of a labourer. A specimen which was found in digging the foundation for a summer-house in a gentleman's garden about two miles from Quebec, in the latter end of May 1787, was " enclosed in a ball of clay, about the size of a cricket ball, nearly an inch in thickness, perfectly smooth within, and about twenty inches under ground. The man who first discovered it, not knowing what it was, struck the ball with his spade, by which means it was broken to pieces, or the ball also would have been presented to me *."

Much stress has been laid upon the *position* which these animals assume, previous to their becoming torpid, on the supposition that it contributes materially to produce the lethargy. In describing this position, Mr CARLISLE observes, " that this tribe of quadrupeds have the habit of rolling up their bodies into the form of a ball during ordinary sleep, and they invariably assume the same attitude when in the torpid state: the limbs are all folded into the hollow made by the bending of the body; the clavicles, or first ribs, and the sternum are pressed against the fore part of the neck, so as to interrupt the flow of blood which supplies the head, and to compress the trachea: the abdominal viscera and the hinder limbs are pushed against the diaphragm, so as to interrupt its motions, and to impede the flow of blood, through the large vessels which penetrate it, and the longitudinal extension of the cavity of the thorax is entirely obstructed. Thus a confined circulation is carried on through the heart, probably adapted to the last weak actions of life, and to its gradual recommencement †." But as none of these effects are supposed to be produced by the same position during ordinary sleep, their existence cannot be admitted in the case of torpidity. Professor MANGILI of Pavia ‡, with greater simplicity of language, says, that the marmot rolls itself up like a ball, having the nose applied contrary to the anus, with the teeth and eyes closed. He also in-

---

* Linn. Trans., vol. iv. p. 156. † Phil. Trans. 1795, p. 17.
‡ *Annales du Museum*, tom. ix.

forms us, that the hedgehog, when in a torpid state, in general reposes on the right side. The bat, however, during the period of its slumbers, prefers a very different posture. It suspends itself from the ceiling of the cave to which it retires, by means of its claws, and in this attitude outlives the winter. This is the natural position of the bat when at rest, or asleep. In short, little more can be said of the positions of all these torpid animals, than their correspondence with those which they assume during the periods of their ordinary repose.

It is also observable, that those animals which are of solitary habits during the summer season, as the hedgehog and dormouse, are also solitary during the period of their winter torpidity; while the congregating social animals, as the marmot, the hamster, and the bat, spend the period of their torpidity, as well as the ordinary terms of repose, collected together in families or groups.

It is generally observed, that animals, previous to their torpidity, have their bodies charged with fat. In the marmot and some others, there are two peritoneal processes, which may be considered as lateral omenta, and which, as well as the great omentum, are filled with fat *. In the dormouse, however, and others, these lateral processes do not exist, the fat being more generally distributed. This store of nourishment enables the animals to support that gradual waste which takes place during the period of their slumbers. By some it has even been regarded as the cause of their lethargy. SPALLANZANI, however, found, among the dormice procured for his experiments, a considerable difference among the individuals in regard to fatness, yet all were equally disposed to become torpid on the application of cold.

---

* CUVIER, Leçons d'An. Com. iv. 91.

Many of those animals, particularly such as belong to the great natural family of *gnawers*, make provision in their retreats, during the harvest months. The marmot, it is true, lays up no stock of food; but the hamsters fill their storehouses with all sort of grain, on which they are supposed to feed, until the cold becomes sufficiently intense to induce torpidity. The *Gricetus glis*, or migratory hamster of PALLAS, also lays up a stock of provision. And it is probable that this animal partakes of its store of food, not only previous to torpidity, but also during the short intervals of reviviscence, which it enjoys during the season of lethargy. The same remark is equally applicable to the dormouse.

Having thus made choice of situations where they are protected from sudden alterations of temperature, and having assumed a position similar to that of their ordinary repose, these hybernating animals fall into that state of insensibility to external objects, which we are now to examine more minutely. In this torpid state they suffer a diminution of temperature; their respiration and circulation become languid; their irritability decreases in energy; and they suffer a loss of weight. Let us now attend each of these changes separately.

1. *Diminished temperature.*—When we take in our hand any of these hybernating torpid animals, which we are now considering, they feel cold to the touch, at the same time that they are stiff, so that we are apt to conclude, without farther examination, that they are dead. This reduction of temperature is not the same in all torpid quadrupeds. It varies according to the species. HUNTER informs us [*], on the authority of JENNER, that the temperature of a hedgehog, in the cavity of the abdomen towards the pelvis, was 95°, and at the diaphragm was 97° of Fahrenheit, in

---

[*] " Observations on certain parts of the Animal Economy," p. 99.

TORPIDITY.

summer, when the thermometer in the shade stood at 78°. Professor MANGILI states the ordinary heat of the hedgehog a little lower, at 27° of Reaumur, or about 93° of Fahrenheit. In winter, according to JENNER, the temperature of the air being 44°, and the animal torpid, the heat in the pelvis was 45°, and at the diaphragm 48½°. When the temperature of the atmosphere was at 26°, the heat of the animal in the cavity of the abdomen, where an incision was made, was reduced so low as 30°. The same animal, when exposed to the cold atmosphere of 26° for two days, had its heat at the rectum elevated to 93°, the wound in the abdomen being so much diminished in size as not to admit the thermometer. At this time, however, it was lively and active, and the bed in which it lay felt warm. As this animal allowed its heat to descend to 30°, when in its natural state of torpidity, and when there was no necessity for action, the increased temperature may in part be ascribed to the wound, which called forth the powers of the animal to repair an injury, which reparation could not be effected at a temperature below the standard heat of the animal. The sources of error in making experiments where the living principle is concerned, are so numerous, that attention ought to be bestowed on every circumstance likely to influence the result.

The zizil (*Arctomys citillus*), according to PALLAS, usually possesses a summer temperature of 103° Fahr. but during winter, and when torpid, the mercury rises only to 80° or 84°. The temperature of the dormouse during summer, and in its active and healthy state, is 101°. When rolled up and torpid, during winter, the thermometer indicates 43°, 39°, and even 35°, on the external parts of the body. When introduced into the stomach, the temperature was found to be 67°, and sometimes 73°. MANGILI found this animal torpid even when the temperature of the air was 66°. Hence he considers it as the most lethargic of animals.

The marmot *(Arctomys marmota)* possesses a summer temperature of 101° or 102°, which is gradually reduced in the torpid season to 48°, and even lower.

Bats have a temperature in summer nearly equal to that of marmots. They are soon affected by the changes of the atmosphere, and they cease to respire in a medium of 43°. In the month of July, the thermometer standing at 80°, the internal temperature of a bat was 101°, which is just the degree of heat in a group of them collected together in summer, and may therefore be considered as the natural standard. Mr CORNISH applied a thermometer to a torpid bat, and found that it indicated 36°. When awakened so much that it could fly a little, he again applied the thermometer, and it indicated 38°. SPALLANZANI found a bat, after being exposed during an hour to a temperature of 43°, to indicate 47°, the bulb of the thermometer being placed in the chest; when exposed to a temperature below the freezing point, the heat of the animal became the same as the surrounding medium, yet it always remained internally higher than the low temperature produced artificially, though the skin did not indicate any difference.

The wood-mouse *(Mus sylvaticus)* became torpid, according to SPALLANZANI, when the thermometer in its cage stood at 43°. The temperature of the belly externally was 45°, but its internal temperature was not much diminished, even by a degree of cold sufficient to induce torpidity.

In these experiments we observe, that the temperature of hybernating quadrupeds is greatly reduced below the summer standard, or the ordinary temperature of the animal in health and activity. Still, however, they continue to maintain a superiority in point of temperature above the surrounding medium, in whatever circumstances they are placed. Even in this torpid state, the energies of life, though feeble, are still sufficient to the production of a certain quantity of heat.

2. *Diminished Respiration.*—In this, as in all the other departments of this curious subject, accurate and varied experiments are still wanting. The following are the principal facts which we have been able to collect.

The hedgehog, according to Professor MANGILI, who has bestowed more attention on this part of the subject than any of his predecessors, respires only from five to seven times in a minute during ordinary repose. When it becomes torpid, the process of respiration is periodically suspended and renewed. Thus a hedgehog, procured after it had revived naturally from its winter lethargy in April, was placed in a chamber the temperature of which was about 54°. It refused vegetable food, and became torpid, and continued in that state to the 10th of May. At first, after every fifteen minutes of absolute repose, it gave from thirty to thirty-five consecutive signs of languid respiration. In the beginning of May, when the thermometer was about 62°, it gave from seven to ten consecutive respirations, after an interval of ten minutes of absolute repose. Upon lowering the temperature, the intervals of repose became greater, while the number of respirations increased to eighteen or twenty.

Marmots, according to the same author, when in health and activity, perform about five hundred respirations in an hour, but when in a torpid state, the number is reduced to fourteen, and these at intervals of four minutes, or four minutes and a half, of absolute repose.

Bats, when kept in a chamber from 45° to 50°, were observed at the end of every two, three, or four minutes of absolute repose, to give four signs of respiration. SPAL-LANZANI, not aware of these periodical intervals of repose, could not discover any signs of respiration. Indeed, when their temperature is reduced to about 47°, this function does not appear to be exercised.

The dormouse, when in a torpid state on the 27th December, exhibited a languid respiration of one hundred and forty times in forty-two minutes. On the tenth of January, the thermometer being at 43°, it respired at intervals in the following manner, according to MANGILI.

Intervals of repose.	Number of consecutive respirations.
5 minutes	16
4	30
3	29
2	29
12	5
9	10
10	6
13	18
12	23
12	8

In some instances, the intervals of repose or suspended respiration lasted sixteen minutes.

MANGILI also found the fat dormouse (*Myoxus glis*), when in a torpid state on the 27th December, and when the thermometer indicated 40°, to respire at intervals. After every four minutes of repose, it respired from twenty-two to twenty-four times every minute and a half. The thermometer being raised one degree of Reaumur, the intervals became only three minutes. The temperature being reduced to 37°, the intervals of repose became four minutes, and the consecutive respirations twenty to twenty-six. The cold increasing, it awoke and ate a little, and then relapsed into torpidity. On the 10th of February the intervals of repose were eighteen or twenty minutes, and then thirteen to fifteen respirations. On the 21st February, the thermometer being 48°, the intervals of repose were from twenty-eight to thirty, and the consecutive respirations from five to seven.

From the observations already made on this important subject, it appears, that respiration is not only diminished, but even in some cases totally suspended. During the severe winter of 1795, SPALLANZANI exposed dormice to a temperature below the freezing point, and enclosed them in vessels filled with carbonic acid and azotic gas over mercury, three hours and a half, without injuring them, and the sides of the vessels were not marked by any vapour. Hence we may conclude that they did not breathe, nor consume any oxygen gas.

MANGILI placed a marmot under a bell-glass, immersed in lime water, at 9 o'clock in the evening. At nine next morning the water had only risen in the glass three lines. Part of the oxygen was abstracted, and a portion of carbonic acid was formed, as a thin pellicle appeared on the surface of the lime-water, which effervesced with nitric acid. SPALLANZANI placed torpid marmots in vessels filled with carbonic acid and hydrogen, and confined them there for four hours, without doing them the least injury, the temperature of the atmosphere being several degrees below the freezing point. But he found, that if these animals were awakened by any means, or if the temperature was not low enough to produce complete torpor, they very soon perished in the same noxious gases. A bird and rat, introduced into a reservoir containing carbonic acid gas, did not live a minute; whereas a torpid marmot remained in it an hour, without betraying the least desire to move, and recovered perfectly on being placed in a warmer medium.

In the exhausted receiver of an air-pump, a torpid bat lived seven minutes, in which another bat died at the end of three minutes. Torpid bats, when confined in a vessel containing atmospheric air, consumed six hundredths of the oxygen, and produced five hundred parts of carbonic acid. Viewing this in connection with his other experiments, this philosopher concluded, that the consumption of the oxygen, and the evolution of the carbonic acid, proceeded from the skin.

The respiration of torpid quadrupeds is thus greatly diminished, and even in some cases suspended; and in general, instead of being performed with regularity as in ordinary sleep, the respirations take place at intervals, more or less remote, according to the condition of the lethargy.

3. *Diminished Circulation.*—From the experiments already detailed, with regard to the reduction of the temperature and the respiration of torpid quadrupeds, we are prepared to expect a corresponding diminution of action in the heart and arteries.

In the hamster the circulation of the blood during its torpid state is so low, according to BUFFON, that the pulsations of the heart do not exceed fifteen in a minute. In its active and healthy state, they amount to 150 in the same space.

It is stated by BARRINGTON in his Miscellanies, that Mr CORNISH applied a thermometer to the body of a torpid bat, and found that it indicated 36°. At this temperature the heart gave sixty pulsations in a minute. When awakened so much as to be able to fly a little, he again applied the thermometer, which now indicated 38°, and the heart beat one hundred times in a minute. As the torpor becomes profound, the action of the heart is so feeble, that only fourteen beats have been distinctly counted, and those at unequal intervals.

Dormice, when awake and jumping about, breathe so rapidly, that it is almost impossible to count their pulse; but as soon as they begin to grow torpid, eighty-eight pulsations may be counted in a minute, thirty-one when they are half torpid, and only twenty, nineteen, and even sixteen, when their torpor is not so great as to render the action of the heart imperceptible.

SPALLANZANI and others are of opinion, that the circulation of the blood is entirely stopped in the remote branches of the arteries and veins, and only proceeds in the trunks of the larger vessels, and near the heart. But it is probable, that however languid the circulation may be, it is still carried on, as the blood continues fluid. He found, that if the blood of marmots be subjected, out of the body, to a temperature even higher than that to which it is exposed in the lungs of these animals, it is instantly frozen; but it is never congealed in their dormant state.

4. *Diminished Irritability.*—The irritability of torpid animals, or their susceptibility of being excited to action, is extremely feeble, and in many cases is nearly suspended. Destined to remain for a stated period in this lethargic state, a continuance of their power of irritability would be accompanied with the most pernicious consequences; as thereby they would be often raised prematurely into action, under a temperature which they could not support, and at a time when a seasonable supply of food could not be obtained. In their torpid state, therefore, they are not readily acted upon by those stimuli, which easily excite them to action, during the period of their activity. Parts of their limbs may be cut off, without the animal shewing any signs of feeling. Little action is excited even when their vital parts are laid open. When the hamster is dissected in this torpid state, the in-

testines discover not the smallest sign of irritability upon the application of alcohol or sulphuric acid. During the operation, the animal sometimes opens its mouth, as if it wanted to respire, but the lethargy is too powerful to admit of its reviviscence.

Marmots are not roused from their torpid state by an electric spark, strong enough to give a smart sensation to the hand, and a shock from a Leyden phial only excites them for a short time. They are insensible to pricking their feet and nose, and remain motionless and apparently dead. Bats are also equally insensible to the application of stimuli.

The most curious experiments on this subject are those of MANGILI. Having killed a marmot in a torpid state, he found the stomach empty and collapsed, the intestines likewise empty, but there was a little fæcal matter in the cœcum and rectum. The blood flowed quickly from the heart, and in two hours yielded a great quantity of serum. The veins in the brain were very full of blood. The heart continued to beat during three hours after. The head and neck having been separated from the trunk, and placed in spirits of wine, gave signs of motion even after half an hour had elapsed. Some portions of the voluntary muscles gave symptoms of irritability with galvanism four hours after death. In a marmot killed in full health, the heart had ceased to beat at the end of fifty minutes. The flesh lost all signs of irritability in two hours; the intercostal and abdominal muscles retaining it longer than those of any other part of the body.

5. *Diminished Action of the Digestive Organs.*—The digestive functions in torpid animals are exceedingly feeble, and in general cease altogether. The situation, and still more the lethargic state of the system, render this process unnecessary. The intestines are in general empty, and in

a collapsed state, and the secretions so small, that a supply of nourishment from the stomach is not requisite. Mr JENNER found a hedgehog, when the heat of the stomach was at 30°, to have no desire for food, nor power of digesting it. But when the temperature was increased to 93°, by inflammation in the abdomen, the animal seized a toad which was in the room, and, upon being offered some bread and milk, immediately began to eat. The heat excited the action of the various functions of the animal, and the parts unable to carry on these actions, without nourishment, urged the stomach to digest.

While many torpid quadrupeds retire to holes in the earth unprovided with food, and in all probability need no sustenance during their lethargic state, there are others, as we have already mentioned, which provide a small stock of provisions. These, we are inclined to believe, eat a little during those temporary fits of reviviscence to which they are subject. This is in part confirmed by the experiments of MANGILI, both on the common and fat dormouse. Whenever these awoke from their torpid slumbers, they always ate a little. Indeed, he is of opinion, that fasting long, produces a reviviscence, and that, upon the cravings of appetite being satisfied, they again become torpid.

6. *Diminished Weight.*—All the experiments hitherto made on this subject, indicate a loss of weight sustained by these animals, from the time they enter their torpid state, until the period of their reviviscence. MANGILI procured two marmots from the Alps, on the 1st of December 1813. The largest weighed 25 Milanese ounces; the smallest only $22\frac{5}{24}$ths ounces. On the 3d of January, the largest had lost $1\frac{8}{24}$ths of an ounce, and the smallest $1\frac{7}{24}$ths and a half. On the fifth of February, the largest was now only $22\frac{21}{24}$th ounces; the smallest 21 ounces. He adds, that they lose

weight in proportion to the number of times in which they revive during the term of their lethargy.

Dr Monro kept a hedgehog from the month of November (1764) to the month of March (1765), which lost in the interval a considerable portion of its weight. On the 25th of December, it weighed 13 ounces and 3 drams, on the 6th of February, 11 ounces and 7 drams, and on the 8th of March, 11 ounces and 3 drams. He observed a small quantity of feculent matter and urine among the hay, although it neither ate nor drank during that period. In this experiment there was a daily loss of 13 grains. According to Mr Cornish, both bats and dormice lose from five to seven grains in weight during a fortnight's hybernation.

Dr Reeves endeavoured to account for the lean state of the marmot when found in the spring, as occasioned by another cause than the slow but uniform exertions of the vital principle. " I have (he says) been repeatedly assured, by men who hunt for these animals in winter, that they are always found fat in their holes on the mountains of Switzerland, and it is only when they come out of their hiding places before provisions are ready for them, or if a sharp frost should occur after some warm weather, that they become emaciated and weak *." This testimony may be received as explaining the emaciated appearance of some marmots, but does not in the smallest degree invalidate the general conclusion, that all torpid animals sustain a loss of weight during the continuance of their lethargy.

From the experiments which we have already quoted, it must appear obvious, that respiration is in general carried

---

* " Essay on the Torpidity of Animals," by Henry Reeve, M. D. 1809, p. 28,

on, although sometimes in a very feeble manner. Carbon, consequently, must be evolved. Accordingly, we find carbonic acid produced in those vessels in which these torpid animals have been confined; and hence must conclude, that a loss of weight has taken place.

Such being the preparatory and accompanying phenomena of this torpid state, let us now endeavour to discover the cause of these singular appearances.

In a subject of this kind, so intimately connected with the pursuits of the naturalist and the physiologist, it was to be expected that numerous hypotheses would be proposed, to explain such interesting phenomena. Unfortunately, indeed, many hypotheses have been proposed, while few, from a connected view of the subject, have ventured to theorise. Perhaps we are not prepared to draw a sufficient number of general conclusions, from the scanty facts which we possess, in order to build any theory. But the following observations may be considered as embracing the principal opinions which have been formed on the subject, and announcing the more obvious causes in operation.

In an investigation of this sort, it was natural to attempt to trace this singularity of habit in torpid animals to some peculiar conformation in the structure of the organs. Accordingly, we find many anatomists assigning a peculiarity of organization, as a reason why these animals become torpid, or, at least, pointing out a structure in torpid animals different from that which is observable in animals that are not subject to this brumal lethargy.

PALLAS observed the thymous gland unusually large in torpid quadrupeds, and also perceived two glandular bodies under the throat and upper part of the thorax, which appear particularly florid and vascular during their torpidity.

MANGILI is of opinion, that the veins are larger in size, in proportion to the arteries in those animals which become torpid, than in others. He supposes, that, in consequence of this arrangement, there is only as much blood transmitted to the brain during summer as is necessary to excite that organ to action. In winter, when the circulation is slow, the small quantity of blood transmitted to the brain is inadequate to produce the effect. This circumstance, acting along with a reduced temperature and an empty stomach, he considers as the cause of torpidity. By analogy, he infers, that the same cause operates in producing torpidity with all the other hybernating animals of the other classes.

Mr CARLISLE, in his Croonian Lecture on Muscular Motion, asserts, that " animals of the class Mammalia, which hybernate and become torpid in the winter, have at all times a power of subsisting under a confined respiration, which would destroy other animals not having this peculiar habit. In all the hybernating mammalia there is a peculiar structure of the heart and its principal veins: the superior cava divides into two trunks, the left passing over the left auricle of the heart, opens into the inferior part of the right auricle near to the entrance of the vena cava inferior. The veins usually called Azygos accumulate into two trunks, which open each into the branch of the vena cava superior, on its own side of the thorax. The intercostal arteries and veins in these animals are unusually large *."

We cannot refrain from observing, that these general views do not appear to be the result of a patient investigation of a number of different kinds of torpid animals, but a premature attempt to theorise from a few insulated particulars. Passing, therefore, from these attempts of the anatomist to illustrate the phenomena in question, let us

---

* Phil. Trans. 1805, p. 17.

TORPIDITY.

attend to those other causes which are concerned in the production of torpidity.

From the consideration, that this state of torpidity commences with the cold of winter, and terminates with the heat of spring, naturalists, in general, have been disposed to consider a *reduced temperature* as one of the principal causes of this lethargy. Nor are circumstances wanting to give ample support to the conclusion.

When the temperature of the atmosphere is reduced, as we have already seen, below 50°, and towards the freezing point, these animals occupy their torpid position, and by degrees relapse into their winter slumbers. When in this situation, an increase of temperature, (the action of the sun, or a fire), rouses them to their former activity. This experiment may be repeated several times, and with the same result, and demonstrates the great share which a diminished temperature has in the production of torpidity. If marmots are frequently disturbed in this manner during their lethargy, they die violently agitated, and a hæmorrhage takes place from the mouth and nostrils.

The circumstance of torpid animals being chiefly found in the colder regions, is another proof that a diminished temperature promotes torpidity. And, in confirmation of this, Dr BARTON states, that, in the United States of America, many species of animals which become torpid in Pennsylvania, and other more northern parts of the country, do not become torpid in the Carolinas, and other southern parts of the continent.

But while a certain degree of cold is productive of this lethargy, a greater reduction of temperature produces reviviscence, as speedily as an increase of heat. MANGILI placed a torpid marmot, which had been kept in a temperature of 45°, in a jar surrounded with ice and muriate of lime, so that the thermometer sunk to 16°. In about half an hour a quickened respiration indicated returning animation. In sixteen hours it was completely revived. It was trembling with cold, and made many efforts to escape. He also placed a torpid bat under a bell-glass, where the temperature was 29°, and where it had free air. Respiration soon became painful, and it attempted to escape. It then folded its wings, and its head shook with convulsive tremblings. In an hour no other motions were perceptible than those of respiration, which increased in strength and frequency until the fifth hour. From this period, the signs of respiration became less distinct; and, by the sixth hour, the animal was found dead. He also exposed a torpid dormouse (from a temperature of 41°) to a cold of 27°, produced by a freezing mixture. Respiration increased from ten to thirty-two times in a minute, and without any intervals of repose. There were no symptoms of uneasiness, and the respirations seemed like those in natural sleep. As the temperature rose, respiration became slower. He then placed it in the sun, when it awoke. Two hours afterwards, having exposed it to the wind, respiration became frequent and painful; it turned its back to the current, without, however, becoming torpid.

That cold is calculated to produce effects similar to torpidity on man himself, is generally known. Those who have ascended to the summits of high mountains, have, by the exposure to cold, felt an almost irresistible propensity to lie down and sleep. Dr Solander, while exploring Terra del Fuego, though perfectly aware of the inevitable destruction attending the giving way to this inclination, nay, though he had even cautioned his companions against indulging it, could not himself overcome the desire. When this feeling is gratified, sleep succeeds, the body becomes benumbed, and death speedily arrives. How long this sleep might continue without ending in death, were the body defended from the increasing cold and the action of the air, will probably never be determined by satisfactory experiments. Partial torpor has often been experienced in the hands and feet, which is easily removed by a gradual increase of temperature. We may add, that, in the case of persons exposed to great cold in elevated situations on mountains or in balloons, there are other causes in operation which may have a tendency to produce sleep. The previous exertions have reduced the body to a very exhausted state—the pressure of the atmosphere is greatly diminished, and the air inhaled by the lungs is rarified.

When the animals now under consideration are regularly supplied with food, and kept in a uniform temperature, it has been observed that they do not fall into their wonted lethargy, but continue lively and active during the winter season. This experiment has often been repeated with the marmot and other animals. But when in this state they are peculiarly sensible to cold. Dr REEVES, in some experiments which he performed, says, " When I was in Switzerland, I procured two young marmots in September 1805, and kept them with the view of determining the question, whether their torpidity could be prevented by an abundant supply of food and moderate heat. I carried them with me to Vienna, and kept them the whole of the winter 1805-6. The months of October and November were very mild. My marmots ate every day turnips, cabbages, and brown bread, and were very active and lively: they were kept in a box filled with hay in a cellar, and afterwards in a room without a fire, and did not shew any symptoms of growing torpid. December the 18th, the weather was cold, and the wind very sharp; FAHRENHEIT's thermometer stood at 18° and 20°. Two hedgehogs died, which were kept in the same room with the marmots; and a hamster died also in a room where a fire was constantly kept, though these animals had plenty of hay and food. The marmots became more torpid than I ever saw them before; yet they continued to come out of their nest, and endeavoured to escape: the food given them in the evening was always consumed by the next morning. In January the weather was unusually mild and warm; my marmots ate voraciously, and were jumping about in the morning; but at four o'clock in the afternoon I examined them several times, and found them not completely rolled up, half torpid, and quite cold to the touch. They continued in this state of semi-torpor for several weeks longer, never becoming so torpid as to live many days without eating, and never so active as to resist the benumbing effects of the cold weather *." SPALLANZANI performed similar experiments with the same result on the dormouse. He found, that although cold to the touch during the day, and completely torpid, that it awoke at night and ate a little, and fell asleep again in the morning. He shewed also that dormice kept in a situation more resembling their wild state, became torpid in the month of November, and remained till the middle of March, without eating the food which was placed near them.

With some animals, at least, a *confined atmosphere* appears to be indispensably necessary to the immediate production of torpidity. This is very strikingly illustrated in the case of the hamster. This animal does not become torpid, though exposed to a cold sufficient to freeze water, unless excluded from the action of the air. Even when shut up in a cage filled with earth and straw, and exposed to cold, he still continues awake; but when the cage is sunk four or five feet under ground, and free access to the ex-

---

* Essay, p. 99.

ternal air prevented, in eight or ten days he becomes as torpid as if he had been in his own burrow. If the cage be brought above ground, in two or three hours he recovers, and will resume his torpid state when again sunk under ground. This experiment may be repeated several times, at proper intervals, either in the day time or during the night, the light having no apparent influence. A confined atmosphere, such as the hamster requires, does not appear necessary to the torpidity of the hedgehog, the dormouse, or the bat. But exposure to the open air seems to be equally hostile to the lethargic state in many animals. MANGILI always found that marmots awoke when taken from their nest, and exposed to the free action of the air. A current of air he always found to have the effect of producing reviviscence, both with dormice and bats. From these circumstances, we perceive the utility of the precautions of those animals in retiring to places where the air is still, and where they may enjoy a confined atmosphere.

Torpidity appears also in some cases to depend on the state of the *constitution*. Thus, in the same chambers, one marmot shall continue awake and active, while the others are in a profound lethargy. A hedgehog, during the winter season, becomes torpid upon the application of cold; but, during the summer season, or after the period of reviviscence, it resists the sedative effects of that agent. MANGILI took a hedgehog, on the 21st of June, and placed it in a temperature of 8° of REAUMUR. It first rolled itself up, afterwards lifted its head and tried to escape. Its respiration became frequent and painful. At the end of the first hour respiration had become feeble; at the end of an hour and a half, it had ceased to respire; and twenty minutes after, it was frozen to the heart. When examined in this condition, the flesh was found white, the veins of the neck were much swollen, and a small quantity of extravasated blood was observed in the brain and the lungs. It appears probable, that, during torpidity, the constitution experiences a change something similar to ordinary sleep, by which its exhausted energies are recruited, and it becomes better able to resist the effects of those ordinary agents with which it has to contend.

There are some circumstances in the history of these animals, which seem to indicate that they possess the power of becoming torpid at pleasure, even in the absence of those disposing circumstances which we have enumerated. SPALLANZANI has seen bats in a torpid state even during summer, and supposes, that as these animals appear to possess some voluntary power over respiration, this torpidity may be some instinctive propensity to preserve life. MANGILI, in spring, when the *Cricetus glis* was awake, and when the temperature of the air was between 66° and 68°, placed it in a vase along with nuts and other food. The animal attempted to escape, and refused to eat. It then became torpid. In this state the number of its respirations diminished. Instead of rolling itself up as usual, before becoming torpid, it lay all the while upon its back, and remained in that state until the 17th of July.

Before concluding our account of torpid quadrupeds, it may be proper to add a few observations on their *reviviscence*. When the hamster passes from his torpid state, he exhibits several curious appearances. He first loses the rigidity of his members, and then makes profound respirations, but at long intervals. His legs begin to move; he opens his mouth, and utters rattling and disagreeable sounds. After continuing this operation for some time, he opens his eyes, and endeavours to raise himself on his legs. All these movements are still unsteady and reeling, like those of a man in a state of intoxication; but he repeats his efforts till he acquires the use of his limbs. He remains fixed in that attitude for some time, as if to reconnoitre and rest himself after his fatigues. His passage from a torpid to an active state is more or less quick according to the temperature. It is probable that this change is produced imperceptibly, when the animal remains in his hole, and that he feels none of those inconveniences which attend a forced and sudden reviviscence.

It is evident, from the situations which some torpid animals occupy, that they must experience, in the course of their lethargy, considerable changes of temperature. It would form a very curious subject of inquiry, to ascertain the superior and inferior limits of this torpid state, with respect to temperature. The *Cricetus glis* has been observed dormant from 34° to 48°; the dormouse from 27° to 66°; the marmot from 40° to 51°; and the hedgehog from 26° to 56°.

It is certainly very difficult to account for the torpidity of those animals, which, like the marmot and hamster, congregate and burrow in the earth. Previous to their becoming torpid, a considerable degree of heat must be generated, from their numbers, in their hole; and besides, they are lodged so deep in the earth, as to be beyond the reach of the changes of the temperature of the atmosphere. Their burrow, during the winter season, must preserve a degree of heat approaching to the mean annual temperature of the climate. If this is the case, how is reviviscence produced in the spring? It cannot be owing to any considerable change of temperature, for their situation prevents them from experiencing such vicissitudes. Is it not owing to a change which takes place in their constitution? and, is not awakening from torpidity, similar to awakening from sleep?

A similar remark may be made with regard to bats in their winter quarters. The caves to which they resort, approach at all times the mean annual temperature. A few individuals, not sufficiently cautious in choosing proper retreats, are sometimes prematurely called into action, at a season when there is no food, so that they fall a prey to owls, and the cold of the evening. But what indications of returning spring can be experienced by those which are attached to the roofs of the deeper caves? Surely no increase of temperature?

There is another very curious circumstance attending the reviviscence of quadrupeds from their torpid state, which deserves to be mentioned. As soon as they have recovered from their slumbers, they prepare for the great business of propagation. This is a proof, that torpidity, instead of exhausting the energies of nature, increases their vigour. It also indicates a peculiarity of constitution, to the preservation and health of which, a brumal lethargy is indispensably requisite.

It appears to be the general practice of modern naturalists, to treat with ridicule those accounts which have been left us, of birds having been found in a torpid state during winter. These accounts, it is true, have in many instances been accompanied with the most absurd stories, and have compelled us to pity the credulity of our ancestors, and withhold our assent to the truth of many of their statements. But are there no authenticated instances of torpidity among birds?

In treating of the torpidity of quadrupeds, we were unable to detect the cause of torpidity, as existing in any circumstances connected with structure, or with circulation, respiration, or animal temperature; nor in the places which they frequent, nor the food by which they are supported. Hence we cannot expect much help from a knowledge of the anatomy, physiology, or even habits of birds, in the resolution of the present question. It has indeed been said, that as birds can readily transport themselves from one country to another, and in this manner shun the extremes of temperature, and reach a supply of food, the power of becoming torpid would be useless, if bestowed on them, although highly beneficial to quadrupeds, that are impatient of cold, and cannot migrate to places where there is a supply of food. This mode of reasoning, however, is faulty, since we employ our pretended knowledge of final causes, to ascertain the limits of the operations of nature. Besides, there are many animals, as we have seen in the class *Mammalia*, which become torpid, and a similar state is well known to prevail among the reptiles. As birds, in the scale of being, hold a middle rank between these two classes, being superior to the reptiles, and inferior to the mammalia, we have some reason to expect instances of torpidity to occur among the feathered tribes.

These remarks have for their object, to prepare the mind for discussing the merits of the question, by the removal of presumptions and prejudices, as we fear preconceived opinions have already exercised too much influence.

In treating of the migrations of the swallow, we endeavoured to point out their winter residence, and even traced them into Africa. We are not however prepared to assert, that in every season all these birds leave this country. If they remain, in what condition are they found?

Many naturalists, such as Klein, Linnæus, and others, have believed in the submersion of swallows during winter in lakes and rivers. They have supposed, that they descend to the bottom, and continue there until the following spring. All the statements which we have yet had an opportunity of consulting, appear either as boyish recollections or hearsay testimony; and although naturalists have been eagerly waiting for an opportunity of obtaining swallows from their subaqueous retreats, for the last fifty years, they have hitherto been disappointed. Indeed, the evidence in support of such voluntary submersion, would require to be of the most unexceptionable kind, as there are the strongest probabilities against the occurrence. Swallows are much lighter than water, and could not sink in clusters, as they are represented to do. If their feathers are previously wetted, to destroy their buoyant power, in what manner can they resist the decomposing effect of six months' maceration in water, and appear in spring as fresh and glossy as those of other birds? Swallows do not moult while they remain with us in an active state, so that if they submerge, they either do not moult at all, or perform the process under water. In the case of the other torpid animals, some vital actions are performed, and a portion of oxigen is consumed; but in the submersed swallows, respiration cannot be performed, and consequently circulation must cease. How, then, is the system able for so many months to resist the destructive influence of its situation, in reference to the abstraction of heat? We may add, that the other torpid animals, in retiring to their winter slumbers, consult *safety*, while the swallow, in sinking under the water, rushes to the place where the otter and the pike commit their depredations, against which it cannot contend, and from which it has no means of escape.

These considerations, joined to the circumstance that migration is in ordinary cases practised by the swallow, lead us to doubt the truth of the reports which have been circulated on the subject. Amidst the endless resources of Nature, it may happen that, from some particular circumstance, migration, with a few, may be impracticable, and these may have been enabled to become torpid. It may likewise have happened, that these birds, taking shelter in holes in the margin of banks, have fallen into the water, in consequence of their retreat giving way, and that they may have been taken, previous to drowning. The general belief throughout Scotland, that they are occasionally found in a torpid state in their nests, could scarcely have originated in any other circumstance than the occurrence of the fact.

But besides the example of the torpidity of the swallow, Bewick relates an instance of the same condition being observed in the cuckoo. " A few years ago, a young cuckoo was found in the thickest part of a close whin-bush. When taken up, it presently discovered signs of life, but was quite destitute of feathers. Being kept warm, and carefully fed, it grew and recovered its coat of feathers. In the spring following it made its escape, and in flying across the river Tyne it gave its usual call."

There is an equally well authenticated example of torpidity in birds recorded by Mr NEILL, as having been observed in the case of the Land-rail, or Corncrake, as it is called in Scotland. " I made," says he, " frequent inquiry, whether corncrakes have been seen to migrate from Orkney, but could not learn that such a circumstance had been observed. It is the opinion of the inhabitants, indeed, that they are not able to undertake a flight across the sea. Mr Yorston, a farmer at Aikerness, further related a curious fact, rather leading to the conclusion that they do not migrate. In the course of demolishing a *hill-dike*, (*i. e.* a mud-wall), at Aikerness, about midwinter, a *corncrake* was found in the midst of the wall. It was apparently lifeness; but, being fresh to the feel and smell, Mr Yorston thought of placing it in a warm situation, to see if it would revive. In a short time it began to move, and in a few hours it was able to walk about. It lived for two days in the kitchen, but would not eat any kind of food. It then died, and became putrid. I do not assert that this solitary instance ought to be regarded in any other light than as an exception to the general rule of migration, till further observation have determined the point *."

---

* " Tour through Orkney and Shetland," p. 204.

These are the only instances with which we are acquainted of actual torpidity having occurred among the feathered tribes. They seem calculated to remove all doubt as to the fact, while they point out to us the numerous resources of nature in extreme cases to preserve existence. Thus, when birds, from disease or weakness, or youth, are incapable of performing the ordinary migrations of their tribes, they become dormant during the winter months, until the heat of spring restores to them a supply of food and an agreeable temperature. But it is only in extreme cases, which happen seldom. The occurrence of torpid swallows is very rare, and the examples quoted of the cuckoo and land-rail are solitary instances, although the last mentioned bird is as abundant in Orkney, as partridges are in the Lothians.

Hitherto we have been considering the torpidity which warm-blooded animals experience. Several cold-blooded animals observe a similar mode of hybernation.

The period of the year at which Reptiles prepare for this state of lethargy, varies in the different species. In general, when the temperature of the air sinks below the 50th degree of Fahrenheit, these animals begin their winter slumbers. They adopt similar precautions as the mammalia, in selecting proper places of retreat, to protect them from their enemies, and to preserve them from sudden alternations of temperature. Those which inhabit the waters sink into the soft mud, while those which live on the land enter the holes and crevices of rocks, or other places where the heat is but little affected by changes in the temperature of the atmosphere. Thus provided, they obey the impulse, and become torpid.

As the temperature of these animals depends, in a considerable degree, on the surrounding medium, they do not exhibit any remarkable variations. When the air is under

50°, these animals become torpid, and suffer their temperature to sink as low as the freezing point. When reduced below this, either by natural or artificial means, the vital principle is in danger of being extinguished. In this torpid state, they respire very slowly, as the circulation of the blood can be carried on independent of the action of the lungs. Even in a tortoise kept awake during the winter by a genial temperature, the frequency of respiration was observed to be diminished.

The circulation of the blood is diminished, in proportion to the degree of cold to which these torpid reptiles are exposed. Spallanzani counted from eleven to twelve pulsations in a minute in the heart of a snake, at the temperature of 48°, whose pulse in general, in warm weather, gives about thirty beats in the same period. Dr Reeves made some very interesting experiments on the circulation of the toad and frog. " I observed," he says, " that the number of pulsations in toads and frogs was thirty in a minute, whilst they were left to themselves in the atmosphere, of which the temperature was 53°; when placed in a medium, cooled to 40°, the number of pulsations was reduced to twelve, within the same period of time; and when exposed to a freezing mixture at 26°, the action of the heart ceased altogether *."

The powers of digestion are equally feeble during torpidity as those of respiration or circulation. Mr John Hunter conveyed pieces of worms and meat down the throats of lizards, when they were going to their winter quarters, and, keeping them afterwards in a cool place, on opening them at different periods, he always found the substances, he had introduced, entire, and without any alteration; sometimes they were in the stomach, at other times they had passed

---

* Essay on Torpidity, p. 19.

into the intestines, and some of the lizards that were allowed to live, voided them, toward the spring, entire and with very little alteration in their structure †.

The immediate cause of torpidity in reptiles has been ascertained with more precision, than in the animals belonging to the higher classes with warm blood. This condition with them, does not depend on the state of the heart, the lungs, or the brain; for these different organs have been removed by SPALLANZANI, and still the animals became torpid, and recovered according to circumstances. Even after the blood had been withdrawn from frogs and salamanders, they exhibited the same symptoms of torpidity as if the body had been entire, and all the organs capable of action.

Cold, with these animals, is evidently the chief cause of their torpidity, acting on a frame extremely sensible to its impressions. During the continuance of a high temperature, they remain active and lively; but when the temperature is reduced towards 40°, they become torpid, and in this condition, if placed in a situation where the temperature continues low, will remain torpid for an unknown period of time. SPALLANZANI kept frogs, salamanders, and snakes, in a torpid state in an ice-house, where they remained three years and a half, and readily revived when again exposed to the influence of a warm atmosphere. These experiments give countenance to those reports in daily circulation, of toads being found inclosed in stones. These animals may have entered a deep crevice of the rock, and during their torpidity been covered with sand, which has afterwards concreted around them. Thus removed from the influence of the heat of spring or summer, and in a place where the temperature continued below the point at which they revive, it is impossible to fix limits to

---

† " Observations." p. 155.

the period during which they may remain in this dormant state.

Since reptiles are easily acted upon by a cold atmosphere, we find but few of these animals distributed in the cold countries of the globe; while in those countries enjoying a high temperature they are found of vast size, and of many different kinds, and in great numbers.

The torpidity of the *Mollusca* has not been studied with care. Those which are naked and reside on the land, retire to holes in the earth, under the roots of trees, or among moss, and there screen themselves from sudden changes of temperature. The different kinds of land Testacea, such as those belonging to the genera Helix, Bulimus, and Pupa, not only retire to crevices of rocks and other hiding places, but they form an operculum or lid for the mouth of the shell, by which they adhere to the rock, and at the same time close up even all access to the air. If they be brought into a warm temperature, and a little moisture be added, they speedily revive. In the case of the *Helix nemoralis*, the operculum falls off when the animal revives, and a new one is formed, when it returns again to its slumbers. The first formed opercula contain a considerable portion of carbonat of lime, which is found in smaller quantity in those generated at a later period. If the animal has revived frequently during the winter, the last formed opercula consist entirely of animal matter, and are very thin. The winter lid of the *Helix pomatia* resembles a piece of card-paper.

All the land testacea appear to have the power of becoming torpid at pleasure, and independent of any alterations of temperature. Thus, even in midsummer, if we place in a box, specimens of the *Helix hortensis, nemoralis* or *arbustorum*, without food, in a day or two they form for themselves a thin operculum, attach themselves to the side of the box, and remain in this dormant state. They may be kept in this condition for several years. No ordinary change of temperature produces any effect upon them, but they speedily revive if plunged in water. Even in their natural haunts, they are often found in this state during the summer season, when there is a continued drought. With the first shower, however, they recover, and move about, and at this time the conchologist ought to be on the alert.

The Spiders pass the winter season in a dormant state, enclosed in their own webs, and placed in some concealed corner. Like the torpid mammalia, they speedily revive when exposed to intense cold, and strive to obtain a more sheltered spot.

Many insects which are destined to survive the winter months, become regularly torpid by a cold exceeding 40°. The house fly may always be found in the winter season torpid, in some retired corner ; but exposure for a few minutes to the influence of a fire recalls it to activity. Even some of the lepidopterous insects, which have been hatched too late in the season to enable them to perform the business of procreation, possess the faculty of becoming torpid during the winter, and thus have their life prolonged beyond the ordinary period. These insects can all be preserved from becoming torpid by being placed in an agreeable temperature, as the following experiments of Mr Gouch (Nicholson's *Journal,* vol. xix.) testify. In speaking of the Hearth Cricket, (*Gryllus domesticus*), he says, " Those who have attended to the manners of this familiar insect will know that it passes the hottest part of the summer in sunny situations, concealed in the crevices of walls and heaps of rubbish. It quits its summer abode about the end of August, and fixes its residence by the fireside of the kitchen or cottage; where it multiplies its species, and is as merry at Christmas as other in-

insects are in the dog-days." Thus do the comforts of a warm hearth afford the cricket a safe refuge, not from death, but from temporary torpidity; which it can support for a long time, when deprived by accident of artificial warmth. " I came to the knowledge of this fact," he says, " by planting a colony of these insects in a kitchen, where a constant fire is kept through the summer, but which is discontinued from November to June, with the exception of a day, once in six or eight weeks. The crickets were brought from a distance, and let go in this room in the beginning of September 1806: here they increased considerably in the course of two months, but were not heard or seen after the fire was removed. Their disappearance led me to conclude that the cold had killed them: but in this I was mistaken; for, a brisk fire being kept up for a whole day in the winter, the warmth of it invited my colony from their hiding-place, but not before the evening, after which they continued to skip about and chirp the greater part of the following day, when they again disappeared; being compelled by the returning cold to take refuge in their former retreats. They left the chimney-corner on the 28th of May 1807, after a fit of very hot weather, and revisited their winter residence on the 31st of August. Here they spent the summer merely, and lie torpid at present (Jan. 1808), in the crevices of the chimney, with the exception of those days on which they are recalled to a temporary existence by the comforts of a fire."

Nothing is known with regard to the hybernation of the intestinal worms. Those which inhabit the bodies of torpid quadrupeds, in all probability, like these, experience a winter lethargy. If they remain active, they must possess the faculty of resisting great alterations of temperature. Among the *infusory animals*, numerous instances of suspended animation have been observed, continued not for a few months, but during the period of twenty-seven years. But such instances of lethargy do not belong to our present subject.

There is another kind of hybernation, in some respects resembling torpidity, which deserves to be taken notice of in this place, and which merits the appellation of Quiescence. The animals which observe this condition, remain during the winter months in an active state, requiring but little food, without however experiencing the change to torpidity.

Of these quiescent animals, the common bear (*Ursus arctos*) is the most remarkable example. Loaded with fat, he retires in the month of November to his den, which he has rendered comfortable by a lining of soft moss, and seldom reappears until the month of March following. During this period he sleeps much, and when awake almost constantly licks with his tongue the soles of his feet, particularly those of the fore paws, which are without hair, and full of small glands. From this source it is supposed that he draws his nourishment during the period of his retirement.

This quiescence appears to differ in its kind from torpidity. This animal is always in season before he retires to his winter quarters, and the female brings forth her young, before the active period of the spring returns, and before she comes forth from her hiding place.

The common badger is supposed to pass the winter in the same manner as the bear, with which, in structure and habit, he is so nearly related. It is also probable, that many species of the genus Arvicola become quiescent, particularly the *amphibia* or common water rat, which always leaves its ordinary haunts during the winter.

It is in this state of hybernation that many of our river

fishes, and even some species that inhabit the sea, subsist, at the season of the year when a supply of food cannot be procured. A similar condition prevails among the fresh water mollusca, and also among many species of Annelides. But we must observe, that accurate observations on this branch of the subject are still wanting.

We have thus endeavoured to trace the various means by which animals are enabled to counteract the baneful effects of the varying temperature of the seasons. This end is accomplished, in some, by a change in the quantity, or the colour, of their dress, by which its conducting and radiating powers are regulated. The inconveniences of excessive heat or cold are avoided by others, by their periodical migrations; while, in a third class, the body, during the cold season, falls into a lethargic state, from which it recovers with the return of spring. But there are other circumstances which influence the distribution of animals, besides those connected with temperature.

II. Food.

Although temperature appears to exercise a very powerful controul over the geographical distribution of animals, yet it is likewise indispensably necessary that they be provided with an abundant supply of suitable food. When this supply is deficient in quantity, both the size and shape of the animals are altered. The frame becomes diminished in stature, its symmetry is marred, and its feebleness indicates the scantiness of its nourishment.

The dependence of the geographical distribution of animals on the supply of nourishment, is most conspicuously displayed in those species which are destined to feed on particular kinds of food. Thus many species of insects are restricted in their eating to one kind of plants, or are parasitical on one species of animal. The distribution of such animals is thus dependent on their food.

The same remark is generally applicable to carnivorous and phytivorous animals. But, in many species, though the restriction is absolute as to the nature of the food, it admits of a considerable range with regard to the variety or kind. Thus, though the lion is restricted to flesh, his cravings are equally satisfied with the carcase of a horse, a cow, or even of man. The hog in general feeds on roots, but it is not confined to those of one kind of plants; hence it can subsist wherever the earth is clothed with verdure.

The seasons exercise a powerful influence on animals, directly, in reference to their temperature, and, indirectly, with regard to the production of their food. Thus, the insect that feeds on the leaves of a particular tree, can only enjoy its repast during that part of the season when this tree is in leaf. How, then, is life preserved during the remaining portion of the year? The resources are numerous. It either exists in the form of an unhatched egg, an inactive pupa, in the imago state, requiring little food, or actually becoming torpid.

The birds which feed on insects in summer, in this climate, are, from the absence of this kind of sustenance in winter, obliged to have recourse to various kinds of vegetable food during that season. Should this change of diet be unsuitable, migration to other districts, where a proper supply can be obtained, becomes indispensably requisite.

In compliance with these regulations, we observe numerous mammalia, birds, and fishes, accompany the shoals of herrings in their journeys; and the grampus and seal enter the mouths of rivers in pursuit of the salmon *. The bats,

---

* The seal sometimes enters fresh water-lakes in pursuit of his favourite repast. In the account of the parish of North Knapdale, by the Reverend

which feed on insects in summer, could not, in this country, obtain a suitable supply of food. Yet the race is preserved, since the same fall of temperature, which is destructive to insect-life, brings on their winter-torpor.

With many quadrupeds, however, and even insects, especially the bee, where migration to more fertile districts is impracticable, and where torpidity is not congenial to the constitution *, there is an instinctive disposition to be provident of futurity,—to subject themselves to much labour, during the autumn, when the bounties of nature are scattered so profusely, in heaping up a treasure for supplying the deficiencies of a winter, of whose accompanying privations, the young, at least, are ignorant. The quadrupeds which possess this *storing* inclination are all phytivorous, and belong to the natural tribe of *gnawers*.

Of all those animals, whose industry in collecting and wisdom in preserving a winter-store, have attracted the notice of mankind, the beaver stands pre-eminently conspicuous. The number of individuals which unite, in cutting down timber to supply their storehouse, and the regularity with which they dispose of it, in order to make it serve the double purpose of food and shelter, are equally calculated to excite astonishment. But as we rather wish to confine our remarks to British animals, wherever the subject will permit, we select as an example of this kind of storing propensity, the common squirrel, (*Sciurus vulgaris*.) This

---

ARCHIBALD CAMPBELL, we are told, that the lake called Lochow, about 20 miles in length, and 3 miles in breadth, " abounds with plenty of the finest salmon; and, what is uncommon, the seal comes up from the ocean, through a very rapid river, in quest of this fish, and retires to the sea at the approach of winter."—*Stat. Acc.* vol. vi. p. 260.

* Some animals, as the field-mouse, even lay up a small stock of provision on which to subsist, previous to torpidity, and in the intervals of reviviscence.

active little animal prepares its winter-habitation among the large branches of an old tree. After making choice of the place where the timber is beginning to decay, and where a hollow may be easily formed, it scoops out with its teeth a suitable magazine. Into this storehouse, acorns, nuts, and other fruits, are industriously conveyed, and carefully concealed. This granary is held sacred until the inclemency of the weather has limited the range of its excursions, and consequently diminished its opportunities of procuring food. It then begins to enjoy the fruits of its industry, and to live contentedly in its elevated dwelling.

III. Situation.

Animals are restricted by instinctive feelings, to particular situations. An attentive examination of all the physical conditions of the situation preferred by each species, and a comparison of these with the arrangement of the organs, will enable us, in many cases, to perceive the propriety, or rather, the necessity of the choice. The animals which reside in *water*, are possessed of organs of motion and respiration, which can execute their functions only in that element, in which also their most suitable food is to be obtained. In some cases, these organs are suited to a residence in salt water, in others, in fresh water, still farther influenced by the temperature and motions which these experience. Again, in the animals which reside on land, we observe the organs of locomotion, respiration, and perception, fitted only for a residence in such a situation. A variety of other circumstances likewise exercise a powerful influence, as temperature, shelter, and moisture.

All aquatic animals, of the vertebral division, which breed and sleep in the water, are restricted to a constant residence in that element, and consequently can exist only in those regions where it occurs. The whale, indeed, could breathe on land, but could not accomplish progressive motion.

But even among the aquatic animals, which are less influenced in their distribution by temperature, and even food, than those which reside on land, there are peculiar arrangements connected with the reproductive system, which exercise a powerful controul over them. With some species, a constant residence in salt water is suitable to every function; but with others, it is indispensably requisite to have occasional access to fresh water. Thus the salmon, smelt, and eel, which perform all the functions of motion, perception, and digestion, in the sea, are impelled to quit it during the season of generation, and repair to fresh water rivers and lakes, where they deposit their spawn, and where the fry sojourn for a short period. To these aquatic animals both fresh and salt water are necessary, so that their distribution is restricted to those regions where both are accessible. The migrations which these animals perform, appear to depend on the condition of the reproductive system, influenced by the obstacles or facilities for changing place, which the varying state of the fresh waters present.

Many quadrupeds and birds, which are able to execute easily the conditions of progressive motion, and obtain an abundant supply of food, in fresh or salt water, are compelled to return to the land at regular intervals, for the purposes of rest. This arrangement is exhibited in seals, otters, and many birds, at all seasons. These animals, consequently, cannot reside at great distances from land, however suited the great ocean may be with regard to temperature and food. In these animals the arrangements of the reproductive system likewise compel them to frequent the land during the breeding season. The term *Amphibious* is usually applied to animals such as are now referred to, which perform their functions partly on the land, and partly on or in the water.

Among terrestrial animals, there are many which execute all the operations of life in one particular kind of situation, influenced, however, by its variable conditions. Such animals are necessarily limited to those countries where such situations occur. There are others, however, which shift their situations at particular seasons, without reference either to temperature or food. The curlew, which can at all times procure a subsistence on the sea shore, and resist or counteract the changes of the seasons, retires, during the breeding season, to the inland marshes to nestle and rear its young. The heron, which is equally successful in procuring food on the shore, is destined to build its nest on trees, and consequently, must betake itself to wooded districts for the purposes of incubation. This bird congregates, during the breeding season, in a few places in Britain, termed *Heronries* or Heronshaws; but in other seasons it is found on all the shores of our remotest islands.

Many terrestrial animals pass the first period of their existence in the water. In compliance, therefore, with the arrangements of the reproductive system, the old animals must seek after that element in which to deposit their eggs, however independent they may be of its presence for their ordinary personal wants.

Animals are thus restrained, in their geographical distribution, to particular places, by a great variety of causes. Each situation possesses peculiar physical characters, each class of organs is accommodated to certain particular external circumstances, and the modifications of these organs are suited to the varieties which these circumstances exhibit. It is, however, one of the most difficult problems in Zoology, upon the situation occupied by any species being given, to determine all the organical and instinctive arrangements which are suited to its conditions.

VI. Foes.

The restraints imposed on particular kinds of animals by the presence of the rapacious tribes, serve to diminish the number of individuals, but do not appear to exercise a very powerful influence on the geographical distribution of the species. When a rapacious animal has committed great ravages in a particular district, a scarcity of food compels it to emigrate, and leave the defenceless kind to multiply in its absence. It is otherwise, however, with animals in the progress of extending the geographical limits of their distribution. They are exposed to a variety of dangers, against which, the organical arrangement of their constitution cannot guard them, and to many foes from which their instincts cannot point out the means of escape.

The restraints, however, imposed on the distribution of animals, by the efforts of the rapacious kinds, bear no comparison to the influence exercised over them by man. Against many species, hostile to his interests, he carries on a war of extermination. Others he pursues for pleasure, or for the necessaries or luxuries of life which they yield. In these conquests, the superiority of his mental powers is conspicuously displayed, and his claim to dominion established. Unable to contend with many species in physical strength, he has devised the pit-fall and the snare,—the lance, the arrow, and fire-arms. Aided by these, every animal on the globe must yield to his attempts to capture. The lion, the elephant, and the whale, fall the victims of his skill, as well as the mouse or the sparrow. Since the use of gunpowder, indeed, the contest is so unequal, that it is in the power of man to controul the limits of almost every species whose stations are accessible.

The views which have been stated in the preceding part of this chapter, pave the way for an examination of the REVOLUTIONS which have taken place in the animal kingdom, as these are indicated in the results of geognosy. It would lead us into details too numerous and extensive for this work, were we to give to this branch of the subject the degree of illustration which its importance and present obscurity require. Besides, we have already touched upon some of its more prominent topics, (vol. i. p. 26.) We shall therefore, at present, restrict our remarks to a few observations of a general nature.

1. The organic remains of the animal kingdom, found in the strata of this country, are generally supposed to bear a very close resemblance to the living races which inhabit the warmer regions of the earth, and to be unlike the modern productions of temperate and cold climates. In the firm belief of the truth of this statement, many naturalists have concluded, either that these organic remains must have been brought into their present situation, by some violent means, from tropical regions, or that our country once enjoyed the warmth of a tropical climate. The tenderness and unbroken state of the parts of these remains, and a variety of circumstances connected with their position, intimate the absurdity of the first supposition, and the truths of astronomy give ample discouragement to the latter. It would have been wiser to have examined all the conditions of the problem before attempting its solution, than rashly suffer the imagination to indulge in speculation and conjecture. Had this examination taken place, we venture to assert that the conclusion, that fossil shells, and other relics, must have been, while recent, the natives of a warm country, would never have been announced as an article of the creed of any geologist.

As the opinion here advanced is so very different from that which is received by modern authors, it seems necessary, before giving the proofs of its truth, to trace those circumstances which have operated in leading into error.

The shells and corals which are found in a fossil state, have probably been quoted more frequently, as proofs of the truth of the popular opinion on the subject, than any other class of relics. Now, here it may be observed, that for upwards of two centuries past, collections of tropical shells and corals have been forming in this country, and on the continent of Europe. During this period the native productions have been examined and collected by few. It therefore happened, that the tropical testacea and zoophytes, were better known than those of temperate regions. An observer, finding a fossil shell or coral, had it not in his power to compare them with the productions of his country, in public collections, or in the descriptions and engravings of books on natural history. He could compare them with the figures or specimens of foreign species only, and if he discovered an agreement in the external appearance of a species, or even a genus, might be led to conclude, that he had found a tropical shell or coral fossil, in Great Britain. A few hasty examinations of this kind, in which remote analogies were suffered to be considered as proofs of identity, and the indications of a *genus* mistaken for the marks of a *species*, could not fail to lead the mind to erroneous results. With the progress of science, however, the geographical limits of species and of genera, have been more carefully investigated, collections of native productions have become more numerous and accessible, and the sources of error greatly diminished. It is perhaps necessary to add, that among all those who have attempted to investigate the history of fossil species, there have been very few at the same time acquainted with the characters of recent animals, and with the details of geognosy. When all these circumstances are duly considered, it need not appear surprising, that the results hitherto obtained are so inconclusive, or rather so remote from the truth.

It may be stated, in opposition to the opinion against which I am contending, that, in general, *there are existing species in this country, belonging to the same genera or natural families, as those which are found imbedded in the strata of the earth in a fossil state.* The Nautilus (of Lin., including those which are spiral, and the straight Orthocera), for example, is found in the oldest and the newest rocks containing petrifactions. Several species from the tropical seas occur in our public collections. But it is not generally known, that, by the labours of Boys, Walker, and Montagu, nearly thirty different species have been detected in a recent state, on our own shores. It is true, that the recent tropical kinds are larger than our indigenous species, and in this character they resemble the fossil species. But this circumstance is of no weight, when considering the geographical distribution of *different* species or genera*. The Anomiæ (Lin.), common in a fossil state, are also represented by many recent kinds in our seas, and the same remark is applicable to the Echini, Madreporæ, and Milleporæ.

It may here be said, that the remains of animals which do not now live in temperate regions, as the Elephant, Rhinoceros, and Tapir, are found in a fossil state in such districts. This objection, however, is of no great weight, and derives its principal support from the prejudices associated with the appellations of the species now noticed.

---

* The Moose-deer, which inhabits countries bordering on the Arctic circle, exceeds the horse in size, while the Guinea-deer, and Meminna, living within the tropics, are not larger than a hare.

When the name Elephant is pronounced in our hearing, the imagination immediately presents us with the picture of an animal browsing in an equatorial forest, guided by a Lascar, or hunted by a Caffre. We are so impatient to speculate, that we do not stop to inquire, whether the bones found in a fossil state belong to the living *species*, or to a member of the same *genus* only, now extinct. Although the lovers of hypothesis have rejected this inquiry as useless, others have submitted to the labour of the investigation. It was ascertained at an early period, by several observers, that many of the large bones dug up from the alluvial soil, in the temperate and cold regions of Europe, Asia, and America, belonged to a species of the genus Elephas; but it was reserved for CAMPER, HUNTER, and CUVIER, to demonstrate, by means of the characters furnished by the bones of the head and the teeth, that these fossil remains belonged to *a species of elephant, different from any of those now living, or known to exist on the globe.* When the specific difference of the fossil species, denominated by the Russians Mammoth, had been determined, it became unsafe to speculate about the climate under which it had subsisted, since *each species of a natural genus is influenced in its geographical and physical distribution, by peculiar laws* *.

RASPE, from an examination of the soil in which the bones were imbedded, and the different countries in which they had been discovered, without the aid of comparative anatomy, arrived at the conclusion, that the elephant, to which these relics belonged, may have been *a native of northern countries,* in which it had lived, propagated, and

---

* The genus Taurus, or Bull, illustrates the truth of this rule, hitherto much neglected, in examining the distribution of petrifactions. The Bos buffalus, or Buffalo, dwells within the tropics; while the Bos moschatus, or Musk Ox, is found within the Arctic circle, and as far north as latitude 74°.

expired *. This important conclusion, which has since been extended to other fossil species, received abundant confirmation, by the fortunate discovery, in 1799, of an entire carcase of this northern elephant, preserved in the ice, at the mouth of the Lena, on its entrance into the Frozen Ocean. The animal had been in good condition at its death, and its flesh was in such perfect preservation, that the Takutski, in the neighbourhood, cut off pieces to feed their dogs with; and the white bears, wolves, wolverines, and foxes, made likewise an agreeable repast. But the most interesting circumstance connected with this individual, was its covering of hair. The elephant, and indeed all tropical animals, as formerly stated, are thinly covered with short hair. This Siberian animal, on the other hand, was thickly covered with long hair. Upwards of thirty-six pounds of the hair were collected: the greater part of which the bears had trod into the ground, while devouring the flesh. Much of the original quantity, therefore, was lost. This quantity of hair indicated not a native of a tropical climate, but an inhabi-

---

* Elephantorum Africæ, Indiæ, et Cylanensium, diversitatem quandam notant, naturæ historici; Cylanenses reliquis omnibus magnitudine et ingenio præcellere prædicantes. Esse igitur potuit elephantorum species in ipso septentrione, reliquis omnibus præcellens frigoris patentia, et à meridionalibus elephantis non magis diversa quam Lappones aut Hurones diversi esse solent a Nigritis et Malabariæ incolis. Sed certe quid pronunciare non ausim. Sufficit, omnium reliquarum hypothesium insufficientiam, naturam situs et ossium ipsorum, tandem etiam exempla extirpatarum specierum persuadere, quammaximè probabile quadam ratione, belluas quarum ossa in septentrione hinc inde latent, non esse peregrinas, non alio advectas sed indigenas in illis terris, ipsis olim natalibus, vixisse, generasse, quamvis hodie non supersint."
Dissertatio epistolaris de Ossibus et Dentibus Elephantum, aliarum Belluarum in America Septentrionali, aliisque Borealibus Regionibus obviis; qua indigenarum Belluarum esse ostenditur. Auctore R. E. Raspe.—Phil. Trans. 1769, p. 136.

tant of a cold region. Its characters served to remove all doubt upon the subject. It consisted of three kinds;—bristles, nearly black, much thicker than horse hair, and from twelve to eighteen inches in length; hairs of a reddish-brown colour, about four inches in length; and wool of the same colour as the hair, but only about an inch and a half long*. These circumstances demonstrate, that this *species* of elephant was suited to reside in the temperate and cold regions, in which its bones are at present discovered, *and that the climate of Siberia, at the time when the mammoth flourished, was the same in temperature, or nearly so, as it is at present.* These facts, viewed in connection with others equally striking with regard to the fossil rhinoceros, indicate the impropriety of speculating about the origin of fossil animals, without having previously determined the *species*, or attended to the laws which regulate the distribution of the existing races.

While there are many genera, containing both extinct and recent species, there are other genera, which have no living examples, as Belemnites, Mastodon, Anoplotherium, and Palæotherium. These facts seem to indicate a former condition of the Earth's surface, very different from that which prevails at present in any latitude.

It has frequently been remarked by British and continental writers, that in the same quarry, or mine, the organic remains contained in one bed often differ from

---

* This interesting skeleton is now put up in the Museum of the Imperial Academy of Sciences of St Petersburgh. The skin still remains attached to the head and the feet. The fore-leg, which was not found, has been restored in plaster of Paris from the other side. The whole is nine feet four inches in height, and sixteen feet four inches in length, from the point of the nose to the end of the tail. The tusks are nine feet six inches in length, measuring along the curve, and, together, weighed 360 lb. avoirdupois; the head alone, without the tusks, weighed 414 lb. avoirdupois.

those in the contiguous beds, and that the same bed, in its course through several miles, may be easily recognised by its petrifactions. WERNER, in attempting to generalise this observation, announced it as his opinion, " That different formations can be discriminated by the petrifactions they contain *." When it is considered that a particular bed of rock can seldom be traced for many miles, the assertion that it may, through this extent, be characterised by its petrifactions, is neither in opposition to observation, nor the laws which regulate the distribution of animals. But when it is meant to be understood, that the same group, or formation of beds, (though occupying the same position with respect to other groups), however remotely connected in geographical position, contain the same petrifactions, it becomes necessary to state, that such an opinion is unsupported by observation, and inconsistent with the laws which *at present* influence the distribution of animals. There is no proof of the same fossil *species* being found in the same formation at the equator, and in temperate and cold regions; and when *genera* only are referred to, the reasoning becomes exceedingly vague, and apt to mislead. The celebrated VON BUCH, a pupil of WERNER, in his Travels through Norway, gives us an example of the Wernerian rule, by stating, that Orthoceratites characterise exclusively the transition limestone formation †. There is, however, no evidence, that the *species* of Orthoceratites were here

---

* See note by Professor Jameson to Cuvier's Theory of the Earth, Edinburgh 1813, p. 225.

† " How great was my joy when, at the steep falls of Aggers Elv, above the lower saw-mills, I discovered the Orthoceratites, which so particularly distinguish throughout all Europe this formation (Transition Limestone), and this formation alone."—English Trans. p. 47.

considered, nor that those which occur in the transition limestones of Norway, Germany, and Ireland, have been investigated with the view of determining the particular species: And I have demonstrated that the rule does not hold true, in reference to the *genus*, by the publication of figures and descriptions of ten species of Orthoceratites, from the beds of the independent coal formation *. Until geologists form more accurate notions with regard to fossil *species*, no reliance can be placed on the observations which they offer respecting their geographical distribution.

We have no reason to doubt, that the laws which now regulate the geographical distribution of animals and plants, did exercise their influence at the period, when the transition and flœtz-rocks were forming. We may, accordingly, expect to find the *fossil* animals and plants of the temperate regions, differing as much from those in tropical countries, as the recent kinds are known to do. Each region may be expected to exhibit a peculiar fossil Fauna and Flora †.

---

* " Observations on the Orthoceratites of Scotland,"—Thomson's Annals of Philosophy, for March 1815 ; where some proofs are likewise given of the same fossil species, occurring in different beds of the same formation, and even in different formations.

† In the article Conchology, in the Supplement to the Encyclopædia Britannica, I have stated, " that in every country there are particular animals and vegetables, which indicate, by their mode of growth and rapid increase, a peculiar adaptation to the soil and climate of that district. Hence we find a remarkable difference in the animals and plants of different countries. Many shell-fish have indeed a very wide range of latitude, through which they may be observed ; but we know, that the same molluscous animals which are natives of Britain, are not found, as a whole, as natives of Spain, while the molluscous animals of Africa differ from both. If the same arrangement of the molluscous animals always prevailed in the different stages of their existence, then we may expect to find the fossil shells of one country differing as

In reference to this view of the subject, we may add, that the laws which regulate the distribution of recent animals, have been, in a great measure, deduced from observations on those which inhabit the countries between the Tropic of Cancer and the Arctic Circle; so that we have much to learn with regard to the characters of those which dwell between the Antarctic Circle and the Tropic of Capricorn. But the observations on the distribution of the fossil species, from having been chiefly carried on in the middle and south of Europe, are more confined. A vast number of fossil species, therefore, remain unexplored in the equatorial and antarctic regions, the characters of which will either confirm the view which is here given, or furnish evidence for that alteration of climate, occasioned by a change in the obliquity of the ecliptic, or in the Earth's axis of rotation, which a few naturalists believe to have taken place. If the fossil animals at the Equator do not resemble recent or fossil Arctic productions but exhibit characters peculiar to themselves, it will be necessary to abandon the idea of great astronomical revolutions, and content ourselves with investigating the changes organised beings are experiencing at present, in order to discover those circumstances which have impressed on the fossil species their peculiar outlandish character.

3. The opinion entertained by WERNER, that the petrifactions of the older rocks, belong to animals of more simple structure and less perfect organization, than those which occur in the recent deposits, is, when considered in a very general

---

much from those of another, as the recent kinds are known to do, so that every country will have its *fossil*, as well as its *recent* testacea. Few observations illustrative of this branch of the subject have hitherto been published." Mr GREENOUGH has since combated the same opinion, in his " Critical Examination of the First Principles of Geology,"—London 1819, p. 287.

point of view, an approximation to the truth. In the beds of the transition class, (the oldest rocks which are known to contain petrifactions), the remains both of radiate and molluscous animals occur; yet the organization of the latter is considered more perfect than that of the former. In the transition class, however, no remains of vertebrose animals have been detected. In the independent coal-formation, (one of the oldest groups of the flœtz class,) in addition to the relics of radiate and molluscous animals, those of several annulose animals occur, as species of the genera Trilobites, Dentalium and Spirorbis; together with fishes, both osseous and cartilaginous. In the newer groups of the flœtz series, relics of amphibia make their appearance, and in the newest groups those of birds and quadrupeds. In the oldest alluvial deposits are found the bones of extinct quadrupeds; in the newer beds, those of such as still survive. From the period, therefore, at which petrifactions appear in the oldest rocks, to the newest formed strata, the remains of the more perfect animals increase in number and variety; and it is equally certain, that the newest formed petrifactions bear a nearer resemblance to the existing races, than those which occur in the ancient strata. The older remains are much altered in their texture, and more or less incorporated with the matter of the rock, while the newer are but little altered. These circumstances lead us to believe, that the strata containing petrifactions were once in a state of *mud*; and that the same process which altered the imbedded relics, communicated to the surrounding matter its present compact or crystalline structure.

These facts, in the history of animals, which have been ascertained by the researches of the geologist, lead the inquisitive mind to investigate those circumstances which have operated in bringing about such mighty revolutions. In conducting the inquiry, it is necessary to impose restraint upon the imagination, and deliberately to examine the existing causes of change in the animal kingdom, in order to comprehend the alterations which have already taken place, or to anticipate those which may yet be produced. What influence has man exerted in producing such changes?

The situation in which we are placed in this world, renders it necessary for us to attempt the destruction of many races of carnivorous animals, to drive them from our dwellings, cut them off in their retreats, and prevent them from living in the same region along with us. When we begin to keep flocks of tamed animals, to plant gardens and sow fields, we expose ourselves to the inroads of a greater number of depredators, and consequently wage more extensive war. The war waged, in the early stages of society, against various animals, is a measure of security. With the progress of civilization this war becomes an amusement; and in the absence of those animals, really destructive to our interests, we make sport of the death of others which are inoffensive. But the employment of the chace is not altogether a measure of safety or amusement. We hunt to obtain food and clothing, and a variety of ornamental and useful articles of life.

The havock which man thus commits in the animal kingdom, has occasioned the extirpation of many species from those countries of which they were formerly the natural possessors. In this island, since the Roman invasion, some species of quadrupeds and birds have disappeared; and others are becoming every year less numerous. Of those which have been extirpated, the bear and the beaver, the crane and the capercailzie, may be quoted as well known examples. The same changes are taking place in every cultivated region of the earth, each having, within the very limited period of history or tradition, lost many of the original inhabitants.

When it is considered, that the business of the chace has ever been keenly followed by man, from the first stages of society to the present day; that its triumphs have been eagerly sought after, and highly prized; and when these circumstances are united, with the recollection of the numbers of the human race dispersed over the globe, all prosecuting the same purpose, for the long period of nearly 6000 years, it does not appear unreasonable to conclude, that man has effected many changes in the geographical distribution of animals. Perhaps, he may have succeeded, in the course of this long period of persecution, in completing the destruction of several species, the memorial of which tradition has failed to preserve, while their remains may yet be traced in the newer and perhaps older alluvial deposits.

However great those changes may have been, in the condition of certain species, brought about by human agency, there are many other revolutions which have taken place in the animal kingdom, over which man could exercise no control. Many corals and shells, the bones of fish, reptiles and quadrupeds, are imbedded in stone, which have belonged to species which do not now exist in a living state on the globe, and which, probably, were extinct before man exercised any control. By what cause, then, have these revolutions been effected?

When we attend to the physical conditions of temperature, food, situation, and foes, which must at all times have exercised their influence over the existence and geographical distribution of animals, it will appear obvious, that a variety of causes, (a change in one or all of these conditions,) may have operated in promoting the increase of some species, and in producing the decay or extinction of others. Have we, then, any proof of such changes in the truths of geognosy, or in the alterations which we witness taking place on the surface of the globe?

When we look at a river after rain, emptying its contents into the sea, we perceive that it has brought along with it a considerable quantity of gravel and mud, which it deposits in the form of bars, sand-banks or deltas. This mud has been obtained from the disintegration and wearing down of the rocks through which it has passed; and contributes to fill up the ocean, by forming land on its borders. The flat ground at the mouths of the Ganges, the Nile, and the Rhine, have derived their origin from this source, as well as the *carses* of Falkirk and Gowrie in Scotland. If the attention is turned from the sea to inland lakes, we observe the same process of *upfilling* going on, with the assistance of other circumstances connected with their condition. Mud is constantly poured into them by the rivulets; the testaceous animals separate lime from the water for their shells, which ultimately go to the formation of marl. Aquatic plants multiply; and, by their annual decay, form layers of peat. The whole mass of foreign matter, by degrees, acquires the altitude of the mouth of the lake, passes into the state of a marsh; and, by the wearing away of the rocks at the outlet, is in part drained, becomes fit for grazing, and, finally, suitable for cultivation. The rivulets now prevented from precipitating their suspended contents in the lake, carry them to some lower pool, or farther on to the sea. Numerous plains, meadows and peat-bogs, indicate the former operations of this process; and in every lake at present, similar changes may be observed taking place.

This obvious tendency of the present order of things to wear down eminences, and fill up hollows, has not been confined to the period of the formation of the alluvial strata, but has exerted its influence during the period of the formation of all those rocks in which organic remains are imbedded. Thus, when the position of the beds of the transition rocks are examined in the great scale, they are found to occupy immense hollows in the primitive rocks. The old red sandstone fills up the hollows of the transition, and occasionally of the primitive rocks. The independent coal formation is found occupying the hollows of the preceding groups. The hollows of these different formations have been still farther filled up, by the numerous series of beds connected with chalk and gypsum; and, at last, we come to the alluvial deposits, which at present are contributing to fill up existing inequalities.

These changes which have taken place, have every where diminished the height of mountains, filled up lakes, and increased the quantity of dry land. We may therefore safely draw the conclusion, that, along with the increase of dry land, there must have been a proportional diminution of aquatic animals and plants, and a corresponding increase of those which inhabit the land *. There is likewise reason to conclude, that, amidst these vast revolutions, so many alterations must have taken place in those physical conditions, on which the life of animals depends, that multitudes must have been annihilated with every successive change. The increase of land, by this process of upfilling, and the reduction of the number of mountains supporting glaciers, must have altered greatly the temperature of the globe; and, in every region, increased the difference between the heat of summer, and the cold of winter, by promoting the intensity of each. This change of temperature may have been somewhat modified by the progress of *vegetation* in the different periods, by the formation of volcanic land, and the heat communicated to the air by volcanic fire. It is impossible to estimate all the effects which these changes may have pro-

---

* This view of the matter is countenanced by the circumstance, that the remains of aquatic plants and animals are more abundant in the rocks of the different ages, than those belonging to the inhabitants of the land.

duced on different species of animals, but little doubt need be entertained that they were of considerable extent.

In consequence of these changes which have taken place on the earth's surface, corresponding alterations must have been produced in its condition, as a residence for animals. Every lake, as it was filled up, would receive the remains of all those of its inhabitants, the locomotive powers of which prevented them from shifting to a more suitable dwelling. If a number of these lakes were filled up nearly at the same time, over the whole, or a large portion, of the globe, (and the universality of many of these upfilling formations justify the supposition,) the total extinction of a race of animals may have taken place; and each succeeding deposition may have been equally fatal to the surviving tribes.

If every physical change which can take place on the surface of the earth, whether it be an alteration of temperature, of the quantity of land or water, of moisture or dryness, is detrimental to some animals, we need not be surprised, that, amidst the vast number which has occurred, many species have disappeared, whole races become extinct, and the general features of the animal kingdom undergone successive changes *.

When we trace the characters of the different depositions which have taken place from the newest alluvial beds to the oldest transition rocks containing petrifactions, we witness very remarkable gradations of character. The newest

---

* Some conception may be formed of the effects of those changes which have taken place on the earth's surface, on animals and vegetables, by observing the alterations which are produced by the drainage of a bog or lake. Plants are destroyed, together with the insects and shell-fish that fed on them. Fish are destroyed, and the worms on which they fed. The frog can no longer find a fit place to deposit its eggs. The food of the water-fowl is destroyed; and their haunts dried up. These ancient inhabitants are succeeded by others suited to the new state of things,—an emblem of the great revolutions of the earth.

formed strata are loose in their texture, and usually horizontal in their position. In proportion as we retire from these, towards the older formations, the texture becomes more compact and crystalline, and the strata become more inclined and distorted. These characters may be traced, by comparing the common loose marl of a peat-bog with the firmer chalk; the compact flœtz limestone with the transition marble; or the peat itself with the older beds of wood-coal, or the still older beds of coal of the independent coal formation. The organic remains in the newer strata are yet unaltered in their texture, and easily separable from the matter in which they are imbedded. In the older rocks, the remains are changed into stone, and intimately incorporated with the surrounding rocks. These facts are of vast importance in a geological point of view, as they make us acquainted with the original condition of the matter with which the organic remains were enveloped, and lead us to believe that the bed now in the form of limestone or marble, was once loose as chalk, or even marl; that coal once resembled peat; and that the strata of sandstone and quartz rock were once layers of sand. They are no less interesting when viewed in connection with the characters presented by the petrifactions of the different æras.

The fossil remains of the alluvial strata, nearly resemble the same parts of the animals which live on the earth at present, and in the newer strata, the remains of existing races are found. As we trace, however, the characters of the petrifactions of the flœtz and transition rocks, we find the forms which they exhibit differing more and more from the animals of the present day, in proportion as the rocks in which they are contained exhibit new characters of texture, position, and relation.

It is impossible to regard these concomitant circumstances as accidental. Their co-existence indicates their relation, and leads to the conclusion, that the revolutions which have taken place in the animal kingdom, have been produced by the changes which accompanied the successive depositions of the strata. According to this view of the matter, the animals and vegetables with which the earth is peopled at present, could not have lived at the period when the transition rocks were forming. A variety of changes have taken place in succession, giving to the earth its present character, and fitting it for the residence of its present inhabitants. And if the same system of change continues to operate, (and it must do so while gravitation prevails,*) the earth may become an unfit dwelling for the present tribes, and revolutions may take place, as extensive as those which living beings have already experienced.

In addition to these circumstances, which must have exercised a powerful influence on the distribution of animals, we must bear in mind, that the universal deluge of NOAH, and the numerous local inundations, the traces of which may be perceived in every country, must have greatly contributed to produce changes in the animal and vegetable kingdom †. To these inundations may be ascribed the occurrence of the remains of supposed land plants, and fresh water animals in strata, alternating with such as contain

---

* The only causes which at present seem to operate in preventing this reduction of the elevated parts of the earth to a level with the sea, are volcanoes and sand-floods. The counteracting effects of the former are of great magnitude; those of the latter are comparatively insignificant.

† In this country, the proofs of these inundations are exhibited in the hills of stratified sand, and the numerous boulder stones which occur in every part of the country. The tempestuous risings of the sea have left traces of their destructive effects, in the heaps of sea shells which occur in our friths. I have described a bed of this kind elevated upwards of thirty feet above the level of the sea, which occurs to the westward of Borrowstownness in the Frith of Forth, in Annals of Phil. August 1814, p. 133.

only marine exuviæ. These appearances occur in the secondary formations of all ages *.

The statements which have now been given, relative to the changes which have occurred on the earth's surface, and in the condition of its inhabitants, point out some of the advantages which the student of zoology may derive from an acquaintance with the truths of geognosy, and how necessary it is to cultivate a knowledge of the history of plants and animals, to qualify for conducting geological speculations. These views may probably be submitted to the public at a future period in a more enlarged form.

An acquaintance with the laws which regulate the geographical distribution of animals, is indispensably necessary in our attempts to NATURALISE exotic species. The temperature most suited to their health,—the food most congenial to their taste, and best fitted to their digestive organs,—the situation to which their locomotive powers are best adapted,—and the foes against which it is most necessary to guard them, are circumstances on which we ought to bestow the most scrupulous attention, in order to insure success. There are many animals which can call forth but few counteracting energies, and consequently, cease to thrive,

---

* The action of gravity on the mud, and its organic contents, at the period of the deposition of the strata, may have given to the whole an arrangement different from that which prevailed at first. This effect of *subsidence* to which I refer, is displayed in every peat-bog. The lower beds are usually formed of sand, or marl. The beds of the latter consist of all the shells which had been formed in the lake before it was filled up. These have subsided to the bottom, and, along with them, the skeletons of deer or oxen which may have perished in the waters. The lighter peat occupies the surface. That the flœtz strata, at their formation, were in a condition to admit of subsidence, is demonstrated by the *complanation* of those trees and other remains which lie parallel to the surface of the bed, appearances which are exhibited in every layer of peat.

upon the slightest alteration taking place in their physical condition. With others, the case is very different, and these we can easily naturalise. They can accommodate themselves to a variety of new conditions, and successfully resist the destructive tendency of the changes to which we subject them.

The change in the condition of the animals we wish to naturalise, should, in all cases, be brought about as slowly as circumstances may permit. In this manner, the first counteracting effects of the system grow into organical habits, before all the evils of the situation are experienced, in which they are destined ultimately to reside. In this gradual manner, man has become fitted to reside in every climate, as well as many of the animals which he has reclaimed.

The details into which we have entered, concerning the distribution of animals, serve to point out the proper method of constructing a FAUNA of a particular country. Among British writers, at least, little or no attention is paid to the geographical distribution of the species. In the list of *native animals*, may be found those species which really live in the country, associated with such as visit it periodically, or only at irregular intervals, and with those which have been extirpated, or which have become extinct, and such as have been naturalised. In consequence of this incongruous assemblage of species, it is difficult to form a correct view of the number or characters of our native animals. Were we to classify them under the following divisions, much ambiguity and even error would be avoided.

1. *Resident Animals.*—Those animals, only, ought to be considered as the genuine inhabitants of the country, which continue in it the whole year, and can accommodate them-

selves to all its annual changes. Such are the fox and the otter, the eagle and the grouse.

2. *Periodical Visitants.*—These arrive and depart at stated seasons of the year. The Equatorial visitants continue during the winter, arriving in autumn and departing in spring. Such are the fieldfare and woodcock. The Polar visitants arrive in spring and depart in autumn. During their abode, they bring forth their young, and differ from the resident animals chiefly in the circumstance of being unable to counteract the changes which winter produces in their condition. Such are the swallow and nightingale.

3. *Irregular Visitants or Stragglers.*—The species here referred to, are either the resident inhabitants or periodical visitants of the neighbouring regions, which have been driven to this country by the fury of storms, or chaced from their native haunts by their foes. In consequence of these causes, and probably many others, a variety of animals have strayed to the British isles, and have been improperly enrolled in our Fauna. Thus, several cetaceous animals, and palmated quadrupeds, have been in this manner, as stragglers, observed and recorded. Among a few of these may be enumerated, Manatus trichecus, or sea-cow; Delphinus albicans, or Beluga; and more recently, Trichecus rosmarus, or Morse. Among birds, and fishes, and insects, the number of stragglers is more considerable. A single instance of the appearance of a single individual of a species on British ground, has hitherto been considered as sufficient to constitute a nominal right to citizenship. How greatly would our lists be reduced, were the names of such foreigners excluded? It is, however, of importance to record the event of their occurrence, and the circumstances

under which it took place, as illustrations of the history of species, and as furnishing marks by which we may trace the changes produced in their geographical distribution.

4. *Extirpated Animals.*—Under this head are included those species which, though they do not now live in this kingdom, still continue to flourish, in a wild state, in other regions more favourable to the continuance of their race. Among the extirpated British quadrupeds, may be enumerated the wolf, brown bear, beaver, boar, fallow-deer, antelope *, ox, and horse. The extirpated birds are fewer in number, being chiefly the capercailzie, crane, and egret. The two last occasionally appear as stragglers.

5. *Extinct Animals.*—These no longer occur in a living state on the globe. Our extinct quadrupeds are few in number,—such as the fossil elephant, rhinoceros, and Irish elk †.

---

* The only notice I have seen of this animal ever having inhabited Britain, is in a paper giving " An account of the Peat-pit near Newbury in Berkshire," by JOHN COLLET, M. D. where there is likewise another proof of the occurrence of the bones of the *beaver* in this country, in addition to those which Mr NEILL has so carefully collected, in the Edinburgh Phil. Journal, vol. i. p. 177. " A great many horns, heads, and bones of several kinds of deer, the horns of the *antelope*, the heads and tusks of boars, the heads of *beavers*, &c. are also found in it; and I have been told, that some human bones have been found; but I never saw any of these myself, though I have of all the others." Phil. Trans. 1757, p. 112.

† Large horns of a stag, differing in size from the present kind, are frequently dug up from marl-beds, but whether they are the relics of a distinct species, or only of a variety, has not been determined. The remains of a fallow-deer and ox likewise occur. The sculls of the latter are so superior in size to those of the present races in the Highlands, which are nearest to a

USES OF ANIMALS. 109

There are no extinct birds known in Britain, but the remains of many extinct species of amphibia, fishes, particularly sharks, crustacea, mollusca, and zoophytes, occur in abundance in the flœtz and transition strata. The fossil species, indeed, probably exceed, in the mollusca, at least, the numbers which live in the present day.

6. *Naturalised Animals.*—These are the natives of other countries, which have, by accident or design, been translated into our own. A few of these continue unreclaimed, as the brown rat, pheasant, and several species of fish; while others depend on our protection, as nearly all our domestic birds. Much, however, remains to be done in this branch of national industry.

CHAP. III.

Economical Uses of Animals.

It is impossible to enter in detail on this part of the subject, which, of itself, might furnish materials for several volumes. We must therefore content ourselves with a very few remarks on the Food, Clothing, and Medicine, together with materials for the arts, which we derive from the subjects of the animal kingdom.

1. *Food.*—Man employs, as articles of food, animals belonging to every class, from the quadruped to the zoophyte. In some cases, he makes choice of a part only of an animal,

---

wild state, as to lead to the conclusion, that they belonged to a well marked variety,—probably to the white cattle which are still preserved in a wild state in a few parks in different parts of the kingdom.

in other cases, he devours the whole. He kills and dresses some animals, while he swallows others in a live state. The taste of man exhibits still more remarkable differences of a national kind. The animals which are eagerly sought after by one tribe, are neglected or despised by another. Even those which are prized by the same tribe in one age, are rejected by their descendants in another. Thus, the seals and porpoises which, a few centuries ago, were eaten in Britain, and were presented at the feasts of kings, are now rejected by the poorest of the people *.

Man, in general, prefers those quadrupeds and birds which feed on grass or grain, to those which subsist on flesh or fish. Even in the same animal, the flesh is not always of the same colour and flavour, when compelled to subsist on different kinds of food. The feeding of black cattle with barley straw, has always the effect of giving to their fat a yellow colour. Ducks fed on grain, have flesh very different in flavour from those which feed on fish. The particular odour of the fat of some animals, seems to pass into the system unchanged, and, by its presence, furnishes us with an indication of the food which has been used.

While many kinds of animals are rejected as useless, there are others which are carefully avoided as *poisonous*. Among quadrupeds † and birds, none of these are to be found, while, among fishes and mollusca, several species are to be met with, some of which are always deleterious to the human constitution, while others are hurtful only at particular seasons.

---

* See some account of the quadrupeds and birds anciently used as food, i PENNANT's Brit. Zool. ii. p. 726.

† The liver of the Arctic bear is stated to be deleterious, by that accomplished navigator, Captain SCORESBY. Arctic Regions, vol. i. p. 520.

2. *Clothing.*—The use of skins as articles of dress, is nearly coeval with our race. With the progress of civilization, the fur itself is used, or the feathers, after having been subjected to a variety of tedious and frequently complicated processes. Besides the hair of quadrupeds, and the feathers of birds used as clothing, a variety of products of the animal kingdom, as bone, shells, pearls, and corals, are employed as ornaments of dress, in all countries, however different in their degree of civilization.

3. *Medicine.*—The more efficient products of the mineral kingdom, have, in the progress of the medical art, in a great measure superseded the milder remedies furnished by animals and vegetables. The blister-fly, however, still remains without a rival; and the leech, which may here be noticed, is often resorted to when the lancet can be of no avail.

4. *Arts.*—The increase of the wants of civilized life, calls for fresh exertions to supply them, and the animal kingdom still continues to furnish a copious source of materials. Each class presents its own peculiar offering, and the stores which yet remain to be investigated, appear inexhaustible.

PART II.

THE METHOD OF INVESTIGATING THE CHARACTERS OF ANIMALS.

In order to render complete the history of any species of animal, it is necessary to examine the peculiar character of all those systems of organs which have been noticed in the first volume of this Work, and to ascertain those laws which regulate its physical and geographical distribution. As this method of investigation is both laborious and difficult, it has been successfully practised by few. The greater number of naturalists have rested satisfied with an examination of the external characters of animals, and have overlooked those which are furnished by their internal structure. In order to form a correct opinion of the merits of these different methods of investigation, we ought to bear in mind, that the history of a species is incomplete, when its external characters only have been determined; that many of these characters are liable to change, and are, consequently, apt to mislead. The characters, on the other hand, furnished by structure are more permanent, yield more certain results, and are more engaging to a philosophical mind. On this important subject, however, it will be necessary to go more into detail, and to consider what those different characters are, and how they are ascertained.

CHAP. I.

EXTERNAL CHARACTERS

NATURALISTS usually employ, as external characters, the marks which are furnished by colour, dimensions, weight, and shape. On each of these we shall now offer a few observations.

1. *Colour.*—This character, the most attractive to the eye, has engrossed a great share of attention. In many description, of birds, for example, it is the only character which is employed, by ornithologists, in the discrimination of species. But this character, we have seen, varies, in many cases, with age, season, and sex; so that an implicit reliance on its indications, unless where the changes have been previously ascertained, must be apt to mislead. Accordingly, we find individuals of the same species exhibiting different coloured plumage, according to circumstances, which have been exalted to the rank of species, and much confusion, in consequence, introduced into the arrangements of the ornithologist. Similar errors have been committed with species of almost every class. Indeed it seems doubtful, whether a species ought to be constituted from characters furnished by colour alone. Even where the colours appear fixed, other marks, of a less suspicious kind, will readily be detected.

It was for a long time a matter of regret among naturalists in general, that no uniform nomenclature for colours was employed, so that it was frequently impossible, in the perusal of the description of a mineral, plant, or animal, to judge accurately of its colour, from the terms in which it was expressed. This evil is in a great measure removed, by the introduction of tables of colours, with examples. The best of these, for the purposes of natural history, and one which, we believe, to be extensively used, is executed by Mr SYME of Edinburgh, an accomplished painter of objects in natural history, and an accurate judge of colours *. It is a work which every naturalist should possess; and it would be of very great advantage to science, were it generally adopted as a standard of colours.

2. *Dimensions.*—In all animals there is a determinate size, which the body, if in health, and placed in favourable circumstances, is destined to attain. A knowledge of the dimensions of an individual is, therefore, of great use in enabling us to ascertain the species to which it belongs. This is a character, however, which ought to be employed with caution, as the circumstances on which its value depends are subject to much variation. Young animals are smaller than old ones, and often exhibit a different proportion of parts. The males, in quadrupeds, are larger than the females; while, among insects, and even some birds, the females exceed the males in their dimensions. But even in the individuals of the same species, climate exercises a powerful influence. When near the limits of their geographical distribution, the growth is counteracted, the relative dimensions of the parts vary, and the ordinary standard of size ceases to be a criterion of the species.

---

* " Werner's Nomenclature of Colours," by PATRICK SYME, flower-painter, Edinburgh; painter to the Wernerian and Horticultural Societies of Edinburgh, 1 vol. 12mo, 1814.

The nomenclature employed to express size, is taken, in each country, from the ordinary standards of length. In Britain, the English inch, and its fractional parts, are universally employed.

3. *Weight.*—Almost every circumstance which operates in varying the dimensions of animals, likewise exercises a powerful influence on their weight. On this account it is necessary to consider the age and sex of the animal, the season of the year in which it has been killed, and its habit of body. The weight, in this country, is usually expressed by grains and other fractions of a pound; but it does not appear that the same kind of weight is generally used. Apothecaries weight is the most convenient in its divisions *.

4. *Shape.*—The characters furnished by colour, size, and weight, are influenced by so many circumstances, that it becomes difficult to avoid all sources of error. It is, fortunately, otherwise with shape. Many of the soft parts of animals, indeed, vary considerably in their form, according to their condition with regard to fatness; but in all the hard parts, or those which are in constant exercise, the shape is subject to little change. The form of the bill of birds, and the teeth of quadrupeds, seldom varies, while change in the size, weight, and colour of the body, are frequent. The breadth of a bird, taken from the tips of the

---

* In taking the dimensions of animals, a small foot-rule is usually employed; but a piece of tape, with the divisions marked upon it, is in many cases more convenient. For the determination of the weight, a tube, with a spiral steel-wire and index, is the most expeditious. Allowance should be made for the loss of blood the animal may have sustained, or the increase to its weight from the shot with which it has been killed. In the case of small birds, the last precaution is indispensably necessary.

extended wings, is not always the same, in the individuals of the same species. But the outline of the wing, where all the feathers are at full growth, and the shape of the feathers, especially the primaries, furnish characters which may be confided in. The more specific the purpose of any organ, or any part of an organ, the fewer variations take place in the forms which are exhibited.

It is a matter of regret, that hitherto so little attention has been paid to the characters furnished by the shape, in the descriptions of animals which have been drawn up. The task is not a difficult one; and the nomenclature, usually derived from well known forms to which definite terms are affixed, is of easy application.

The naturalists who exclusively employ, in the classification of animals, the marks furnished by the external characters, usually take some notice of the places to which animals resort, denominated the Station, or *Habitat.*

The examination of this character can only be carried on successfully in the natural haunts of the species. In those places, the instincts of their nature are exhibited in action; and we can trace the motions which are executed, and the ends which these are destined to accomplish. When circumstances do not permit us to conduct this examination in the situations which animals ordinarily frequent, we form what is termed a MENAGE; and endeavour, in a confined state, to study those manners, which, in a wild state, were precluded from our observation. The aquatic animals are the most difficult to preserve in a living state; they have, consequently, presented so many obstacles to an examination of their manners, that naturalists remain comparatively ignorant of their history.

The terms which are usually employed to express the relations of the different parts of the animal frame, are seldom so much restricted in their signification as to prevent ambiguity. Dr BARCLAY, inreference to the anatomy of the human body, has succeeded, in his work on Nomenclature, which we have had occasion to quote, in removing much of this ambiguity, by the employment of restricted terms, to express the conditions of position, aspect, and connection. Were this Nomenclature generally adopted by zoologists, much confusion, which at present prevails, might in future be avoided. The following remarks may be of some use to the reader, until an opportunity presents itself of consulting the valuable Work of this distinguished anatomist.

The *Mesial Plane* is supposed to be perpendicular to the horizon, when a horse, for example, is standing in a natural position, and it divides the body, from back to breast, and from the head to the opposite extremity, in two equal parts. The edge of this plane, passing along the back, is termed the *Dorsal*, while the opposite edge, passing along the belly, is termed the *Ventral* edge. Dr BARCLAY terms this last edge the *Sternal*. Strong objections occur to the use of this term, in such an extended sense. The terms, back and belly, which the terms dorsal and ventral express, are used in their most extended sense, to denote the whole extent of both edges of the mesial plane, and are applicable to nearly every animal. The term sternal, from *sternum*, expresses only a small portion of one of these edges, and is alone applicable to the vertebral animals.

If we restrict the phrases, dorsal and ventral, to express the *edges* of the mesial plane, By what terms shall we designate its extremities? The terms *coronal* and *sacral*, which sufficiently mark the terminations of the mesial plane in man, would indicate, in the lower animals, places considerably removed from either end. It would be of importance, therefore, to avoid the use of phrases so very limited in their application, or, when extended, too apt to mislead.

Perhaps the terms *anteal* and *retral* may appear appropriate, the former expressing the extremity at the head, the latter its opposite.

The *lateral plane* is supposed to intersect the mesial plane at right angles. Its extremities are designated by the same terms; but its edges are termed *dextral* or *sinistral*, according as they are situated on the right or left sides, the human body being the standard.

When the body is flattened along the edges of any of these planes, there are particular terms used which express this condition. Thus *depression* refers to a flattening on the dorsal aspect, while *compression* indicates the flattening to be on the side.

The length of the body is usually measured along the mesial, and the breadth across the lateral, plane.

Instead of the terms internal and external, those of central and dermal, or peripheral, have been substituted with propriety. The term *dermal* applies to the skin, or exterior covering of the body, or any organ; while the term *peripheral* expresses the surface of any included organ.

All these terms, which thus end in *ar* or *al*, are adjectives, and are restricted in their signification by the addition of the terms *position*, *connection*, or *aspect*. Thus, the orifice of an organ, with a central aspect, will indicate its *direction* towards the centre. An organ having a *sternal position*, is one situated at or near the sternum; while *sternal connection* would intimate its union with the sternum.

As these terms frequently occur in composition, where their use as *adjectives* would render circumlocution necessary, the learned author of the work on Anatomical Nomenclature already alluded to, recommends their conversion into *adverbs*, by uniting to them the termination *ad*, (or rather *ade*, which is similar). This, however, is a termination in the English language appropriated to substantives, and invariably expresses repetition of the same quantity, or continuation. Its use is by no means necessary, as many of these adjectives, having the termination *ly*, are already employed in the English language as adverbs, as *centrally*, and *laterally;* and all the others can admit easily of a similar change. The termination *en* has likewise been proposed, to limit these different terms, to express connection. In the English language, this termination, when not a sign of the plural, usually expresses composition. Besides, it is seldom that, in description, we are called upon to express simply *connection*. It is necessary that we express the kind of connection. Thus, a *radien* muscle may be regarded as one particularly connected with the *radius*, but the kind of connection is not stated. On the other hand, a muscle of *radial origin* or *insertion*, is one, the connection of which is thus shortly and distinctly expressed.

In many cases, when speaking of the extremities of an organ, we may use the terms *central*, *baselar*, or *proximal*, to express the one which is situated nearest to the body, or its centre; and peripheral, dermal, or distal, to express the other which is most remote. The *base* and *apex*, the *fixed* and *free* extremity, are terms which may sometimes be substituted with advantage.

It must appear obvious, that, with all the knowledge which external characters can convey, there is yet wanting a great deal of information respecting those conditions of structure which influence the manners of species, limit their actions, and give indications of their natural alliances.

CHAP. II.

INTERNAL CHARACTERS.

The examination of the internal characters of animals has hitherto been chiefly carried on by those who have practised the art of dissecting, as connected with medical studies. Among many naturalists, the examination of the structure of animals is considered unnecessary, with a view to their classification, however useful it may be to the physiologist. There has resulted from this neglect a superficial method of examining, describing, and arranging animals, observable in many works on zoology; great importance being attached to characters drawn from organs remotely connected with those most essential in the functions of life. Sounder views, however, are now entertained by the most distinguished supporters of the science The internal structure of animals is now investigated, in order to explain their functions; and the methods of classification which are employed, owe their stability to the justness of the views entertained of their peculiar organisation. The employment of the external characters requires little labour or reflection, and the accuracy of the results is at all times doubtful. The use of the internal characters presupposes considerable exertion and patience, but the results are to be depended upon. Manual dexterity is indeed requisite, but it can be acquired by practice.

In order to examine the internal structure of an animal with any degree of success, some attention is required to its condition at death. In many molluscous animals, the organs are found in such a state of contraction, that it is difficult to determine either their form or structure; the feelers are withdrawn; the entrance to the pulmonary cavity is closed, and the whole body distorted. These evils may be in some measure removed, by suffering the animal to die slowly in water; by dropping it, when in an expanded state, into boiling water, or by removing the shell. In many cases, where immediate dissection is impracticable, the animal is immersed and preserved in spirits of wine. In this state, however, every part becomes corrugated, and the dissection, at a subsequent period, requires the cautious maceration of the parts in water, and furnishes results which are not always to be depended upon. Previous to the dissection of animals, with soft external parts, variable in the forms, it is of great importance to observe their characters in a living state, the forms which they assume, and the motions they are capable of performing. The errors which have resulted from the absence of this previous knowledge, are, we fear, numerous.

In the course of dissection, it is necessary to keep the subject under examination in a steady position. In the larger animals, the methods of accomplishing this end are obvious. In the smaller kinds, which are soft, it is convenient to place them on a board, covered with cork or soft wax, to which they can be fixed by long pins; when hard, it is usual to employ a vice or forceps.

In examining the character of the soft parts of animals, in *air*, it is frequently impracticable to prevent the different portions from pressing upon one another, and interrupting the progress of dissection. This evil is easily obviated, by performing the operation in *water*. The surrounding fluid suspends the different parts, and keeps them in a natural position, and enables the eye more readily to discover their structure and connection. In tracing the viscera of molluscous animals, insects and worms, it is impossible, without the use of water, to obtain any satisfactory views of their position. The subjects may be kept steady in the water, by having the bottom lined with cork or wax to hold the pins. Where very transparent objects are examined, the bottom of the vessel may be of glass, covered with wax, except a circular spot in the middle, on which the parts to be examined may be placed, and light transmitted through them from below.

The instruments employed in dissection, differ in their kinds and form, according to the views or ingenuity of the operator. Scalpels of various forms are used for the mere cutting part of the operation. Fine pointed sharp scissars, make a cleaner cut than any knife, and leave the parts more nearly in their natural and relative position. In tracing the course of vessels or muscular fibres, where the object is not to cut, but to unravel or separate, a needle is usually employed, or, in some cases, where the surrounding matter is very soft, the point of a hair-brush forms an advantageous substitute.

Besides the preparatory circumstances which have been thus shortly noticed, a variety of processes of a chemical and mechanical nature, are employed in the course of dissection, which, in many cases, are indispensably necessary to ensure success. The following merit some notice in this place.

1. *Maceration.*—Water is the fluid most extensively used in this process. When the soft parts of animals are immersed in it for a time, they swell and exhibit more clearly their intimate structure. If the immersion is prolonged, the more soluble parts are abstracted, and the framework of the organ left more fully exposed to view. It is sometimes necessary to press out the water gently, after it has

remained for some time, and substitute a fresh portion in its stead. If the object has remained a sufficient length of time, the different layers of which it consists will separate easily, and a structure will be developed, which its original appearance would not have led one to expect.

The principal use of maceration in water is to separate the softer parts, and to expose the denser framework of the organs. It is, however, in many cases expedient to reverse this process, to abstract the hard parts, with the view of observing the distribution of the softer parts. The macerating fluid employed in this case is an *acid*, usually the muriatic, diluted with water, as this menstruum exercises but a feeble action on the soft parts. When bones, shells, or corals, are steeped in a fluid of this kind, the earthy matter is abstracted, and the cartilaginous basis is left behind, exhibiting in its arrangement its mode of growth and intimate structure.

Frequently, the organs are so much covered with fat, as to be concealed from the view of the observer, and to be in a great measure protected from the action both of water and acids. In such cases, oil of turpentine is usually employed, in which the fat readily dissolves. This method was successfully employed by SWAMMERDAM, in his examination of the viscera of insects.

2. *Coagulation.*—Many small animals, and some of the parts of larger animals, are so soft and tender, as scarcely to admit of examination in their natural state. In some cases they are exposed to cold and suffered to freeze. In general, however, the requisite degree of firmness is communicated to them, by immersing them in alcohol or vinegar. In many cases, in attempting to trace the course of vessels, their walls are found so thin, and their contents

so fluid, as to render the examination extremely difficult. Where the contained fluid is of an albuminous nature, as in the ordinary circulating fluids, its coagulation is easily effected by alcohol or vinegar. Sometimes sudden immersion in boiling water is more convenient. Where the contents of the vessel are chiefly gelatine, as in the dorsal vessel of insects, *tannin* has been successfully employed in its coagulation.

3. *Injection.*—Although the veins are found filled, after death, with the blood of the animal, which, by being coagulated, enables us readily to trace their course, yet the arteries are frequently in an empty state. It is usual to inject these and other empty vessels, with *air* from a blowpipe or glass tube, by which their dimensions and distributions are more distinctly displayed. Instead of air, *water* is sometimes used, either in its natural state, or charged with glue, which gelatinises upon cooling, or with different colours, to point out more clearly the direction of the different parts. *Quicksilver* is likewise frequently employed, where the coats of the vessels are of sufficient strength to resist its weight. Another kind of injection is formed, of a mixture of bees-wax, resin, turpentine, and colouring matter, which, when cold, is sufficiently firm to exhibit the ramifications of the vessels in their natural position *. In some cases, the exterior vessels are corroded by diluted nitric acid, for the purpose of exhibiting more distinctly the wax casts of their interior. The cavities of the bones of the ear, of

---

* Those who wish to become acquainted with the art of injecting, and preparing the organs of animals, may consult, with advantage, " The Anatomical Instructor," by Thomas Pole, 12mo, London 1813.

shells and corals, are filled with wax, or even metal, and the external parts then abstracted by acids.

In many animals of small size, it is impossible to obtain satisfactory information concerning their internal structure. The relations of such, therefore, with other animals, can only be guessed at, under the guidance of *analogy*. Since the examination of the external appearances of animals has been conducted with care, and extended to the minuter distinctions of form, naturalists are now better qualified, by a knowledge of the degree of resemblance, to determine what value is to be attached to external characters, as indifications of internal structure. Formerly, a very few points of resemblance were considered sufficient to warrant the inference of similarity in internal structure, and justify the insertion of the species thus compared, in the same genus. Such hasty combinations are frequent with Linnæus, especially among the mollusca. From a slight resemblance in the form of the aperture of the shell, he included in the same genus, animals which breathe by means of gills and lungs, which are oviparous and ovoviparous, which have the sexes separate or united in the same individual. The repeated exposure of such incongruous conformations has led naturalists, before inferring a resemblance of internal structure, to be convinced that there is a very strong resemblance externally; and even with all this caution, it may be questioned if a resemblance of external form warrants the conclusion of a similarity of internal structure. Yet, perhaps, nine-tenths of the species of animals which are known, rest their claims to their present place in the system, to the adoption of such a rule. Every year, however, demonstrates its fallacy, and produces corresponding revolutions in the systems of arrangement.

In the case of small animals, the external characters and internal structure of which cannot be investigated by any of the methods already pointed out, another instrument is employed, *the Microscope*—scarcely more remarkable for the extent of discovery which it has yielded, than for its fruitfulness of error. In using this instrument, so much depends on the dexterity of the observer, the condition of the eye, and the distribution of the light, that it is exceedingly difficult to avoid error. We have already pointed out, in the case of the shape of the globules of the blood, a very remarkable example of discordant results, in the microscopical examination of the same object by different observers. Numerous examples of the same kind might be referred to, in the highly magnified representations of the parts of the same animals, as given by different naturalists. Indeed, no great confidence can be reposed in the accuracy of such designs, where the sources of deception are so numerous, and where the imagination is often permitted to guide the hand, while delineating appearances but obscurely perceived by the eye.

In the examination of objects with the microscope, it is of great importance to employ, at the first, a small magnifying power, to enable the eye to take in at one view a considerable extent of surface, and perceive the relative position of the different parts. Higher magnifiers may then be applied in succession. Single lenses are seldom used, where the focus is less than one-fifth part of an inch. At this focal distance, it becomes exceedingly difficult to avoid touching the objects, and if their surface is unequal, the relation of the small portion which is distinctly seen, cannot be traced to the surrounding parts, the image of which is obscure. It is of importance, therefore, to endeavour to gain a *distinct,* rather than a highly magnified image of an object.

The *compound* microscope is chiefly employed where amusement is the object in view. In using it, the eye is not so much fatigued, as the image of the object appears under a greater angle, and of an agreeable softness. But when we attempt to employ this instrument for the purposes of discovery, there is reason to complain of the want of distinctness in the image, and the ill-defined appearance of its outline. *Single* microscopes have, therefore, obtained the preference among naturalists, and have yielded all the most valuable information concerning minute objects which we possess. They were employed by Leeuwenhoeck, Spallanzani, Ellis and Muller. The most convenient form of the instrument, is the one which was employed by Ellis, (with the addition of rackwork to move the stage and arm of the lens), and described by him in his essay on Corallines. The pocket-glass or *Hand Megaloscope*, is the simplest form of the single microscope, and constitutes a necessary part of the travelling apparatus (or rather daily dress) of the zoologist.

In the examination of objects with the microscope, it is of importance so to manage the light, as to cause the rays which illuminate the object proceed from one point. This, in the case of the sun, is easily effected by making the rays enter through a hole in the window-shutter, or, when a lamp is used, the rays may be received through a hole in a sheet of pastboard, and all the superfluous ones excluded. The intensity of the light may be increased by a condensing lens, or diminished by the interposition of plates of mica or glass.

When animals are examined in water, it is frequently a matter of considerable difficulty, to keep the lens from touching the surface of the fluid, displacing the object, and deranging the investigation. If the lens, however, be fixed in its socket by means of wax, so as to prevent the water reaching to the upper surface, it may be kept immersed in the fluid, and much practical advantage obtained from its situation. There is, indeed, a considerable loss of magnifying power, which may be supplied by the addition of an eye-glass *, or second lens.

The preceding remarks on the methods of investigating the external and internal characters of animals, indicate the means used in acquiring accurate zoological knowledge. It is now necessary to enquire into the different methods employed in communicating this information to others, and to point out their advantages or defects. Descriptions, drawings and preparations are those in common use.

1. *Descriptions.*—In attempting to communicate a distinct representation of scenery, it is, in general, necessary to avoid a minute detail of circumstances, and rather to make choice of those prominent characters which have made the deepest impression on our own minds. The use of description in this case, is merely to convey an outline, which the imagination is afterwards to fill up. The end in view, in giving descriptions of objects in natural history, is widely different. Each description is intended to serve as a *standard of comparison*, and its excellence, consequently, depends on the accuracy and minuteness of details. If we record merely the more obvious appearances of the indivi-

---

* Dr BREWSTER, in his valuable " Treatise on New Philosophical Instruments," gives the following directions for fitting up a microscope for inspecting objects in water. " The object-glass of the compound microscope, should have the radius of the immersed surface about nine times the focal distance of the lens, and the side next the eye, about three-fifths of the same distance. This lens should be fixed in its tube with a cement which will resist the action of water or spirits of wine ; and the tube, or the part of it which holds the lens, should have an universal motion, so that the axis of the lens may coincide to the utmost exactness with the axis of the tubes which contain the other glasses."

duals of a species, we shall probably fix upon those characters which distinguish the genus, and which are, therefore, common to all the species which it includes.

The descriptions in Natural History are daily becoming more laboured in their details. This arises from the increase of species, and the necessity of determining the characters on which their claim depends. Parts are now studied with care, as furnishing specific marks which were formerly overlooked as useless, and the characters by which species were designated, are now employed to distinguish families or orders.

In Great Britain, during the latter half of the last century, descriptions of animals were usually drawn up in a very superficial manner. The internal structure was in a great measure overlooked, and the more obvious varieties of colour were selected, rather than the more characteristic appearances of the shape. Such, generally, are the descriptions of PENNANT, SHAW, DENOVAN, and even MONTAGU. This is the more surprising, as the eminent naturalists who flourished towards the end of the seventeenth and beginning of the eighteenth centuries (the golden age of British Zoology), excelled in the minute details with which their descriptions abounded. The writings of LISTER, WILLOUGHBY, RAY and ELLIS, furnish very striking examples.

It would contribute greatly to the progress of the Science of Zoology, if the descriptions of species were drawn up in reference to all the different organs, and in the order in which they have been already detailed, with additional notices of their physical and geographical distribution. The observations which have already been made, could, in this manner, be readily classified, and the blanks which require to be filled up would become more apparent.

VOL. II.

2. *Drawings.*—It is often impracticable to convey a correct idea of the characters of an animal, by any description, however minute in its details. Many relations of parts, and many gradations of form, may be perceived by the eye, which words are unable to express. Drawings, therefore, have been resorted to, and have largely contributed to the progress of the science.

In the execution of drawings of zoological objects, the greatest attention should be bestowed on the delineation of those characters which are most intimately connected with organization, and which are, consequently, most permanent. The shape of the body should be carefully studied, and the peculiar form of the component organs. When these are faithfully delineated, as they have appeared in the progress of the investigation, a perfect representation of the object may be communicated.

Coloured drawings of animals are eagerly sought after. They please the eye, and frequently furnish, at a glance, the most marked peculiarities of a species. They are attended, however, with great expence, and can only be purchased by those whose wealth usually leads them to seek after other pleasures than science affords. Fortunately for zoology, they are not necessary to its progress. Colour is the least important character which is employed, and can safely be dispensed with, where peculiarities of form have been faithfully delineated. Indeed, we fear that coloured drawings have retarded, rather than promoted, the science of zoology, by diverting the attention from the form and structure of the organs, which are more immediately connected with the functions of life.

So much attention is required, in executing drawings of animals, to the minute peculiarities of form, that even an experienced artist is apt to overlook the most essential characters. The closest inspection of the drawings should therefore be practised by the eye that has regulated the investigation. The young zoologist ought to study the art of drawing himself, by which his progress would be greatly facilitated, and more accurate results obtained.

Even after the drawings of the objects have been executed with fidelity, there is some risk of errors being introduced in the course of the engraving, in order to suit professional taste *.

3. *Specimens.*—The exhibition of a collection of objects in zoology, well preserved, is calculated to excite an interest in the Science, to refresh the memory, and furnish standards with which newly discovered objects may be compared. Such a collection is usually termed a *Museum*, and the specimens which it contains, are either exhibited in a dried state, or immersed in spirits of wine.

Many of the hard parts of animals, as corals and shells, require little trouble in their preparation. They ought to be freed from extraneous matter, and made clean by maceration in water, aided by the use of a soft brush. Maceration in fresh water is absolutely necessary for the preservation of marine objects. When this is neglected, the salt which they retain becomes moist in damp weather, and never fails to injure the specimen. In the preservation of bones, it is necessary to remove all the adhering flesh, and to extract the fat. They may then be articulated by wires in their natural position †.

---

* Many of the figures copied into the " British Zoology," of PENNANT, have been greatly spoiled by alterations of this kind. The figure of the Blunt Headed Cachalot, for example, which, he says, was copied from the one by ROBERTSON, Phil. Trans. 1770, vol. LX. p. 321. Tab. ix., is as unlike the original as can well be supposed.

† The following directions for cleaning and preparing bones, are given by Mr POLE, and may serve to guide the student in all similar operations.

Many of the parts of animals, which can be dried without losing their form, as the skins of many quadrupeds and birds, are afterwards subject to be attacked by insects, and

---

" As much of the fleshy parts should be taken from bones intended for preparation, as can conveniently be done; but it is not necessary they should be separated from each other, more than is required for the convenience of placing them in a vessel for the purpose of maceration, as in this process it will readily take place. The bones are to be laid in clean water, of such a depth as entirely to cover them, which water should be changed every day for about a week, or as long as it becomes discoloured with blood; then permit them to remain without changing, till putrefaction has thoroughly destroyed all the remaining flesh and ligaments; this will require from three to six months, more or less, according to the season of the year, or temperature of the atmosphere, &c. In the extremities of the large cylindrical bones, holes should be bored about the size of a swan's quill, to give the water access to their cavities, and a free exit to the medullary substance. As by evaporation the water will diminish, there should be more added, from time to time, that none of the bones, or any part of them, may be suffered to remain uncovered, as by exposure they would acquire a disagreeable blackness, and lose one of the greatest ornaments of a skeleton,—a fine, white, ivory complexion. It will be necessary, in order to preserve the skeleton as clean as possible, especially in London, and other large cities, where the atmosphere abounds with particles of soot and other impurities, to keep the macerating vessels always closely covered; as from neglect of this, the water will acquire so much of it, as to blacken the bones. When the putrefaction has destroyed the ligaments, &c. the bones are then fit for cleaning; this is done by means of scraping off the flesh, ligaments and periosteums; afterwards, they should be again laid in clean water for a few days, and well washed; then in lime water, or a solution of pearl-ash (two ounces of pearl-ash to a gallon of water), for about a week, when they may be taken out to dry, first washing them clean from the lime or pearl-ash. In drying bones, they should not be exposed to the rays of the sun, or before a fire, as too great a degree of heat brings the remaining medullary oil into the compact substance of the bones, and gives them a disagreeable oily transparency; this is the great objection to boiling of bones, for the purpose of making skeletons, as the heat applied in that way has the same effect, unless they are boiled in the solution of pearl-ash, which some are of opinion is one of the most effectual methods of whitening them, by its destroying the oil. Bleaching is, of all

speedily destroyed. Various methods have been resorted to, in order to guard against these depredators. If the specimens are occasionally baked in an oven, the eggs and larvæ of many insects will be destroyed, but it is scarcely possible by heat alone to kill them all, without injuring the specimens. The protecting method which has been resorted to, consists in incorporating with the substance of the specimen, some poisonous ingredient, which shall prove fatal to any animal that ventures to feed upon it. The two substances which are now universally employed for this end, are corrosive sublimate or oxymuriat of mercury, and the common white oxide of arsenic.

The corrosive sublimate is usually kept unmixed, and in a state of solution in water or alcohol. This solution is applied to all the parts of the preparation with a brush. Although an effectual preventive of the attacks of vermin, it scarcely possesses sufficient antiseptic powers to resist the

---

methods, the most effectual, where it can be done to its greatest advantage, that is in a pure air; and more especially on a sea-shore, where they can be daily washed with salt-water." *Anatomical Instructor*, p. 99. In order to make natural skeletons of small animals, he adds, " Mice, small birds, &c. may be put into a box, of proper size, in which holes are bored on all sides ; and then burried in an ant-hill, when the ants will enter numerously at the holes, and eat away all the fleshy parts, leaving only the bones and connecting ligaments: They may be afterwards macerated in clean water for a day or two, to extract the bloody matter, and to cleanse them from any dirt they may have acquired; then whitened by lime or alum-water, and dried in frames, or otherwise as may be most convenient. In country places, I have sometimes employed wasps for this purpose, placing the subject near one of their nests, or in any empty sugar cask where they resort in great plenty; they perform the dissection with much greater expedition, and equally as well as the ants. I have seen them clean the skeleton of a mouse in two or three hours, when the ants would require a week." P. 105. Anointing the animals with honey, after flaying them, encourages these dissectors to begin their work.

progress of putrefaction until the specimen is dried. To remedy this evil, the corrosive sublimate is kept dry, and mixed with equal quantities of burnt alum and tanners' bark, or coarse snuff, reduced to a fine powder. The application of this powder facilitates the drying of the specimen, and at the same time powerfully retards the tendency to putrefaction. When the drying is completed, it will be safe from insect foes. Quantities of musk and camphor are sometimes added, which conceal any disagreeable smell; but they may safely be omitted, as they are not necessary to the future preservation of the object, while they add to the expence attending its preparation.

Arsenic has, in a great measure, usurped the place of the corrosive sublimate, in the *preserving powders* which are employed at present, as it does not injure the wires which are occasionally employed in preserving the forms of the specimens. But neither of these powders can be used in the preservation of animals, without incurring the risk of some of the parts drying too quickly, and proving inconvenient to the operator. These evils, however, are removed by the use of the *arsenicated soap*, which is usually made as follows : White oxide of arsenic and soap, two pounds each; sub-carbonat of potash, twelve ounces; quicklime, four ounces; and camphor five ounces; the whole beat up into a uniform mass. This mixture lathers with water like common soap, and may be applied by means of a brush. It does not contract and dry up the parts like the powders, but keeps them in a soft pliable state, counteracts the tendency to putrefaction, and effectually protects them against the attack of insects. It appears to have been first used in the Public Collections at Paris.

Many entire animals, and the parts of others, are of too soft a consistence to suffer being dried, without the loss of form and even texture. These are preserved in widemouthed phials filled with alcohol. When the objects are large, it is proper, before immersing them, to inject a quantity of the spirits into their interior, to prevent incipient putrefaction, and consequent discoloration of the spirits. In order to suspend the specimen in the fluid, it is affixed by a thread to a float of clean cork, or, which is much better, of glass, and sometimes to a slip of wood fixed across. In many cases, the introduction of a bubble of air into the specimen itself, by means of a blowpipe, will effectually secure the object in view.

The greatest inconvenience that attends the use of objects preserved in alcohol, arises from the evaporation of the spirits, and the risk of having the specimen destroyed by putrefaction before the change is observed. When the glass is closed by cork merely, the evaporation sometimes takes place very rapidly, by its capillary attraction, and this effect is sometimes accelerated by a thread from the object passing through the mouth of the vessel along with the cork. The evaporation of the spirit may be retarded by giving it a thin covering of fixed oil; or, it may be altogether prevented, by covering the mouth of the vessel with two or three folds of bladder bound round the edges tightly with pack-thread. If the layers of bladder are well coated with mucilage of gum-arabic, glue, or the white of an egg, the utmost security will be obtained. If a piece of tin-foil be coated on the under side with glue, and then tied closely over the mouth of the vessel, and again coated with glue on the outside, and a slip of bladder tied closely over it, every risk may be avoided. The surface of the bladder may now be coated with coloured varnish, to improve the appearance of the preparation. The advantage attending a covering of glue or mucilage, over the common varnish frequently used, arises from their insolubility in the alcohol. Objects preserved in alcohol may be taken out for the purpose of examining their structure, after carefully macerating them in water.

PART III.

ON NOMENCLATURE.

The ultimate object which the zoologist has in view, in the employment of the preceding methods of investigation, is to complete the History of Species. For the full accomplishment, however, of this end, it is not only necessary to acquire a knowledge of their structure and functions, but likewise of all their mutual relations. This last task can only be executed, by calling to our aid the principles of arrangement, and by distributing animals into divisions or classes, according to the characters which they exhibit. Attempts of this kind have been made by numerous observers; and the various *systems* which have been proposed, differing from one another in the characters employed, and the divisions recognised, intimate very plainly the difficulties inseparable from the subject.

The methodical investigation and distribution of Animals, would be comparatively easy, if the forms and modifications of the different systems of organs exhibited constant mutual relations. Thus, if we consider the organs of any system to be in their most perfect state, when they admit into their structure the greatest variety of combinations, and execute the greatest number of motions or functions, does it happen, that, when we have discovered in any species, one system of organs in its most perfect state, all the other systems may be expected to be in the same condition. The whole history of the animal kingdom contradicts such expectations of *co-existing* characters, and justifies the conclusion, that, in the same species, one or more of the systems of organs may be in a perfect state, in co-operation with others which may be considered as imperfect *.

---

* It is truly surprising to find such an observer as Cuvier, in the face of observations and his own experience, asserting the existence of this mutual dependence of the different organs; or, as he is pleased to term them, the *necessary conditions of existence.* In " his view of the Relations which exist amongst the Variations of the several Organs," (Comp. Anat. vol. i. p. 47.), he says, " It is on this mutual dependence of the functions, and the aid they reciprocally yield to one another, that the laws which determine the relations of their organs are founded,—laws which have their origin in a necessity equal to that of the metaphysical or mathematical laws; for it is evident, that a suitable harmony between organs which act on one another, is a necessary condition of the existence of the being to which they belong; and that if any one of the functions were modified in a manner incompatible with the regulations of the others, that being could not exist." That such harmony prevails in every species is evident; but instead of being always produced by the same agents in the same state of mutual dependence, it is maintained in the midst of a diversity of combinations, by a variety of *compensating* means, which display in the most astonishing manner, the endless resources of the wisdom and power of the Great Creator.

In illustration of the same views, he adds, " an animal, therefore, which can only digest flesh, must, to preserve its species, have the power of discovering its prey, of pursuing it, of seizing it, of overcoming it, and tearing it in pieces. It is necessary, then, that this animal should have a penetrating eye, a quick smell, a swift motion, address, and strength in the claws and in the jaws. Agreeably to this, necessity, a sharp tooth, fitted for cutting flesh, is never coexistent in the same species, with a foot covered with horn, which can only support the animal, but with which it cannot grasp

In order to exhibit the mutual relations of animals, by distributing them into groups, the naturalist may employ two different methods, which will not, however, furnish the same results, or lead to the same combinations.

If we attend to any system of organs, we speedily discover that it exhibits in the different species a variety of modifications. In some, it exists in its simplest, in others, in its most complex form; and between these extremes, there are many conditions distinguished by the presence or absence of particular parts, or by equally obvious variations of form or structure. In reference to the cutaneous system, we have already seen that it possesses hair in some, feathers in others, or scales, shells, or crusts, while in many, it is destitute of these appendices. We have here both positive

---

any thing; hence the law by which all hoofed animals are herbivorous, and also those still more detailed laws which are but corollaries of the first, that hoofs indicate dentes molares with flat crowns, a very long alimentary canal, a capacious or multiplied stomach, and several other relations of the same kind." P. 55. This specious reasoning, would certainly lead to the admission of these necessary laws of co-existence, were the statements advanced correct in all their bearings. But the operations of Nature are not restrained by such trammels. Quadrupeds possessing the common quality of being carnivorous, have not all the same number of teeth, nor of the same shape, neither the same kind of stomach or intestines. Again, all herbivorous animals are not hoofed, for many of them are digitated as the hare. All hoofed animals have not flat crowned teeth, like the bull, nor pointed teeth like the boar, nor a simple stomach like the horse, nor deciduous horns like the stag, nor a reservoir for drink like the camel, nor digestive organs that do not require any like the sheep. Indeed, the number of varieties included under one species, the number of species belonging to a genus, and the number of genera in an order, intimate the variableness of the conditions of co-existence, and the absence of those supposed laws of relation, the belief in the *mathematical necessity* of which, has contributed to augment the clumsy fabric of modern Materialism."

and negative characters, with which to construct the primary divisions. In the first division, subordinate groups may be formed, distinguished by the terms Pilose, Plumose, Squamous, Testaceous and Crustaceous. Each group might be farther subdivided by the numerous modifications which these cuticular appendices exhibit. If we pursue the same course with any of the other systems,—the motive, sentient, nutritive or reproductive, each will exhibit peculiar classes and orders. If we attempt, therefore, to classify animals in this manner, we shall obtain as many *Natural Methods* of classification as there are *Systems of Organs*.

The advantages attending the classification of animals, by the systems of organs they possess, are numerous. We are able, in this manner, to comprehend all the modifications of form, structure, and function, of which an organ is susceptible, and consequently to become better acquainted with its relations and its uses. But, at the same time, it would be inconvenient to have so many different primary divisions, and to have the characters of a species so divided and separated, that for the right understanding of the whole, it would be necessary to consult the divisions of eight or ten different classes. This mode of classifying animals by their different systems of organs, though the only one entitled to the appellation of a *Natural Method*, has scarcely ever been practised by zoologists, unless in detached physiological disquisitions.

If we employ, exclusively, as the basis of our classification, any one of the systems of organs, we shall find, that having formed a *natural genus* from the obvious marks which it furnishes, it shall contain species differing greatly from each other in the characters exhibited by one or more of the remaining organs of the other systems. The genus Helix of Linnæus, is a natural genus in reference to a portion of the cutaneous system, the shell, but it exhibits in the characters of sentient, nutritive, and reproductive systems of the species, *artificial* combinations. Even when there is a very close agreement among the species, in regard to several systems of organs, remarkable differences prevail in the functions of others. The genus Lepus, universally considered as a natural one, may serve as an illustration, by contrasting two of the best known species, the hare and rabbit. These animals so nearly resemble each other in form and structure, that it has puzzled the most experienced zoologists to assign definite distinguishing marks*. Yet there are many circumstances in which they differ (besides the colour of the flesh, when boiled, and their manner of escaping from their foes), in reference to the reproductive system. The nest of the hare is open, constructed without care, and destitute of a lining of fur. The nest of the rabbit is concealed in a hole of the earth, constructed of dried plants, and lined with fur which is pulled from its own body. The young of the hare at birth, have their eyes and ears perfect, their legs in a condition for running, and their bodies covered with fur. The young of the rabbit at birth, have their eyes and ears closed, are unable to travel, and are naked. The maternal duties of the hare are few in number, and consist in licking the young dry at first, and supplying them regularly with food. Those of the rabbit are more numerous, and consist of the additional duties of keeping the young in a suitable state of cleanliness and warmth. The circumstances attending the birth of a hare are analogous to those of a horse, while those of the rabbit more nearly resemble the fox. This illustration furnishes a very striking example of

---

* There is a paper by the Honourable DAINES BARRINGTON, containing an "Investigation of the specific Characters which distinguish the Rabbit from the Hare." Phil. Trans. 1772, p. 4.

the danger of trusting to analogy, since very remarkable differences may prevail, even where the points of resemblance are very numerous, and justifies us in concluding, *that in a natural genus there are artificial combinations.*

In the formation of classes and orders from the characters furnished by one system of organs, arbitrarily selected, we are in danger of being misled, by considering all the parts of that system of equal importance to its functions. This, however, is very far from being the case. There is frequently an intermixture of the organs of different systems, so that we are apt to be led away from the one we resolved to employ, and to use the characters of the organs of another system, which, even in their most perfect form, we had deliberately rejected. Thus, in examining the gills of fishes, we may confine our observation to the number, extent of surface, and mode of attachment of the red parts, which are subservient to the purposes of respiration. But if we extend our observations to the central surface of the arches of those with free gills, we enter the sphere of the organs of deglutition. Errors of this kind are frequently committed even by experienced observers.

The *Natural Method* of classifying animals by the different systems of organs they possess, being inconvenient in its application, and the *Artificial Method*, consisting in the exclusive employment of any one system of organs, producing incongruous combinations, a *Mixed Method* has been adopted by many naturalists, which appears to answer every useful purpose. The characters which are employed in the construction of this last method, are derived from all the systems of organs, and the subordinate parts of these, assigning as the test of their importance, the *extent* of their occurrence and of their *influence.*

The characters which are of *universal* occurrence, are to be considered as the distinguishing marks of the subjects of the *animal kingdom*. These are few in number, and have already come under our consideration *. If we lay aside these, and attentively examine the remaining characters, we shall find some of them common to extensive groups, while others make their appearance in a small number of species. If we take a hundred species, for example, it will be practicable, in many cases, to discover characters which are common to nearly fifty of these, but which are either absent or incorporated with other characters in the remainder. We thus obtain two classes of fifty each, the one distinguished by a *positive*, the other by a *negative* mark. If we procure an animal, and wish to determine to which of these hundred species it belongs, we first determine its class, from its positive or negative characters, and we are thus saved the trouble of comparing it with the fifty species of the class to which it does not belong. If we subject each of these classes to a similar examination as that which led to their discovery, we may be able to effect a similar twofold division, and thus facilitate, in an equal degree, the labour of the comparison. If we proceed in this manner, halving every successive division, we shall ultimately reach that limited number of species with which the individual we possess can be readily compared.

It must appear obvious, that, by this mode of classification, a great deal of tedious and useless repetitions will be avoided in the descriptions of species, and the enumeration of the marks of genera. The characters of the class will comprehend all the properties common to its orders, genera, and species. The characters of the order will include all the properties common to its genera and species. The characters of the genus will include all the properties common to the species; and in the description of a species, it will only be necessary to enumerate

---

* Vol. i. chap. 3.

those modifications of character which constitute the *peculiarities* of its vital principle. In giving the characters of a species, we are apt to introduce those which are co-extensive with the genus or order to which it belongs. In the enumeration of the characters of what are falsely termed by modern zoologists *natural genera,* such repetitions prevail to a very inconvenient extent, and in some measure retard rather than promote the interests of science.

The employment of the twofold method of division, by positive and negative characters, is so obvious in its expression, and so easy of application, that the reluctance which many naturalists seem to display in using it, may well excite our surprise. It would keep in check that rage for innovation, which so peculiarly marks many modern cultivators of the science, by assigning the limits within which it might be exercised, and it would direct their efforts to fill up the blanks which remain, instead of amusing themselves with new systems, fabricated one day, and destroyed another. This method, however, may be considered as yet in its infancy, but its adoption appears to me to be the only remedy hitherto devised for checking those fluctuations which have deservedly exposed the systems of natural history to ridicule; and for rendering the progress of zoology steady and triumphant.

In selecting the characters by which we are to be guided in the formation of classes and orders, there will be some room for difference of opinion. This, however, may be considerably diminished, by considering that those characters which are of easiest detection, and which produce the most equal division of species, in all cases, deserve the preference. Judging by these rules, the twofold division of animals into vertebral and invertebral, is preferable to their division into such as have the nervous matter apparent, and such as have it concealed or disseminated. The one set of characters is easily recognised, and keeps together a great number of animals, in which the systems of organs exhibit the same general structure. So apparent, indeed, are the characters of the former groups, that they have been recognised by Aristotle, Ray, and many modern authors.

When we employ a threefold method of primary division, as Linnæus, or a fourfold as Cuvier, we are obliged to make use of more characters than one in their construction, and depart from that unity of principle to which the twofold method can exclusively lay claim; in other words, it becomes impossible to assign any limits to the number of the primary classes. If we employ one character, it must either be positive or negative. The negative is indivisible, but the positive is susceptible of inferior arrangement, from the numerous modifications of the organs from which it has been derived. If, instead of making divisions, from the modification of the positive character, or from its negative, we make each of our *primary* divisions from positive characters, in that case, their numbers must necessarily equal the number of systems of organs which animals possess, and the same animal must have a place assigned it in each class (or in several of them,)—a method of classification which, we have already said, is *natural*, but which has been declared inadmissable.

The primary divisions of Linnæus, which are three in number, may serve to illustrate some of these remarks.

1. Cor biloculare, biauritum; *sanguine* calido, rubro. (Mammalia, Aves, .)

2. Cor uniloculare, uniauritum; *sanguine* frigido, rubro. (Amphibia, Pisces.)

3. Cor uniloculare, inauritum; *sanie* frigida, albida. (Insecta, Vermes.)

---

* Systema Naturæ. Holmiæ, 1766, i. p. 79.

It is here assumed, that a *heart* exists in all animals, and that the modifications which it exhibits, are best adapted for the construction of the primary divisions. It was, however, well known to LINNÆUS himself, that the existence of a heart in many vermes, had not been demonstrated, and that its occurrence was not even probable. If he had resolved to employ the heart as the basis of his classification, he ought to have formed his divisions from the circumstance of some animals having a heart, while others have nothing analogous to that organ. But, even after having made choice of its modifications, we still find him departing from that unity of principle which alone is compatible with precision. The heart having two auricles and two ventricles, distinguished one class sufficiently from the other, in which one auricle and one ventricle only were present. In the formation, however, of his third class, he has recourse to a repetition of the term *uniloculare*, by which it becomes obvious that it is merely a modification of his second class. But it may be said, that this third class is distinguished from the second by a negative character *inauritum*. It is obvious, however, that the characters *uniauritum* and *inauritum*, occupy a subordinate rank to *uniloculare*, and should mark orders, not classes. It is likewise unfortunate, that this character should be false in reference to several of the tribes which it includes. There is a new character introduced into the third class, which shews very plainly the obstacles which presented themselves to the author in its formation. The circulating fluid of the two first classes, he terms *sanguis*, that of the second *sanies*. It appears, however, that this character is subordinate to the one which should have formed an order, as it depends not merely on the unilocular heart, but on the absence of an auricle. The coldness of the sanies is similar to the coldness of the sanguis of the second class, and it was not unknown to the author, that the white colour was not common to all species of the class. The institution of the third class was not here necessary, as the animals which it included obviously appeared deserving of holding the rank of an order only in the second class. The system of Cuvier, though free from the error of assigning characters to classes which do not apply to all the orders included under it, (although in the subordinate groups this error is of frequent occurrence), is still faulty in the employment of modifications of positive characters, along with positive and negative ones, in the construction of classes of the first degree.

In the application of the mixed method, by the employment of positive and negative characters, it will frequently be found difficult to obtain a division into equal numbers of species; one division probably including twice as many as the other. This inequality, however, in many cases, proceeds from the attention bestowed on the examination of one tribe in preference to another, and perhaps from the extinction of some species which belonged to the present order of animals. Yet, whether a natural, mixed, or artificial method be employed, there is no numerical equality among the species included under each of the divisions.

The various subdivisions which occur in the mixed, or indeed in any other method hitherto proposed, can never be regarded as co-ordinate, although the terms by which they are designated intimate an equality of rank. The same characters which form an order or genus among birds, are not employed, or rather do not exist, in fishes; the orders and genera of which are founded on characters derived from other sources. There never, therefore, can exist that relative subordination of groups, in the different divisions, which many naturalists seem to acknowledge, by the anxiety they display to have the same terms placed in the same rank of succession, even where there is no necessity for their employment. Zoologists have not yet ventu-

NOMENCLATURE. 147

red to form species without individuals, or genera where there are no species; but in a valuable work on fishes, fifteen *orders* may be observed, instituted and named, which have no representatives in nature *.

In consequence of adopting a single character, derived from a particular organ, in the construction of subdivisions, these will be found variously related to the different categories of animals in the system. Even the positive character of a particular organ on which a division depends, will, in many cases, be found so modified in some of the subdivisions, by the intermixture with the characters of other organs, as to intimate a resemblance to other groups, having a rank depending on very different relations. Among the Mammalia, the whale, by swimming, intimates its resemblance to fishes, and the bat, by flying, makes an approach to a bird, although vast differences exist in other characters. However numerous the relations of a group may appear to be with many others, it is not practicable to make these follow one another, in reference to such gradations or transitions. When a group is constituted by the help of a positive character, it is seldom expedient to constitute the subdivision from its modification, a positive character, furnished by some other organ, being preferable. In like manner, a group with a negative character, will be subdivided most suitably by the employment of a positive one.

Besides these remarks, which are of a very general nature, it is necessary here to take notice particularly of those subdivisions which are employed in Zoology, the names by which they are designated, and the rules which are observed in their construction.

---

* " Histoire Naturelle des Poissons," par le C.^{en} Lacepede, Paris, 5 vols. 4to. 1803.

I. Species.

This term is universally employed to characterise a group, consisting of individuals possessing the greatest number of common properties, and producing, without constraint, a fertile progeny.

The number of individuals belonging to a species, bears no constant ratio to the numbers in other species, even of the same genus. In the same species, the number is subject to remarkable variations, at different times. These changes are either produced by alterations in the physical circumstances of their station, or by the diminution or increase of their foes, their food, or their shelter.

Differences of character prevail, to a limited extent, among individuals of the same species. These tend greatly to embarrass the student in his investigations, and have led to the introduction of many spurious species into the Systems of Zoology. Some of these differences necessarily prevail, as the indications of the sex; others are accidental, and constitute what are termed Varieties.

A. *Sexual Differences.*—These occur, more or less, in all those species in which the male organs are seated on one individual, and the female organs on another. In all cases, the male is considered as the representative of the species. While the female, in some species, differs remarkably from the male in external characters, there is still an agreement in structure, with the exception of the organs of the reproductive system, and the modifications of some parts subservient to their functions. When a female individual comes under notice, it is frequently very difficult, if not impossible, to determine the species to which she belongs, where external characters alone are employed. This difficulty, in the case of birds, meets the student at every step;

but it in a great measure disappears, when the internal characters are chiefly relied on.

In a domesticated state, it sometimes happens, that female birds assume the plumage of the male, when they have ceased, from old age, to be capable of producing eggs. This was first distinctly determined by JOHN HUNTER, in the case of the pheasant and peacock *. Sir E. HOME likewise records an example of the same change taking place in a duck, at the age of eight years †. Mr TUCKER, and more recently, Dr BUTTER, have recorded similar occurrences in the common fowl ‡. The females of the last species in the absence of the cock, may frequently be observed assuming the attitudes of the male, and even, according to ARISTOTLE, acquiring small spurs. The same illustrious observer adds, that the cock has been known to make up the loss of the mother to the brood, by devoting himself entirely to the supply of their wants ‖.

---

* " Account of an extraordinary Pheasant." Phil. Trans. 1780, and republished in An. Econ. p. 63.

† Phil. Trans. 1799, p. 174, where some cases of sexual monstrosity in quadrupeds are noticed.

‡ Mem. Wern. Soc. vol. iii. p.183. In addition to all these species which thus change, BLUMENBACH (Phys. p. 369.) refers to the Pigeon, Bustard, and Pipra rupecola; and BECKSTEIN records the Turkey.

‖ Hist. Anim. ix. 49. The following singular change of character in the Turkey-cock, came under my own observation four years ago. While two female birds, his companions, were sitting on their eggs, he was in the habit of frequenting the nests occasionally. Even after the females were removed, along with their young, he resorted to one of the nests, and continued sitting for several days in succession, scarcely moving away to take food. At length a dozen of hens' eggs were put into the nest, and on these he sat regularly until they were hatched. When, however, the young chickens began to make a noise, and to break open the shell, he endeavoured to kill them with his bill, and throw them out. Only one, indeed, was saved from his fury. After performing this tedious task of incubation, he returned to his ordinary habits.

These instances of monstrosity, as they occur but seldom, and chiefly in domesticated animals, present few obstacles to their systematical arrangement. It is frequently, however, a matter of considerable difficulty to determine the sexes at an early age, and the young from mature individuals of the same species.

B. *Varieties.*—Although the vital principle of every animal is restrained, in all its operations, within certain limits, peculiar to each species, these are not so very confined as to prevent slight alterations of character from taking place, without disturbing the harmony of the whole.

Individuals of the same species, living in a wild state, and exposed alike to the same physical circumstances, rarely exhibit variations of character to any extent. When, however, the individuals of the same species are compelled to live in different countries, they are not placed in the same physical circumstances. Constitutional efforts will be made by each of these groups, to accommodate themselves to the conditions of their station, and these will display themselves in variations of colour, size, and even the form of certain parts.

As domestication subjects animals to the greatest variety of changes in their physical circumstances, so it operates most powerfully in the production of varieties. These take place in obedience to the tendency to accommodation, or redundant parts are produced in consequence of excess of nourishment.

Many of these variations are peculiar to individuals, while others are permanent, and capable of being transmitted to the offspring. These permanent varieties in the human species, now generally limited to five, the Caucassian, Mongolian, Ethiopian, American, and Malay, have given rise to the belief, that there are several species of the genus Homo. This opinion, however, is now generally abandoned.

When it is intended to give a correct and complete delineation of the character of a species, it is necessary to attend to the following circumstances:

1. *Specific Name.*—This is intended to be used along with the name of the genus, to distinguish it from the other species with which it may be combined *. It is, therefore, usually an adjective, in concord with the name of the genus as a substantive; and expresses some circumstance connected with colour, form, habits, station, or distribution. Size is seldom resorted to as furnishing a name, in consequence of the absence of a fixed standard, and the terms *majus*, *minus*, *parvus*, and others of the same kind, are therefore seldom employed.

In some cases, the specific name is a substantive, and occurs either in the nominative or genitive case, and without reference to the gender of the name of the genus. This want of concord happens, when a species has been long known by a distinct appellation, and when this is employed in science as its specific name. Thus, in the genus *Turdus*, while one species is distinguished in the ordinary manner, viz. *T. torqua-*

---

* The names of the different divisions, as Species, Genera, and Classes, are expressed, by universal consent, in the Latin language. They are either derived directly from that language or from the Greek, in which latter case they obtain a Latin termination. Provincial names are excluded as barbarous. These are frequently difficult to pronounce, harsh in their sound, and can scarcely admit of a Latin termination. Where they are not liable to such objections, they may be occasionally suffered, yet, for the sake of *uniformity*, they should be used sparingly.

tus, another *T. Merula*, furnishes an example of the exception here referred to. When the specific name is in the genitive case, it is always derived from the proper name of the zoologist who discovered it, or who contributed to illustrate its characters. Thus, *Liparis Montagui*, was so named by DONOVAN*, in honour of the late GEORGE MONTAGU, Esq. who first detected it on the Devonshire coast. The application of the proper names of zoologists, to the construction of the specific names of animals, ought to be restricted to those who have illustrated the species. Of late years, however, this honour has been bestowed on observers to whom the species has even been unknown; and not contented with using the names of zoologists, those of wives, friends, or patrons, have been extensively employed. To bestow zoological honours on those who are not interested in the progress of science, is ridiculous; and to neglect the original discoverer in order to do this, is base †. It were better, perhaps, to proscribe the practice; or if it is to be persisted in, the termination ought not to be in the genitive, but the nominative case. In all the exceptions noted in this paragraph, it is customary to distinguish them from the ordinary specific names, by making the first a capital letter.

Before concluding these remarks on specific names, it is necessary to state, that the discoverers of species have the undoubted right of imposing the names, and that these

---

* British Fishes, Tab. lxviii.

† The motive might be misinterpreted, were I to point out instances in which such names have been imposed from weak and selfish purposes. It may be of more use to quote the opinions of LINNÆUS, in reference to such honours, as connected with the generic names of plants. " Unicum Botanicorum præmium, hinc non abutendum est."—" Hoc unicum et summum præmium laboris, sancte servandum et caste dispensandum ad incitamentum et ornamentum Botanices."—Ph. Bot. 170.

ought never to be altered. They may have a harsh sound, be barbarous, or even absurd, yet all these objections are as nothing, when compared with the evils accompanying the multiplication of synonymes. Even without any good reason, many naturalists have presumed to change the names which the discoverer of the species imposed upon them, in order to obtain what appeared to them uniformity of nomenclature, or rather for the purpose of increasing their own importance. The period is probably not very remote, when this mischievous spirit of innovation shall receive an effectual check, in consequence of credit being attached only to those who develope new *characters*, and not to those who disturb science by the fabrication of unnecessary names. Where synonymes have unavoidably been created, in consequence of the want of communication between distant observers, the rule universally known, but not equally extensively observed, is to give the preference to *the name first imposed.*

2. *Specific Character.*—When a genus contains only one species, it is not necessary, even for the sake of uniformity, to employ a specific character. Its use is merely to assist in discriminating readily the different species which a genus may contain. For this purpose, the species, or their descriptions, are compared together, in order to discover their most striking differences. These essential marks are expressed in one sentence, which is added immediately after the specific name. Before the introduction of the Linnæan system, this specific character was frequently employed instead of the specific name. Its inconvenient length, however, was generally acknowledged; so that the substitution of the specific name by the illustrious Swede, in references and conversation, was universally acceptable to naturalists. The specific names were termed by him *nomina trivialia;* because they frequently had a reference to the accidental, instead of the essential, characters of animals.

LINNÆUS was disposed to restrict his specific characters to twelve words, an example which succeeding naturalists have usually imitated. In general, however, all that is necessary to be included in a specific character, particularly where all the species which are known are accessible to our inspection, may be expressed, even within the limits of the Linnæan rule. Where the characters of a species are subject to variation, from sex, age, or season, it is of importance to employ those marks only which are permanent, and common to all the individuals.

Specific characters necessarily vary with the addition of new species. If a genus consists of two species, the one constantly black, and the other white, the colour will furnish the marks for the specific character. Should a third species be discovered likewise of a black colour, it will not only be necessary to frame a specific character for the third species, but to reconstruct a new one for the first.

3. *Specific References.*—It is usual to add, immediately after the specific character, a reference to those works in which the species has been described, under whatever name, and to arrange the whole in the order of time. It does not appear to be worth while to quote those authors who have not contributed any thing towards the elucidation of the species, but who have merely copied from others, unless in the case of those systematical works which are generally consulted. When references are made without selection, much room is unprofitably occupied.

4. *Specific Description.*—This includes every circumstance in the history of the species, not previously recorded in the characters of the groups in which it is included. The mature male, during the season of love, is described as the standard of the species, with which all the differences arising from sex, age, or season, are compared.

II. Genus.

This division, which is designated by a peculiar title, serves as a surname to the species which it includes.

If we are to consider as a *genus*, every division by which *species* are distributed into groups, and regard it as only one step higher in the scale of our classifications, the number of species which it can include will in general be few in number. Indeed, if we employ the twofold method of division, by positive and negative characters, the lowest groups will seldom contain more than two species. When genera occur greatly exceeding in species this limited number, we may safely conclude, that the structure of the species is imperfectly understood, and that the distinguishing characters are chiefly derived from the modifications of one organ.

1. *Generic Name.*—The term employed to designate a genus, is always a substantive in the singular number, and nominative case. In order to avoid all grounds of confusion, and that the term may lead the mind to no other image, primitive words, and such as are appropriated to other objects, are carefully avoided. For similar reasons, all words indicative of comparison or resemblance, or those ending in *oides*, are rejected. Neither hybridous nor barbarous terms are employed. Those in common use are almost exclusively derived either from the Greek or the Latin [*].

It is somewhat remarkable, that, in mineralogy and botany, generic titles should, in many cases, be derived from the names of those naturalists who have contributed to the

---

[*] The " Philosophia Botanica," 8vo, 1751,-6. of LINNÆUS, may be perused with advantage, on the subject of Nomenclature. Many of his rules, however, are empyrical, and have never been attended to by naturalists.

advancement of these sciences, while the generic titles of animals are never derived from the same source, nor similar honours bestowed on zoological observers. We have never heard a satisfactory reason assigned for this adopted course of nomenclature, though the practice of LINNÆUS has been quoted by the slaves of authority in its justification. In perusing the work of LAMOUROUX, " Histoire des Polypiers Coralligènes Flexibles," we could not avoid being struck with the absurdity of commemorating the Mythology of Greece and Rome, by bestowing upon his new-formed genera, the titles of the Nymphs of the Springs and of the Sea, instead of the names of PEYSSONNEL, TREMBLEY, ELLIS, and PALLAS, and a host of other observers, whose labours did honour to the age in which they lived.

That the author had no particular dislike to the use of proper names as generic titles, appears obvious from his genera Elzerina and Canda, in which, however, they are applied without judgment *.

When a generic name has once been bestowed on a well defined group, it ought not to be changed, even although a

---

* " J'ai donné à ce genre le nom agreable D'Elzerine, parce que celle qui le portait, fille de Neas, Roi de l'ile de Timor ou se trouve ce Polypier, est citée honorablement dans le voyage aux terres Australes, de PERON et LESUEUR," p. 122.—" J'ai donné à ce genre le nom de Canda ; ainsi, s'appelait une jeune Malaise citée dans le voyage de PERON et LESUEUR ; ces naturalistes ont raporté cette elegante cellarieé des cotes de Timor," p. 131. Did it not occur to this author, that the claims of the two French naturalists PERON and LESUEUR, the discoverers of the species, for the reception of which these genera were formed, were preferable to the two Asiatic girls whom he has attempted to immortalize in science ? Why are none of the new species discovered by these observers, named in honour of them, but employed to celebrate the author's living friends? The names of species which have survived half a century, are, in some cases, wantonly changed, and PALLAS is robbed to flatter SAVIGNY.—See p. 212.

more appropriate term has been discovered. This rule, which the most judicious zoologists have uniformly respected, was underrated and violated by several French naturalists, during the period of the Republic. It was their object, at that time, to make every thing appear to originate in their own nation. The names both of genera and species were therefore changed, to suit their purposes. In Mineralogy, M. Haüy has contrived, by degrees, to change nearly all the generic names; and it is still more surprising, that many of these new names are coming into use even in this country. In the writings of Lacepede and Lamark, such innovations are but too frequent. I would be deficient in candour, however, were I not to state, that M. Cuvier has, with more propriety, in general, avoided any obvious exhibitions of this national appropriation. Where useless changes are thus produced in nomenclature, their authors, and their names, should be overlooked.

It has become customary, of late years, to add, immediately after the generic name, the name of the author who first established it. This is a very useful plan, as it facilitates research, and guards against the inconvenience of synonymes. A few naturalists, in addition to the author of the genus, add likewise the names of those who have adopted it, and even enrol their own in the number. Such quotations, however, are, to say the least, unnecessary.

2. *Generic Character.*—As this is intended to distinguish the different groups of a higher division from one another, in the most expeditious manner, the more obvious and essential marks only are employed. Where the previous subdivisions have been executed with care, the generic character may usually be expressed in a very short sentence, not exceeding the Linnæan limits of twelve words.

3. *Generic Description.*—As the generic character is intended solely to facilitate investigation, its requisite brevity precludes the possibility of enumerating all the properties which the species possess in common. This defect, however, is supplied by the description, which is better known among naturalists under the denomination of *natural character*. When this description is full, and constructed with due regard to the characters of the higher groups, it saves a great deal of repetition in giving the description of the species; and, during its composition, unfolds the peculiar marks by which they are to be distinguished. It powerfully exercises the judgment, and seldom fails to unfold new views of the natural affinities of animals, and the modifications of their several organs.

In the Linnæan System, the divisions of a higher kind than genera were limited to two in number, viz. orders and classes. Classes constituted the primary divisions, and included orders: these last included the genera and species. The modern improvements of science, and the vast additions of new species, have multipled the number of divisions in the system to an extent greatly beyond the five originally employed by the Swedish naturalist. In many cases, they exceed twenty in number. In order to give to each of these groups an appropriate title, naturalists have denominated them divisions, classes, orders, tribes, legions, families, sections, subdivisions, &c. We have already stated the want of co-ordination between these groups, and are therefore disposed to prefer distinct appellations for each, expressive, if practicable, of their essential character, rather than to designate them by terms, which, while they occur frequently, have never the same equivalent expression.

In the construction of *Families*, which consist of genera, related to each other by certain common properties, the introduction of new terms is easily avoided, by denominating them by the name of the oldest established genera, and bestowing upon them a patronymic termination, as *des* or *dæ*. The latter is generally preferred as a termination, since it never occurs in generic names; while the former, being of frequent occurrence, may, if employed, occasion mistakes.

In all these different divisions, it is of importance to express briefly, the essential distinguishing marks, in order to facilitate research; and to enumerate all the common properties, to avoid the necessity of repetition.

The observations which have now been offered, lead to the conclusion, that Species alone, in the divisions of zoology, are permanent, all the others being subject to change. The higher divisions are of our own creation, and are altered occasionally, to make them correspond with our increasing knowledge. The discovery of a few new species, compels us to form sections in a genus, or, instead of these, to construct new genera. The old genus, then, becomes a family, and a similar advancement of rank takes place in all the higher divisions. Many of the Linnæan genera are now at the head of eight or ten inferior divisions, and the genera of modern systems, in consequence of future discovery, will, in their turn, experience a similar elevation.

PART IV.

CLASSIFICATION OF ANIMALS.

If we employ the twofold method of classification recommended in the preceding part of this work, we shall be at no loss to obtain the first divisions of the animal kingdom, VERTEBRAL and INVERTEBRAL, however great difficulty may be experienced in the construction of the subordinate groups. These two divisions depend, the one on a positive, the other on the negative character, and possess the advantage of being easily recognized. In this respect, and indeed in every other, they have the decided superiority. Any other basis of division hitherto employed is faulty, in not including a number of common properties, in effecting unnatural separation among kindred tribes, or in being founded on characters which are merely modifications of some positive quality. Without wasting the time of the reader, in dwelling on the defects of these different systems, we shall proceed at once to an exposition of the characters of the method employed.

I.
VERTEBRATA.
Vertebral Animals.

CHARACTER.—ANIMALS FURNISHED WITH A SKULL AND VERTEBRAL COLUMN FOR THE PROTECTION OF THE BRAIN AND SPINAL MARROW *.

DESCRIPTION.—The properties which the vertebral animals possess in common, are numerous, and clearly indicate the unity of the plan according to which they have been constructed. In reference to the nervous system, indeed, a conformity of character here prevails, which is not observable in that, or any other system of organs among the invertebral tribes.

1. *Cutaneous System.*—The skin here appears in its most perfect state. It exhibits the cuticle, the mucous web, and the corium, in all cases; nor are the muscular and cellular webs often wanting.

2. *Osseous System.*—The bones, which are destined for the support of the body, always abound in phosphate of lime, and they increase in size, by the expansion of their cartilaginous basis. The basis of the skeleton may be regarded as the skull, to which are joined the bones of the face, on one quarter, and the spine on another. The spine consists of vertebræ, with a groove or canal on the dorsal aspect, for the reception of the spinal marrow. The limbs, or

---

* It may be objected to this primary division of animals into Vertebral and Invertebral, that the characters which are employed in their discrimination, are derived from *modifications* merely of the nervous system, instead of depending on the presence and absence of particular organs. It would certainly appear to be more consonant with the principles of classification which have been recommended, to divide animals into such as had apparent nerves, and such as were destitute of apparent nerves. Still, however, as there is no animal without a nervous system, it would only be employing a different modification of the same organs, and one, the limits of which, are less distinctly defined.

extremities, where present, never exceed four in number, and are more or less intimately articulated to the vertebral column.

3. *Muscular System.*—The muscles in this group of animals, are principally supported by the osseous system. Those which are destined for the motion of the extremities, have their origin and insertion in the bones by which these are supported, or in the portion of the skeleton to which they are articulated. In this peculiar character, they differ widely from the invertebral animals.

4. *Nervous System.*—The peculiar characters of the nervous system of vertebral animals, have been already given, in considerable detail, in the preceding volume,—p. 150.

5. *Digestive System.*—The opening of the mouth is always transverse, the motion of the jaws being performed in the same plane. The liver is always present.

6. *Circulating System.*—Lacteals, lymphatics, and veins, occur as absorbents, and are aided by a pulmonic auricle and ventricle. Where a systemic heart is present, it is always in contact with the pulmonic. The blood is uniformly of a red colour. The kidneys always occur for the secretion of the urine.

7. *Reproductive System.*—The sexual organs are usually placed on different individuals, and the females are viviparous, oviparous, or ovoviparous.

The subdivision of the vertebral animals is easily accomplished, from the characters furnished by the Circulating System. In one group, there is a perfect systemic and and pulmonic heart, with warm blood; while, in the other, the heart is more or less defective, and the blood is cold.

Vertebral Animals with Warm Blood.

As the temperature of the body is usually higher than the surrounding medium, the skin is either furnished ex ternally with appendices, or internally with a layer of fat, to act as non-conductors of the heat which is generated. The regulation of the temperature, to suit the varying external circumstances, is effected by the quantity produced, by the quantity or colour of the non-conducting substances, regulating its escape by transmission or radiation.

The presence of ribs, and a sternum, for the protection of the cavity containing the viscera, is universal. The anterior extremity is likewise furnished with scapular and clavicular bones. The muscles contain a greater quantity of fibrin than in the other classes. The brain occupies the whole cavity of the skull, and exhibits several markings on its surface *. The eyes are furnished with eye-lids. The passage of the nose communicates with the windpipe. The cochlea of the ear is present. In many species, whiskers or long hairs are placed about the mouth. The intestinal canal is distinguished into small and great guts, and supplied with gastric, pancreatic, and biliary glands.

The heart is double, and the circulation is complete. All the blood must pass through the lungs before entering the systemic heart, to be distributed through the aorta to the body. Free air is respired through a windpipe furnished with a larynx. The lungs are double. The sexes are uniformly separate, on different individuals, and internal fecundation is requisite.

The warm blooded animals are divided into Quadrupeds and Birds.

QUADRUPEDS.

Char.—Ovarium double.

Des.—The appendices of the skin of quadrupeds consist either of hair or scales, and differ remarkably from the

---

* Vol. i. ib.

feathered covering of birds. The thorax is separated from the abdomen by a muscular diaphragm. The pancreatic and biliary ducts enter the intestine by a common opening. The omentum is always present. There are two occipital condyles *. The lower jaw only moves.

---

* The spine of quadrupeds, though exhibiting many common properties, is subject to considerable variations in the number of the vertebræ of which it consists. In the three-toed sloth, the cervical vertebræ amount to nine; while, in the other quadrupeds, their number is limited to seven. In the ternate bat there are no caudal vertebræ. The following Table, extracted from Cuvier's Comp. Anat., exhibits the number of vertebræ in different species of British quadrupeds.

SPECIES.	Dorsal ver.	Lum ver.	Sacral ver.	Caudal ver.
Man,	12	5	5	4
Common Bat,	11	5	4	12
Great Bat,	12	7	3	6
Horse-shoe Bat,	12	6	3	12
Hedge-hog,	15	7	4	12
Shrew,	12	7	3	17
Mole,	13	6	7	11
Badger,	15	5	3	16
Otter,	14	6	3	21
Martin,	14	6	3	18
Weasel,	14	6	3	14
Cat,	13	7	3	22
Dog,	13	6	3	22
Fox,	13	7	3	20
Hare,	12	7	4	20
Rabbit,	12	7	2	20
Common Mouse,	12	7	4	24
Field Mouse,	13	7	3	15
Black Rat,	13	7	3	26
Norway Rat,	13	7	4	23
Harvest Mouse,	12	7	3	23
Water Rat,	13	7	4	23
Dormouse,	13	7	4	24
Hog,	14	5	3	More than 4
Stag,	13	6	3	11
Goat,	13	6	4	12
Sheep,	13	6	4	16
Ox,	13	6	4	16
Horse,	18	6	2	17
Seal,	15	5	2	12
Dolphin,	13 }		In all	66
Porpoise,	13 }			

There is usually an obvious difference between the two sexes. The neck of the male is thicker, his voice louder, his size and strength greater than in the female.

The more recent examinations of the Marsupial and Monotrematous tribes, render the more common divisions of quadrupeds insufficient for the purposes of systematical classification. The term Mammalia may still be the appellative of those quadrupeds which suckle their young; but it must necessarily exclude the Monotremata, the young of which, like those of birds, are either left to shift for themselves, or are guided by the mother to suitable food.

MAMMALIA.

Char.—The young, after birth, suckled by the mother.

Descr.—The female is furnished with teats, and the mouth of the young is fitted for sucking. There is an uterus for the perfection of the fœtus. The external orifice of the seminal canal is single.

It appears to be necessary, in this place, to take notice, more particularly, of the *Teeth*, as the characters furnished by these organs, are extensively employed in the classification of the mammalia.

1. *Structure of the Teeth.*—The grinding surface of a tooth is termed its *Crown*, or summit; the portion which is concealed by the gum its *Root*, or *Fang;* and the intervening part, its *Body*. These three parts are seldom constructed of the same materials, but are usually formed from the different substances which are termed Ivory, Enamel, and Cement.

The *Ivory* is harder than common bone, of a denser texture, and contains less animal matter. It seems to consist of different layers of delicate fibres intimately united, exhibiting remarkable peculiarities in different species. It is susceptible of a good polish, and is extensively used in the arts. It is principally obtained from the different species of elephant and hippopotamus. Some teeth, as the tusks of these species, are entirely composed of ivory; while, in other species, it is intermixed with the enamel or cement.

The *Enamel* is still harder than the ivory; contains less animal matter; and consists of minute fibres, uniformly arranged perpendicular to its surface. It never occurs alone, but always as a coating to the ivory.

The *Cement* is of a softer substance than any of the preceding. It contains more animal matter, and is nearly similar to ordinary bone. It occurs as a coating to the ivory and enamel, or filling the interstices of their folds.

When a tooth, consisting of these three substances, is subjected to the action of the fire, the cement blackens first, and then the ivory, while the enamel retains a good deal of its original whiteness. These effects result from the charring of the animal matter *.

2. *Formation.*—The teeth are formed in sockets in the jaw-bones, which are termed Alveoli. In each socket, there is a pulpy substance, consisting of gelatine and albumen, and liberally supplied with bloodvessels and nerves. This pulp is surrounded by a capsule, to which, however, it does not adhere, except at the base. This capsule enters inio all its sinuosities. From the pulp the ivory is secreted, the

---

* On the surface of the tooth, there is sometimes deposited a substance termed the *Tartar* of the teeth. It frequently assumes a yellow colour with a smooth surface in the ox and the sheep, and has been ignorantly considered as gold derived from the pasture. It is merely a precipitation from the saliva. BERZELIUS found it to consist of earthy phosphate, 79.0; mucus, not yet decomposed, 12.5; peculiar salivary matter, 1.0; and animal matter, soluble in muriatic acid, 7.5 = 100.0.—An. Phil. vol. ii. p. 381.

external layer being first formed, and the others deposited beneath, in succession. The socket of the tooth is left at first hollow, owing to the presence of this pulp; and, where there are more fangs than one, each encloses a portion of the base of the pulp. In proportion as the ivory is secreted, the pulp diminishes in size, and at last nearly disappears.

The capsule with which the pulp is surrounded, is membranaceous, and consists of two layers. The central layer secretes the enamel. In some cases, the enamel is spread over a uniform surface; while, in other cases, it penetrates the body of the tooth, in the form of tortuous bands, extending from the crown to the root, or descending to a limited depth into the body In the first and last cases, the enamel is deposited after the ivory has acquired its form, while in those teeth where the enamel extends from the base to the crown, both the ivory and enamel appear to be of nearly contemporaneous formation. In such teeth, exhibited in the rabbit, the growth may be indefinitely extended; so that the wearing down of the crown of the tooth may be compensated by the increase at the base. Such teeth are of equal thickness throughout; all the others terminate in narrow fangs.

Considerable difference of opinion prevails with respect to the formation of the cement. According to M. TENON, it is to be regarded as originating from the ossification of the membrane of the capsule. BLAKE viewed it as a deposition from the peripheral surface of the capsule; while M. CUVIER thinks that it is deposited from the same surface of the capsule as the enamel. As the cement only occurs in those teeth where the walls of the capsule are tortuous, and where the enamel penetrates the ivory in bands, the formation of the cement appears to be connected with this complex structure, and to obviate an imperfection which would necessarily result, if the membrane which secreted the tortuous bands of enamel did not become ossified, as the space which it occupied would be left empty by its absorption or decay. Instead of this, we find the place of the membrane occupied by this osseous cement. In some cases, the external surface of this membrane encloses spaces which are not filled up, as in the case of the horse. This view of the matter will be confirmed by a view of the position of this cement, wherever it occurs *.

While the crown of the tooth is forming, the capsule is covered with the gum. Towards the termination of the process, however, the crown bursts through the gum, and is elevated to its proper place, by the farther growth of the body and fangs.

The teeth of the mammalia exhibit very remarkable differences with regard to their duration. The *Milk Teeth* are produced in early life, and fall out before the animal reaches maturity. These are replaced by others, which continue during the vigour of the animal to old age. Some of these *Permanent Teeth* are not preceded by temporary ones, as the back teeth in the jaw, but occupy those spaces which have been produced by the growth and lengthening of the jaw. These castings and renewal of the teeth take place at different times, according to the species. In the elephant, wild boar, and a few other species, the grinding teeth are shed and renewed in succession, throughout the life of the animal. While the last formed tooth is wearing down, another, of a larger size, and suited to the increase of the jaw, is forming beneath, to supply its place. These

---

* Sir EVERARD HOME, when speaking of this cement, says, " The ligamentous structure on which the third substance is formed, is divisible into layers, shewing that it is made up of separate membranes; and, between these, small ossifications, in different places, are readily detected."—Comp, Anat. vol. i. p. 201.

successive renewals take place at intervals of several years.

The teeth are usually divided into three kinds, from the position which they occupy in the jaw. The front teeth, or *Incisores*, are four in number in each jaw, in Man; and, in the lower animals, however numerous, they are distinguished by their insertion in the intermaxillary bone in the upper jaw; or, in the lower, by their opposition to the intermaxillary teeth or bone. The *Tusks* are situated immediately behind the incisors, and, in man, are four in number, or two in each jaw. They are likewise termed *Dentes canini*, or cuspidati. The grinders, or *molares*, occupy the back part of each jaw. These different kinds of teeth usually exhibit well-marked peculiarities. The incisors have, in general, an even sharp summit, suited for cutting. In order to preserve their sharpness, the outside only, in some species, as the hare, is covered with enamel, which, being less subject to wear than the inner layer of ivory, is kept constantly fit for use. The tusks are in general conical. The crown of the grinders is more extended. It is uneven, and wholly covered with enamel, in the carnivorous kinds; while, in the herbivorous, it is flat, or the enamel partially distributed.

The mammiferous animals admit of subdivision, from the circumstances connected with the condition of the fœtus. In one extensive tribe, the fœtus is nourished directly by the aërated blood of the mother, received through a placenta; while, in the other, the fœtus derives its nourishment from the absorption of the glairy fluid with which it is surrounded.

PLACENTARIA.

Char.—Uterus furnished with a placenta.

The nature of the fœtal connection with the mother has

already been illustrated *. There is not in this tribe any pouch connected with the mother, into which the expelled fœtus is deposited, as occurs in the marsupial animals. The young of the placentular animals are more perfect at their birth, and consequently less intimately dependent on their mother.

The animals of this division may be subdivided into terrestrial or aquatic; two groups which are characterized by remarkable conditions of their limbs.

PEDATA.

Char.—Skin with appendices in the form of hair, or spines, or scales.

Desc.—The posterior extremities are always developed, and attached to the pelvis. Rest and sleep are enjoyed on land, even in those species which seek their food in the water. These pedate mammalia are either fingered or hoofed.

UNGUICULATA.

Char.—The four extremities terminating in fingers furnished with nails or claws.

This condition of the extremities, observed in the fingered mammalia, gives them a capability of grasping objects, of climbing, and burrowing, which the hoofed tribes do not possess. In many of the unguiculata, the incisors exist in one or both jaws, while in a small tribe they are wanting.

1. Furnished with incisors.

Those animals which have tusks are separated from such as are destitute of them.

---

* Vol. i. p. 402.

a. Furnished with tusks.

The teeth appear to be unfit to furnish characters for farther subdivision, until we come to the construction of genera. The organs of motion, however, supply their place.

1. *Thumbs fitted to act in opposition to the fingers.*

All the animals of this group have the eyes directed nearly as in man. The orbitar and temporal fossæ are separated from each other. The anterior ventricles of the brain have digital cavities. The cerebellum is covered by the posterior lobes. This division of animals has long been subdivided into two orders; the first containing Man, the second the Monkey.

BIMANA.

Posterior extremities formed for walking.

This order contains one genus, consisting of one species, Man.

1. Homo *sapiens.*

Man is peculiarly distinguished from all those animals which make an approach to him in bodily configuration, by his *erect position.* His head is attached by the middle of its base to the vertical spine, and is destitute of those ligaments which serve as its support in the species in which it is attached to a horozontal spine. The pelvis is wide, and enables the legs to support the trunk more steadily. The toes are short, and the great toe is uniform in its position with the others; circumstances which, while they unfit the feet for grasping or climbing, qualify them for supporting the body by resting on the ground. For this purpose, the sole of the foot and heel are broad and flat, and their surface placed at right angles to the vertical direction of the body. The muscles are likewise arranged, so as to preserve the leg in a state of extension, and give to the calf and thigh their peculiar enlargement and form. If the foot is not adapted for laying hold of objects, neither is the hand fitted for walking. It is placed in the same direction as the arm, and is incapable of remaining in a horizontal position, suited for supporting the body on the ground. It is, however, well adapted for laying hold of objects by the length of the thumb and fingers, and the facility with which these oppose each other in the act of grasping. These indications of the erect posture of man are common to all the varieties of the race. The advantages attending this position may readily be comprehended, by an examination of his intellectual and instinctive powers.

QUADRUMANA.

The four extremities are formed for grasping. For this purpose, the thumb on the feet, as well as the hands, is capable of acting in opposition to the toes. Both feet and hands are placed nearly in the direction of the members to which they are attached. Neither of them, however, are fitted for walking, but both are suited for grasping. The manner of life corresponds with this arrangement. The species live in woods, and are accustomed to climb.

Simiadæ.

Four approximate incisors in each jaw.

The animals of this group have the transparent point on the retina, as in Man. In the Lemuridæ, a slight fold of the retina may be observed, but it is destitute of spot or transparent point *.

---

* BLUMENBACH, in reference to this point, observes, " As I have discovered this central aperture in the eye of no animal besides Man, except the *quadrumana*, the axes of whose eyes are, like the human, parallel to each other, I think its use connected with this parallel direction of the eyes. As, on the one hand, this direction of the eyes renders one object visible to both at the same time, and, therefore, more clearly visible ; so, on the other,

Grinders twenty in number.

A. Nails of the fingers and toes, flat and rounded.

(A.) Destitute of a tail.

a. No cheek pouches. The liver, cæcum and os hyoides, resemble those in Man.

2. MIMETES (of Dr LEACH,) *Chimpanse.* There is no intermaxillary bone; the last joint of the great toe is perfect, and the thigh-bone has the ligamentum suspensorium. The Simia troglodytes of authors, is the type of the genus.

3. SIMIA. Orang-Outang. There is here an intermaxillary bone; but the last joint of the great toe, and the suspensatory ligament of the thigh-bone are wanting. The Simia Satyrus is the type *.

b. With cheek-pouches.

4. PONGOS. This genus contains only one species, from the Island of Borneo.

(B.) Furnished with a tail. With cheek-pouches, and callous buttocks.

a. The last grinders in the lower jaw, with four tubercles.

5. CERCOPITHECUS. This genus, which contains many species, will probably require to be subdivided. The Kahau of Bornea, *C. nasicus,* appears to constitute a genus apart.

b. The tubercles in the last grinders of the lower jaw unequal.

6. PITHECUS. This includes the Magot and Macaques of CUVIER, and is distinguished from the following genus, by the oblique position and dorsal aspect of the nostrils.

---

this foramen prevents the inconvenience of too intense a light, if it is probable that it expands and dilates a little, and thus removes the principal focus from the very sensible centre of the retina."—Institutes of Phys., Note, p. 148.

* Annals of Phil. xvi. p. 104.

7. CYNOCEPHALUS. Snout lengthened, with terminal nostrils.

B. Nails compressed, pointed. Thumbs of the hands scarcely separate from the fingers.

8 HAPALE. Ouistiti. Head round; face flat; buttocks hairy; destitute of cheek-pouches.

Grinders, twenty-four in number.
A. Tail prehensile.
a. The part of the tail used to grasp objects naked.
9. MYCETES. Head pyramidal. The os hyoidis has a singular vesicular enlargement communicating with the larynx. Simia seniculus and Beelzebub of LIN., belong to this genus.
10. ATELES. The thumb of the hand is concealed under the skin. This generic distinction observable in the Simia paniscus of LIN.; ought probably to be employed to form the primary divisions of the quadrumanous animals.

b. Tail hairy.
1. CEBUS. Thumbs distinct. Simia appella of LIN.

B. Tail not prehensile.
12. CALLETRIX. Simia Pithecia of LIN.

Lemuridæ.

Incisors either not four in number in each jaw, or placed on a divided line.
A. Incisors in the lower jaw six in number.
a. Furnished with a tail.
13. LEMUR. Macauco. The Lemur Macaco is the type of this genus.
14. OTOLICNUS. The tarsi are here greatly elongated and disproportionate. The motions are slow.

b. No tail.

15. STENOPS. The summits of the grinders are more pointed than in the two preceding genera.

B. Incisors in the lower jaw fewer than six.

16. LICHANOTUS. Lower incisors four in number. No tail.

17. TARSIUS. Lower incisors two in number. Furnished with a tail.

2. Thumbs incapable of acting in opposition to the fingers.

The animals which belong to this division, denominated by CUVIER *Sarcophaga*, are incapable of grasping objects with the same dexterity or firmness as those which have separate thumbs. They retain the hold of their prey, however, by means of their long claws, aided by the pressure of the foot upon the object. The eyes are not directed ventrally, as in Man. The cerebellum is destitute of the covering of the posterior lobes of the cerebrum. The orbitar and temporal fossæ are united. Their sense of smell is, in general, acute. They are chiefly supported by animal food. Their jaws have a very limited lateral motion, and their intestines are narrow. They admit of subdivision from their organs of motion.

CHEIROPTERA.

Furnished with wings.

The Cheiroptera, consisting of the bats, are peculiarly characterised by a naked expansion of the skin uniting the anterior and posterior extremities. By means of this expanse of membrane, the act of flying can be easily performed. They walk awkwardly. Their tusks are large, and their incisors frequently fall out at an early age. They usually bring forth two young at a birth. They have pectoral mammæ, and a pendulous penis. They are crepuscular feeders, and become torpid during the winter. They are more closely related to the quadrumana than to any of the other groups.

A. Claws on all the fingers as well as the toes.

The fingers are not elongated, as in the true bats; and the membrane which extends on each side, from the head to the tail, and embraces the legs and arms, is incapable of supporting continued flight. The tusks are notched. · The incisors in the upper jaw are two in number, notched and widely separated. Those in the lower jaw are six, and so regularly and deeply divided by the notches, as to become pectinated.

18. GALEOPITHECUS. This is the only genus, and the *Lemur volans* of LIN., the only known species.

B. Some of the fingers destitute of claws.

This division includes the true Bats, which have the bones of the fore-arm and fingers elongated, for the purpose of giving a greater expanse to the membrane of flight. The fingers, indeed, have here lost their power of seizing objects, having become ribs for the support of the wing[*]. The thumb is small and free, with a claw as a hook, by which the body is suspended during repose in the roofs of caverns. The pectoral muscles are strong, and the sternum has a mesial ridge like that of birds, to give to them a greater extent of base. The hind legs are feeble; and the toes, which five in number, are of equal height, and armed with claws. They have no cæcum.

---

[*] This form is exhibited in the *Plecotus auritus*, or Long-eared Bat. Plate 1. Fig. F.

a. Grinders with flat summits. Pteropusidæ.

19. PTEROPUS. Fore-finger with a claw. The membrane between the extremities is deeply notched. The tail is either wanting or very short. The fore-finger, which is half as short as the middle one, has three phalanges and a small claw. The other fingers have only two phalanges and no claws. The nose is simple. The ear has no auricle. Tongue rough, with reversed prickles. This genus has been divided into those without a tail, and those that have a very short one. In both sections there are four incisors in each jaw.

20. CEPHALOTES. Fore-finger destitute of a claw.

This genus is likewise characterised by the lateral membranes of the wings coalescing on the back with a vertical and longitudinal partition. The only known species is the *C. Peronii* of GEOFF. from Timor.

b. Summits of the grinders with conical points. No claw on the fore-finger.

(A.) The middle finger with three bony phalanges; the other fingers with only two.

Nose with appendices.

a. Nose with warts.

21. NOCTILIO. Incisors four above, two below. Tail short, and disengaged from the interfemoral membrane. *Noctilio Americanus* of LIN.

b. Nose, with a leaf-like process.

22. PHYLLOSTOMA. The incisors are four in each jaw. On the nose is a transverse membrane. The tongue is rough with papillæ towards the end, and is capable of considerable extension. The lips are furnished with tubercles. CUVIER gives the indications of three sections in this genus.

1. Such as have no tail, as *Vespertilio spectrum of* Lin. 2. Such as have the tail united with the interfemoral membrane, as *Ph. hastatum*. 3. Such as have a tail free from the interfemoral membrane, as *Ph. crenulatum*. To these may be added the Glossophaga of Geoffroy, represented by the *Vesp. soricinus* of Pallas.

Nose simple.

(*a.*) With a tail.

23. Molossus. Two incisors in each jaw. Upper lip simple. The species of this genus, which are numerous, were confounded under the *Vesp. molossus* of Gmel.

24. Nyctinomia. Four incisors in the lower jaw. Upper lip deeply notched.

(*b.*) No tail.

25. Stenoderma. The interfemoral membrane interrupted at the coccyx. Two incisors in the upper jaw, and four below.

(B.) One bony joint in the fore-finger, and two in the rest.

Nose with appendages.

(*a.*) With a tail.

Tail extending beyond the interfemoral membrane.

26. Rhinopoma Geof. Forehead with a shallow pit. Nostrils with an obscure plate above. Ears reunited. *R. microphyllus.*

Tail included in the interfemoral membrane.

27. Rhinolophus. Nostrils with a complicated membrane like a horse-shoe. Incisors two above, in a cartilaginous intermaxillary bone, and four below. *Rh. ferrum equinum.*

28. Nycteris. Forehead with a pit. Nostrils with a circle of projecting plates. Incisors, four above and six below. *N. hispidus.*

(*b.*) Without a tail.

29. MEGADERMA. Ears reunited on the crown of the head. The interfemoral membrane is entire. *Ves. Spasma*, LIN.

Nose simple.

Tail free.

30. THAPHOZUS. GEOF. Forehead with a depression. Two incisors above, and four below.

Tail united with the interfemoral membrane.

31. VESPERTILIO. Ears separated. Incisors, four above, and six below.

32. PLECOTUS. Ears reunited on the head.

The arrangement of the Cheiroptera is still imperfect. M. GEOFFROY has contributed much important information regarding their classification, in his various papers in the Annales du Musuem. The following species are natives of Britain.

Rhinolophus
    1. ferrum equinum.
    2. hipposideros.

Vespertilio
    1. murinus.
    2. noctula.
    3. emarginatus *.

Plecotus
    1. auritus.
    2. barbastella.

---

* This speciee was first found in England, near Dover, by M. A BRONGNIART, who communicated it to M.GEOFFROY. Annales du Museum, vol. viii. p. 198. Tab. xlvi. In the summer of 1820, I ascertained it to be a native of Fifeshire.

FERÆ.

Destitute of wings.

The absence of wings, confines the progressive motion of the Feræ to the land or water. Those which are destined to live chiefly in the water are incapable of using their hind-feet to walk with on the land, and their fore-feet can only aid them in crawling. Taking advantage of these modifications of the motive organs, this tribe may be divided into such as have the hind-legs formed for walking, and such as have the hind-legs incapable of walking.

The feræ chiefly subsist on animal food, which in some species, is mixed with vegetables. Their intestines are slender; and neither colon nor rectum are greatly enlarged. The teeth are formed with prominent points, or so to lock into each other, as to be capable of chewing the soft food, or of bruising the bones with which it is intermixed. Where much vegetable food is used, the summits of the grinders are much flattened; but where carrion alone is the food used, the summits of the grinders are full of sharp eminences.

The grinders vary in character so much, as to be capable of division into three sorts. The *tearing grinders* or *fausses molaires* of M. F. CUVIER, correspond in some degree to the *bicuspides* of the human subject. They occur immediately behind the tusks, and vary in number according to the species. They are compressed, with elevated points, and are used along with the incisors and tusks, in tearing the food. The *chewing grinders*, or *carnassieres* of M. F. CUVIER, are placed immediately behind the preceding. They are four in number, one on each side in each jaw, and are usually the largest in the row. Their summits are flatter, and more extended than the tearing ones. They are subservient to the purposes of chewing. The *bruising*

QUADRUPEDS. 181

grinders, the *tuberculeuses* of M. F. Cuvier, do not exceed two in number, and occupy the back-part of the jaw. Their summits are flatter than the chewing grinders, and they are usually of a smaller size. The larger they are, the less carnivorous the species to which they belong *. The teats are ventral.

Toes separate, and the feet suited for walking. Hind legs fully developed.

PLANTIGRADA

Walk on the soles of the feet.

The toes are five in number †; and the entire sole, which is bare, rests on the ground. There is no cœcum. They feed chiefly during the night; and many of them become torpid in winter.

1. Middle incisors produced; lateral ones and tusks short.

a. Two small incisors between the produced ones in the lower jaw. No external ear ‡.

Five fingers.

33. Mygale. (Cuvier.) Snout produced and flexible. The Musk Rat, or Desman of Lapland (*Sorex moschatus*), Gm.) is the type.

34. Scalops. (Cuvier) Snout pointed. *Sorex aquaticus,* Lin.

---

* These descriptions will be better understood, by consulting Plate I. Fig 2., where a delineation of the grinders of the badger is given. *a* is an inside view of the teeth of the upper jaw. The tusk is followed by one small, and two large tearing grinders. To these succeed one chewing grinder, followed by a large bruiser. In the lower jaw b_2 the bruiser is small; the chewer large, and there is an additional tearer.

† There are likewise five fingers, unless in the genus Chrysochloris.

‡ The toes are webbed, enabling the species to swim. In this character they approach the palmated division.

Three fingers.

35. CHRYSOCHLORIS. (Lacep.) Hair of a metallic brilliancy. *Scalpa Asiatica*, LIN.

b. Produced incisors without intermediate small ones.

36. ERINACEUS. Hedgehog. Two middle incisors of the upper jaw cylindrical. Back covered with prickles. *E. Europæus.*

37. SOREX. Two middle incisors of the upper jaw bent and notched at the base. *Sorex araneus* *.

2. Incisors nearly equal, tusks large.

Incisors in a regular row.

38. CENTENES. Tanrec. Back covered with prickles like a hedgehog. The body, however, is incapable of rolling up like that animal. There is no tail. *Erinaceus ecaudatus*, Gm.

39. TALPA. Mole. Back with hair. A tail. Six incisors above, and eight below. The sternum is furnished with a mesial crest, to give support to the base of the pectoral muscles.

The second incisor on each side in the lower jaw, placed a little behind the others.

All the plantigrada already noticed, have the summits of the grinders covered with conical points, and feed chiefly on insects and worms. In those now to be noticed, these three kinds of grinders, which, in the others, can scarcely be traced, are very obvious. In the former, the

---

* It is probable, that the mole described by BARRINGTON, from North America, will constitute a genus belonging to this section. " There are two very long and large cutting teeth in the centre, calculated to fill the vacancy in the lower jaw, which contains only two short cutting teeth, followed immediately by two long canine ones."—Phil, Trans., vol. lxi. p. 292.

clavicles are developed, in those that follow they are imperfect.

(*a.*) The three last grinders in each jaw, with tubercular or flattened summits.

Tail about the length of the body.
Tail prehensile.
40. CERCOLEPTES. Snout short, tongue slender and extensile. Two tearing grinders. *Viverra caudivolula*, Gm.

Tail simple.
41. PROCYON. Raccoon. Nose pointed. *Ursus lotor*, LIN.
42. NASUA. Coati. Snout elongated. *Viverra nasua*, LIN.

Tail very short.
43. URSUS. Bear. Two recent species are known, the Brown and White; and two extinct species occur in the limestone caves of Germany *.

(*b.*) The two last grinders in each jaw, with tubercular or flattened summits.
44. MELES. Badger. A transverse glandular pouch or scent bag, under the tail. *M. Taxus.*

---

* Mr SCORESBY, when describing the use of the White Bear as food, adds, "The liver, I may observe, as a curious fact, is hurtful, and even deleterious; while the flesh and liver of the seal, on which it chiefly feeds, are nourishing and palatable. Sailors, who have inadvertently eaten the liver of bears, have almost always been sick after it: some have actually died; and the effect on others has been to cause the skin to peel off their bodies. This is, perhaps, almost the only instance known of any part of the flesh of a quadruped proving unwholesome."—Arct. Reg., vol. i. p. 520.

45. Gulo. Glutton or Wolverine. A fold under the tail instead of a scent-bag. *Ursus Gulo,* Lin.

The greater number of the plantigrada burrow in the ground, have their fore-legs remarkably strong, and the nails of their fingers produced, with a groove below.

DIGITIGRADA.

Support themselves in walking on the extremities of the toes.

1. Bruising grinders in each jaw.

a. Two bruising grinders in the upper jaw.

Two bruising grinders in the lower jaw. Furnished with a small cœcum [*].

46. Canis. Pupil circular, diurnal. This genus includes the common dog and its numerous varieties, the wolves and Jackalls.

47. Vulpes. Fox. Pupil linear, nocturnal. The species of this genus are more numerous than the preceding. The tail is bushy, the nose more pointed, and the scent stronger. The upper incisors are not so much notched. They burrow in the ground.

One bruising grinder in the lower jaw.

Tongue rough. Claws in part withdrawn in walking. Viverradæ.

Scent-bag pierced by the anus.

48. Ichneumon. Five fingers and four toes. The scent-bag is large and simple. *V. icheumon,* Lin.

49. Ryzæna. (Iliger.) Four fingers and toes. *V. tetradactyla* of Gm.

---

[*] There are usually four tearers above, and three below, on each side.

Scent-bag separate.

50. VIVERRA. Civet. Scent-bag placed between the anus and sexual orifice, divided into two. *V. civetta,* LIN.

51. GENETTA. Genet. Scent-bag simple and shallow.

(*b.*) One bruising grinder in the upper jaw. Destitute of a cœcum. Body not much thicker than the head. Musteladæ, including the species of the genus Mustela of LINNÆUS.

Two tearing grinders in the upper, and three in the lower jaw.

52. PUTORIUS. Polecat. Bruising grinder in the upper jaw broader than long. Chewing grinder below destitute of a tubercular surface. *Putorius vulgaris.*

53. MEPHITIS. Skunk. Bruising grinders as long as broad. Chewing grinders below, with two tubercles on the inner side. Nails of the toes produced.

Three tearing grinders in the upper jaw.

54. MUSTELA. Martin. Four tearing grinders in the lower jaw. *M. martis.*

2. No bruising grinder in the lower jaw.

55. HYÆNA. Three tearing grinders in the upper, and four in the lower jaw, on each side; scent-bag large. This genus includes two species,—the Spotted and Striped Hyæna.

56. FELIS. Cat*. Two tearing grinders in both jaws. The sharp eminences of chewing grinders, and the di-

---

* The Felis catus, or Wild Cat, which still frequents the remote woods of Britain, is probably a different species from the domestic cat, of which it has usually been regarded as the *stock*. The tail of the domestic cat is tapering, of the wild cat nearly cylindrical. The weight and size of the latter

minished number and size of the bruising ones, intimate the most sanguinary of animals. The species are numerous, but ill defined, as the characters have chiefly been taken from the colour markings of the fur.

Dr FORSTER proposes a very natural division of this genus unto *Jubatæ*, including the Lion : *Aelures*, consisting of tigers, cats, &c. and *Lynces*, or those with a brush of hairs on the tips of the ears, as the Lynx, caracal, serval, &c *.

Feet webbed, hind-legs adapted for swimming.

PALMATA.

The legs are short, and much enveloped in the skin, so as to execute on land the motion of crawling, rather than running. The pelvis is remarkably narrow, and the hind feet approach the tail, and are spread out horizontally †. The fingers and toes are connected by webs. The chief motions are executed in the water, although sleep and parturition are performed on shore. All their food is from the water, and consists of fishes.

1. Incisors and tusks in both jaws. Condyles of the lower jaw retained in their sockets.

---

are much larger than the former. The high value which was set upon domestic cats in the ninth century, as appears from the Welch laws of HOWEL the Good; the price of a kitten, before it could see, being a penny; until it caught a mouse, twopence; and when it commenced mouser, fourpence; militates against the commonly received opinion. It is probable that the domestic kind is originally from Asia.

* Phil. Trans. 1781. p. 2.

† The position of the feet indicates an approach to the tail of the whale, as may be seen, Plate I. Fig. 3., in which there is a representation of the common seal, with the feet and tail apart.

a. With the three kinds of grinders.

57. LUTRA. Otter. Six incisors on each jaw. Tail nearly the length of the body. Two anal scent-bags. *Lutra vulgaris.*

58. ENHYDRA. Sea Otter. Six incisors above, and four below. Tail much shorter than the body. No anal scent-bags.

I have ventured to remove the Otters from the Polecats, and unite them with the Seals, with which they so nearly agree in their manner of life. The separation of the Sea Otter from the fresh-water one appears to be justified by the characters which it exhibits. The communications of Sir E. HOME and Mr MENZIES (Phil. Trans. 1796, p. 385.), unfold several facts illustrative of its anatomy.

All the grinders nearly uniform in their appearance. Incisors, six above, and four below.

59. PHOCA. Seal. Without external ears. Summits of the grinders flattened. Outer incisor on each side in the upper jaw large. *P. vitulina* *.

60. OTARIA (Peron). Ursine Seal. With external ears. Summits of the grinders conical. External incisors small. *O. ursina.*

2. Without incisors or tusks in the lower jaw.

61. TRICHECUS. Walrus. The tusks of the upper jaw greatly produced, and directed ventrally. These aid the animal in climbing upon the rocks and ice-bergs. Only one species is known. *T. rosmarus.*

---

* Some seals, as Ph. monachus, are said to have four incisors in each jaw. Such will probably be constituted into a new genus, under the title Monachus.

The animals of this tribe will probably be held in higher estimation at a future period than they are at present. The seal yields a considerable quantity of oil, the skin is valuable, and the carcase can be converted into manure, or adipocire. The last species might even be used as an article of food, if modern fastidiousness did not prevent the imitation of ancient manners. The seal, which is thus so useful, is easily tamed, and herds of them might be kept in a partially domesticated state, on many parts of the coast, with evident advantage. The otter, too, is easily tamed, and, in a domesticated state, might be found useful in catching fish. We pursue the beasts of the field with dogs, and the birds of the air with hawks, nor would the watery element protect the fishes from the chase, were the otter enlisted into our service.

b. Destitute of tusks.

GLIRES.

The animals of this tribe, in the absence of tusks, are ill fitted for tearing their food to pieces. The incisors, usually reduced to two in each jaw, have their edges sharp, placed so as to act against each other, and qualified to nip off small portions of the substances against which they are directed. Their jaws, too, are feeble, and they accomplish their object by efforts, which are frequently repeated. The hind-legs are the longest and strongest, in nearly all the species, reducing their progressive motion to a series of leaps, rather than steps. The clavicles are seldom fully developed. The modifications in this respect, which they exhibit, are employed by CUVIER, in the primary distribution of the genera, although they exhibit no definite line of distinction.

The intestines are long, and, in many species, there is a very large cœcum. The brain presents fewer convolutions than in the preceding tribes; it is indeed nearly smooth.

1. Summits of the grinders, with conical processes, covered with enamel. The species subsist on animal food, or are omnivorous.

Incisors with pointed summits.

Hind feet palmated.

62. HYDROMYS. The summits of the grinders are obliquely quadrangular, with a spoon-shaped hollow. *H. leucogaster* of GEOFFROY.

Hind feet simple.

(*A.*) Three grinders in each jaw.

With cheek-pouches.

63. CRICETUS. Hamster. Tail short and hairy. *Mus cricetus*, LIN.

Without cheek-pouches.

64. DIPUS. Jerboa. Tail long, ending in a bush. Hind-legs remarkably long, so that the animal moves by leaping. *Mus jaculus*, LIN.

65. MUS. Tail nearly naked. Hind legs of moderate length. This restricted genus includes the common mouse and rat.

(*B.*) Five grinders on each side above, and four below.

66. ARCTOMYS. Marmot. Social, becoming torpid. *Mus marmota*, LIN.

Incisors with chisel-shaped summits; the Dentes scalprarii of GREW.

Grinders three on each side in both jaws.

67. SPALAX. Zemni. Fives toes before and behind,

with flat nails. Tail and ears short. The eyes are concealed under the skin. *Mus typhlus.*

Grinders four on each side in both jaws.

(*a.*) Five fingers.

Five toes.

68. CHEIROMYS. Aye-Aye. Great toe capable of acting as a thumb, by opposing the other toes. Incisors greatly compressed. *Sciurus Madagascariensis*, GM.

69. BATHYERGUS. No opposable thumb. Tail very short. Eyes minute. *Mus maritimus*, GM.

Four toes.

70. PEDETES. Hind legs remarkably long, like the Jerboa. Nails of the fingers long, those of the toes broad. *Mus cafer.*

(*b.*) Four fingers and five toes. Incisors greatly compressed.

71. PTEROMYS. Skin expanded into a wing on the sides, between the fore and hind legs, enabling the animal to support itself for a short time in the air. *Sciurus volans.*

72. SCIURUS. Squirrel. No wing. There is in youth the rudiments of a fifth grinder in the upper jaw*.

2. Summits of the grinders flat, and the enamel appearing partially on the surface.

The animals of this division are herbivorous. They may be still farther subdivided, from the structure of their

---

* It is probable that the genus Condylura of ILIGER, including the Sorex cristatus, ought to form a new section in this division of the Gnawers, with which it is connected, by the absence of tusks. The number of incisors, however, and grinders, indicates an affinity with the Feræ. According to DESMARETS (Journ. de Phy. ii. 89, p. 225.), it has six incisors above, and four below; fourteen grinders in the upper jaw (in all), and sixteen in the lower.—Ann. Phil. xvi. p. 105.

grinders. In some, these teeth seem, from the distribution of the enamel, to consist of two or more teeth joined together, as the entire tooth is constructed of transverse and vertical plates of enamel and ivory. In the second, the enamel is distributed more superficially, covering merely the salient and entering angles of the sides, or penetrating but a short way into the body of the summit.

(A.) Enamel in vertical plates.

Subsidiary incisors in the upper jaw. This division includes the genus Lepus of LINNÆUS, and is peculiarly distinguished by two small incisors, placed immediately behind the ordinary ones in the upper jaw. The grinders, which are prismatic, are five in number in the lower jaw, and six in the upper, on each side. The inside of the lips and cheeks hairy. There are five fingers and four toes. The cæcum is uncommonly large, with an internal spiral plate.

73. LEPUS. With a tail. Ears large, hind-legs long, clavicles imperfect, and a suborbitar space, with the bone reticulated. *L. timidus.*

74. LAGOMIS. No tail. Ears and hind-legs of moderate size. Clavicles nearly perfect. Suborbitar hole simple. *Lepus alpinus,* PALLAS.

No subsidiary incisors in the upper jaw.

(*a.*) Roots of the grinders with fangs. Grinders four on each side.

75. ECHIMYS. Body covered with spines. *Hystrix chrysurus.*

76. MYOXUS. Body covered with hairs. *Mus avellanarius,* LIN. *

---

* It is probable that these two genera should be associated with the Castor and Histrix, in the last division of the genera, with flat-crowned grinders.

(*b.*) Roots of the grinders simple, prismatical.

Grinders three on each side in both jaws.

Tail compressed, scaly.

77. FIBER. Ondatra. Feet webbed. Tail long. *Castor zibethicus*, LIN. Has much the habit of the beaver.

Tail round and hairy.

78. ARVICOLA. Campagnol. Tail about half the length of the body. *Mus amphibius*, LIN.

79. GEORYCHUS. Leming. Tail and ears very short. Toes formed for burrowing. This genus differs but little from the preceding. M. CUVIER assigns to the former a length of tail nearly equal to the body. In the A. amphibius, he says, " La queue de la longueur du corps." In the specimens of A. amphibius and A. agrestis, which have come under our notice, the tail did not exceed one-half of the length of the body.

Grinders four on each side, in both jaws· Clavicles imperfect. Caviadæ.

Five fingers and toes.

80. CÆLOGENUS. Immediately under the eye, in the jaw, on each side, there is a remarkable slit or cavity *Mus pactra*, LIN., or Spotted Cavy.

Four fingers and three toes.

Feet webbed.

81. HYDROCHOERUS. Capibara. Claws remarkably large and strong. *Cavia capibara*, GM.

Feet simple.

82. CHLOROMYS. Agouti. With a tail. *Mus aguti*, LIN.

83. CAVIA. Cavy. No tail.

(B.) Enamel superficial. Four grinders.

84. HISTRIX. Porcupine. Body covered with prickles. Tongue covered with spines. Five fingers and four toes. The species *H. prehensilis* and *fasciculata* having long prehensile tails, may be separated from the *H. cristata* and *dorsata*.

85. CASTOR. Beaver. Body covered with hair. Tail oval, depressed and covered with scales. Five fingers and toes, the latter webbed. One opening for the sexual orifice and anus, thus making an approach to the monotrematous animals. *C. Fiber*, LIN.

2. Destitute of incisors.

The animals of this division are termed Edentés by CUVIER, although, in some of the genera, both tusks and grinders exist. Their toes and fingers are armed with very strong claws. They are slow in their motions.

Furnished with grinders.

Furnished with tusks. Tardigrada. The toes are concealed in the skin.

86. BRADYPUS. Sloth. Face short. Grinders cylindrical. Tusks pointed and long. The stomach is divided into four pouches, with simple walls. Intestine short, and without a cœcum. There are two species, *B. tridactylus* or Ai, and *didactylus* or Unau, both natives of South America.

87. PROCHILUS of Iliger. Snout lengthened. Six grinders in each jaw. Five nails on each foot. Lips and tongue extensile. There is only one species known, *P. ursinus*, an accurate representation of which is given in BEWICK's Quadrupeds, immediately after the figure of the brown bear.

No tusks. Snout produced.

a. Grinders simple.

88. DASYPUS. Armadillo. Grinders cylindrical, separated from one another. Tongue smooth, stomach simple, and the intestine without cœcum. Five toes. Body covered with large plates or scales, not imbricated. There are several species, distinguished by the disposition of the scales.

89. MEGATHERIUM. Grinders contiguous. This is a fossil genus, instituted by CUVIER for the reception of two species, the skeletons of which have been dug up in America. The first is the Megalonoix, of the size of an ox. The second, or Megatherium primitivum, is equal to the rhinoceros.

b. Grinders penetrated by canals.

90. ORYCTEROPUS. Grinders cylindrical, traversed longitudinally by a number of small canals. The stomach muscular towards the pylorus. Cœcum small. Body covered with hair. *O. capensis.*

Destitute of grinders, tusks and incisors. Genuine Edentata. No cœcum.

91. MYRMECOPHAGA. Ant-eater. Body covered with hair. Tongue filiform, extensile, moist. Stomach simple. *M. jubata.*

92. MANIS. Pangolin. Body covered with imbricated scales. Stomach slightly divided. *M. pentadactyla.*

UNGULATA.

HOOFED QUADRUPEDS.

The fingered quadrupeds are all capable, more or less, of seizing objects with their extremities, so that their fingers and toes are not only subservient to the purposes of progressive motion and protection, but of the digestive system. The extremities of the hoofed tribes are exclusively employed to support and to move the body. They have no clavicular bones. They are all herbivorous.

PECORA.

Ruminate.

The genera of this tribe possess many common properties. Each foot consists of two toes, covered with strong hoofs, which are flattened on the opposing faces. The two metacarpal and metatarsal bones are united, in those genera without incisors in the upper jaw, to form the canon bone. The vestiges of lateral toes exist.

The incisors are placed only in the lower jaw, except in the camel, and are opposed to an indurated gum. There is a space between the incisors and grinders in some species containing canine teeth. The grinders are hollowed transversely, with the enamel distributed in irregular circles. The stomachs are four in number, and so disposed, that the food from the gullet can enter two of them at pleasure. This structure is necessary for the performance of the act of ruminating, or *chewing the cud*.

The first stomach (venter, rumen, or ingluvies), is called the *Paunch*. It is of the largest capacity, and receives the food directly from the gullet. It is imperfectly subdivided internally into compartments, by muscular bands, and its lining is beset with small blunt processes. The food in this stomach undergoes little change, and appears to be retained merely until the second stomach is fit for its reception.

The second stomach (reticulum), *honey-comb*, *bonnet*, or *king's-hood*, is less than the paunch, and peculiarly distinguished by the great number of cells on its central surface, like a honey-comb. In this second stomach, the food, which passes into it by degrees from the paunch, is mixed with the drink of the animal, which does not enter the first stomach, but is conveyed directly to the second. From the reticulum, the food is returned in small portions through the gullet to the mouth, to be subjected to mastication by the grinders, as when first taken into the mouth, it is transmitted, without chewing, to the paunch. When properly reduced, it is now conveyed through the gullet into the opening of the third stomach.

The third stomach (omasum), *manyplies*, *tripe* or *feck*, is distinguished by numerous longitudinal folds of its internal coat, varying alternately in breadth, and having their surface closely covered with glandular grains. The food in this stomach now begins to change its character, and emits an offensive odour.

The fourth stomach (abomasum), *read* or *red*, has an internal villous coat, with longitudinal folds. Into this stomach the gastric juice appears to be poured, as it is here that the milk is found curdled in a calf, and this is the only part used as rennet. The food passes from the third into the fourth stomach, by a projecting valvular orifice *.

The two first stomachs are analogous to the cheek-pouches of some of the apes, and belong exclusively to mastication. In the young animal, in which mastication is unnecessary, these stomachs are of small size. The third and fourth are merely full developments of the cardiac and pyloric portions of the stomachs of other quadrupeds.

---

* The configuration of the gullet, and mouths of the three first stomachs, by which the food is guided in these different motions, is well exhibited in a drawing of the stomachs of a cow, by Mr Cliff, published by Sir E. Home, Phil. Trans. 1806, Tab. xv. xvi., and re-published in Comp. Anat. Tab. xxi. xxii.

Incisors in the upper jaw.

This natural family of Camelusidæ have tusks in both jaws, two incisors above, and six below. Hoofs small, and covering the upper extremities of the two toes. The cells of the second stomach are large, and capable of retaining water *.

93. CAMELUS. Camel. Each foot united below by a common sole. A dorsal haunch. There are two species, *C. bactrianus* and *C. dromedarius*; the former with two haunches, and the latter with one.

94. AUCHENIA. Lama. Toes divided. No dorsal haunches. *Camelus glama*, LIN.

Without incisors in the upper jaw.

With horns. Grinders, usually six in number on each side in both jaws.

A. Horns with a cone. The horns consist of an osseous cone, or elongation of the frontal bone on each side, and covered with horn. The horns are permanent. The structure of the cone, which in one is solid bone, and the other is full of sinuses, furnishes characters for subdivision.

a. Cone porous. Eight incisors.

Horns bent anteriorly.

95. Bos. Ox. The horns are smooth, and bent anteriorly and laterally. There are several species of this genus, the distinguishing characters of which have been very imperfectly ascertained.

---

* LINNÆUS says, " Ventriculus secundus cellulosus pro aqua pura diutius asservanda per siticulosa diserta."—Syst. Nat. 90. That the cells are thus employed, has been demonstrated by Sir E. HOME, Phil. Trans. 1806, p. 357, Tab. xvii. xviii. xix., and Comp. Anat. I. 1651, II. Tab. xxiii, xxiv, xxv.

Horns bent posteriorly, angular and rough.
96. CAPRA. Goat. With a beard. *C. hircus.*
97. OVIS. Sheep. Without a beard. *O. aries* *.

b. Cone solid.
98. ANTILOPE. Antelope. Horns annulated and spirally twisted.
99. RUPICAPRA. Chamois. Horns smooth. *Capra rupicapra,* LIN. To this genus must be provisionally added the Nyl-ghaw (Phil. Trans. 1771, p. 170. tab. v.) and Gnu. The ruminating animals all possess, in an exquisite degree, the sense of smell. This has frequently been observed of cows, sheep, goats, and the nyl-ghaw.

B. Horns simple, formed of bone.
100. CERVUS. Deer. Horns deciduous. The horns are produced under a soft velvety skin, which ultimately dries up, and is rubbed off by the animal. In the spring season they are annually shed, a natural separation forming at the base. They are renewed in the course of a few months after.

In many kinds of deer and antelope, there is a bag on each side of the head, situated between the eye and the nose, which some have considered as destined to hold the tears. The French naturalists, indeed, term them *Larmiers.* They have, however, no connection with the tears, or lachrymal ducts; but their glandular walls secrete a

---

* CUVIER says, ' Ils méritaient si peu d'etre séparés generiquement des chèvres, qu'ils produisent avec elles des metes feconds."— Reg. Ann. I. 267. In the Statistical Account of the parish of Urr, by the Rev. JAMES MUIRHEAD, it is asserted, that the hybrid produce of the sheep and goat is fertile, black-faced, and differs little in form from the black-faced sheep. The intercourse is stated to be common. Vol. xi. p. 66.

matter similar to the wax of the ears. (See Phil. Trans. 1804, p. 73.) For want of a more appropriate term, these bags may be denominated, in our language, *Crumens*, from the Latin crumenæ. In some species, these crumens are lined with a hairy cuticle, while in others the cuticle is naked.

101. CAMELOPARDALIS. Camelopard or Giraffe. The horns are permanent, as well as the hairy skin by which they are enveloped. There is only one species known of this singular animal, a native of Africa.

Destitute of horns.

102. MOSCHUS. Musk. Tusks in the upper jaw long, and bent ventrally and posteriorly. The males have the bag in which is contained the valuable musk.

BELLUÆ.

Do not ruminate.

The animals of this division have few characters in common. They have usually a clumsy shape, and a thick hide. This last character induced CUVIER to term them, *Pachydermata*.

With tusks.

Hoof entire. In the inside of the hoof are the vestiges of two toes.

103. EQUUS. Horse. Six incisors in each jaw, and six grinders. The stomach is simple, cœum long. CUVIER has enumerated five species *E. caballus, hemionus, asinas, zebra,* and *quagga*. The second of these is probably nothing more than a variety of the first.

Hoof divided.

Among the genera of this division, some have the principal hoofs arranged so as to make the foot bifid, the lateral hoofs, where such exist, being useless as supporters of the body. In the others the hoofs are irregular, and the foot is not bifid.

A. Foot bifid.

a. Tusks produced. Six incisors in the lower jaw. Suesidæ.

With a tail.

104. Sus.—Hog. Six incisors above. Grinders simple, with tubercular summits. Tusks prismatic.

105. Phacocherus. Two incisors in the upper jaw. Grinders formed by the union of cylindrical pieces, united by cement. Tusks rounded. A fleshy lobe under each eye. *Sus Æthiopicus* *.

Without a tail.

106. Dicotyles. Peccary. Incisors and grinders like the hog. The tusks do not protrude beyond the mouth. In the back of the loins, there is a scent-bag with a small orifice which pours out a fetid odour. They have a canon bone like the ox. Stomach divided into pouches. The two species are natives of South America.

b. Tusks short. The three kinds of teeth forming an uninterrupted line, as in Man.

107. Anoplotherium. Six incisors in each jaw. Twenty-eight grinders in all. No canon bone. This is an extinct genus. The remains of five species have been determined by Cuvier, from the gypsum quarries in the neighbourhood of Paris.

---

* The structure of the teeth of this species is well displayed by Sir E. Home, Comp. Anat. II. Tab. xxxviii. xxxix. That of the wild-boar, Tab. xxvii.

B. Feet not bifid. The primary hoofs, which exceed two in number, equally serve for support.

With a snout. Six incisors and two tusks in each jaw.

108. TAPIR. Four hoofs before, and three behind. For a long time, one species only, *T. Americanus*, was known. But, of late years, another species, *T. Malayanus*, has been identified as a native of the larger islands of the Indian Seas. It was first noticed by Mr MARSDEN, as a division of Summatra. There is a good figure of the animal in Dr HORSFIELD's "Zool. Researches in Java;" and an interesting specimen may be seen in the Royal Museum of Edinburgh.

109. PALEOTHERIUM. Three hoofs on each foot. An extinct genus, determined by CUVIER. Many species have been established.

No snout.

110. HIPPOPOTAMUS. Four incisors in each jaw, and twelve grinders. The three foremost grinders conical, the three last with tubercles, the summits of which become bare of enamel by detrition. Four hoofs on each foot. *H. amphibius*.

111. HYRAX. Daman. Two incisors in the upper, and four in the lower jaw. Fourteen grinders in each jaw. No tail. Stomach divided into several pouches. Cœcum large, and the colon has cœcal appendages. *Hyrax capensis*.

No tusks.

A snout or proboscis. The upper jaw is furnished with two large incisors, improperly termed tusks, as they are seated in the intermaxillary bones; in the lower jaw there are none. Bones of the feet divided into five toes.

112. ELEPHAS. Elephant. Grinders with flat summits. There are two distinct recent species, *E. Indicus*, and *E. Africanus*.

113. MASTODON. Summits of the grinders with tubercular processes, the points of which become flat by the wearing down of the enamel and ivory. This is an extinct genus, the remains of the species of which are found in the debris of large rivers.

Without a snout.

114. RHINOCEROS. Three hoofs on each foot. Nose horned. There appears to be some reason for considering the three recent species as the types of as many genera. CUVIER assigns twenty-eight grinders to all of them, but from the observations of Mr BELL, the *R. sumatrensis* has only twenty-four *. This species has two incisors in each jaw. In the *R. Indicus*, there are two incisors above, and four below, while in the *R. Africanus*, there are none. The skin of the *R. Indicus* is folded, that of the others is nearly smooth.

APODA.

Hind-feet united with the tail in the form of a fin.

The skin on the body is smooth, and nearly destitute of appendices. The fingers are all enveloped in skin, without a protruded arm, and form a fin or *swimming paw*. They are destitute of separate hind-feet and tail, but their place is supplied by a broad horizontal expansion, posteriorly emarginate, which, though differing in direction, serves the same purpose as the caudal fin in fishes, and is the principal organ of progressive motion. The horizontal position of this fin is probably connected with the vertical motion of the animal, in coming to the surface, at intervals,

---

* Phil. Trans. 1793, p. 3. Tab. ii. iii., where there is a good figure of the animal and the cranium.

to respire. This fin is usually termed the Tail. There are two small bones embedded in the muscles, near the anus, which are the only vestiges of a pelvis. They have no external ears, the aperture being a small hole.

HERBIVORA.

Nostrils terminating in the snout.

The animals of this group, established by CUVIER, were formerly arranged with the Walrus, to which they bear some resemblance. Their skin is sleek, and has here and there a few hairs. The mouth is furnished with teeth, having flat summits fitting them to brouse on sea-weeds. The teats are pectoral.

Swimming paws, with the rudiments of nails.

115. MANATUS. Lamantine. The grinders are eight on each side, with two transverse ridges. No tusks. In youth there are two small incisors in the upper jaw, which speedily fall out. The stomach is complicated, the cœcum branched, and the colon swollen. *Trichecus manatus* of LINNÆUS, is the only well established species.

It is probable that this animal, or some of the other species of the tribe, with pectoral teats, may have given rise to the belief in the *mermaid*. The lamantine is said to carry its young between its paws, and, when viewed in this attitude, it would furnish materials for the imagination to form those exaggerated pictures, which have from time to time been communicated *.

---

* The Lamantine occupies a place in the British Fauna, as a straggler. Mr STEWART, in his Elements of Nat. Hist. I. p. 124., states, that ' The carcase of one of these animals was, in 1785, thrown ashore near Leith. It was much disfigured; and the fishermen extracted its liver, and other parts,

Without the rudiments of claws.

116. HALICORA. Dugong. Grinders, twelve in number, three on each side. Two incisors above, and the vestiges of several small ones below. The structure of *H. Indica*, the only well established species, has recently been ably unfolded by Sir E. HOME, in the three papers which he has communicated in the Philosophical Transactions for 1820, Part. II., with finished delineations by Mr CLIFF.

117. RYTINA. Syren. One grinder on each side. Lips whiskered. The only ascertained species of this genus inhabits the north-western coasts of America.

This singular group of herbivorous apodal mammalia probably consists of many more species than those which have hitherto been described.

CETACEA.

Nostrils opening on the crown of the head.

The skin is smooth and glossy. The cuticle resembles a piece of oiled silk cloth, and the corium is thick, and consists of vertical fibres. The stomach is complicated. The larynx forms a tabular projection across the pharynx into the canal of the nose *. The bronchial termina-

---

from which a considerable quantity of oil was obtained." Mr STEWART has since informed me, that it came ashore at Newhaven, in the harvest season. Though it had been dead for some time, and was in a putrid state, he was able to satisfy himself with regard to the species.

* When whales come to the surface to respire, they produce a hissing noise, and a column of vapour arises, sometimes to a considerable height. When expiration happens to take place, before the head has actually reached the surface of the water, some spray may also be thrown up. This, however, is seldom the case; hence the term *blow-holes*, instead of spout-holes, to express the nostrils, is more appropriate. The term nostril, indeed, may

tion of the trachea is not bifurcated. The teats, two in number, are situate on each side of the vulva.

Palate covered with baleen.

The baleen, improperly termed *whalebone*, is too well known in commerce to require any description. It occurs in the form of sub-triangular plates, with the free edge fringed towards the mouth, the fixed edge attached to the palate, the broad end fixed to the gum, and the apex to the middle of the arch. These plates form a series on each side, and are placed transversely at regular distances. Their use is to strain the water, which the whale takes into its large mouth, and to retain the small animals, on which it subsists. There are no teeth, and the throat is narrow. External opening of the blow-hole double, with an internal septum.

Back furnished with a protuberance or fin. Piked whales *.

118. BALÆNOPTERA. Pectoral skin folded longitudinally, and capable of inflation. The species of this genus are very imperfectly determined. The *B. musculus,* or

---

be objected to, on the supposition that whales have not the sense of smell, neither olfactory nerves, nor holes for their passage. This, however, is perhaps not strictly true. On the 26th July 1816, being on board a sloop in the Irish Channel, a large flock of grampuses were sporting immediately around us. The master asserted, from his former experience, that a little *bilge-water* would put them to flight. The experiment was tried, and the pump had made but a few strokes, when the whole disappeared, I may almost say, instantaneously, and rose to blow at a considerable distance. In the Dugong, according to Sir E. HOME, " there are orifices in the crebriform plate of the skull, for the olfactory nerves."—Phil. Trans. 1820, p. 153, and HUNTER assigns the sense of smell to the Baleen whales, Phil. Trans. 1778, where some judicious observations on the structure and economy of whales may be found.

* So named by SIBBALD, on account of the pike or process on the back.

round-lipped whale (SCORESBY's "Arctic Regions," tab. xiii. fig. 2.), and *B. rostrata*, or sharp-nosed whale (HUNTER, Phil. Trans. 1778, tab. xx.), may be regarded as distinct.

119. PHYSALIS. Finner. Without pectoral folds. The longest of the whales reaching to 100 feet. The most difficult to capture.

No dorsal fin.

120. BALÆNA. Common whale. Under lip whiskered. Head very large.

Destitute of baleen. One external opening of the nostrils.

A. Nostrils double within, being divided in the skull by a bony septum.

Teeth in the mouth.

The following interesting remarks are made by HUNTER (and copied by Sir E. HOME into his Comp. Anat. i. 261.), on the dentition of whales. " The situation of the teeth, when first formed, and their progress afterwards, as far as I have been able to observe, is very different in common from those of the quadruped. In the quadruped, the teeth are formed in the jaw, almost surrounded by the alveoli, or sockets, and rise in the jaw as they increase in length; the covering of the alveoli being absorbed, the alveoli afterwards rise with the teeth, covering the whole fang; but in this tribe the teeth appear to form in the gum, upon the edge of the jaw, and they either sink in the jaw as they lengthen, or the alveoli rise to inclose them; this last is most probable, since the depth of the jaw is also increased, so that the teeth appear to sink deeper and deeper in the jaw. This formation is readily discovered in jaws not full grown; for the teeth increase in number as the jaw lengthens, as in other animals. The posterior part of the jaw becoming larger, the number of teeth in that part increases, the sockets becoming shallower and shallower, and at last being only a slight depression.

" It would appear that they do not shed their teeth, nor have they new ones formed similar to the old, as is the case with most other quadrupeds, and also with the alligator. I have never been able to detect young teeth under the roots of the old ones; and, indeed, the situation in which they are first formed, makes it, in some degree, impossible, if the young teeth follow the same rule in growing with the original ones, as they probably do in most animals.

" If it is true that the whale tribe do not shed their teeth, in what way are they supplied with new ones, corresponding in size with the increased size of the jaw? It would appear, that the jaw, as it increases posteriorly, decays at the symphysis, and while the growth is going on, there is a constant succession of new teeth, by which means the new formed teeth are proportioned to the jaw. The same mode of growth is evident in the elephant, and in some degree in many fish; but in these last, the absorption of the jaw is from the whole of the outside along where the teeth are placed. The depth of the alveoli seems to prove this, being shallow at the back part of the jaw, and becoming deeper towards the middle, where they are the deepest, the teeth there having come to the full size. From this forwards they are again becoming shallower, the teeth being smaller, the sockets wasting, and at the symphysis there are hardly any sockets at all. This will make the exact number of teeth in any species uncertain *."

---

* Phil. Trans. 1788, p. 398.

The examination of the teeth of two individuals of the common porpess, before I had perused the preceding observations of HUNTER, had induced me to adopt a different opinion with regard to the growth of the jaw, and to conclude, that it enlarges at the symphysis as well as at the base. In one of the individuals, which was only about three feet in length, there were twenty-five teeth apparent, and one before and another behind, underneath the surface of the gum, in each side of the upper jaw, and twenty-four in the under, and one not come through the gum in front. In the other specimen, which was a pregnant female, five feet three inches in length, there were on the one side of the upper jaw twenty-three perfect, and two uncut ones in front, and on the other side, twenty-six perfect, and three imperfect ones before. In the lower jaw there were twenty-one perfect, and three imperfect ones on the one side, and nineteen perfect and four imperfect ones on the other. There are two incisors in the intermaxillary bones.

In these examples, the uncut teeth at the symphysis, exactly resembled the uncut teeth at the base of the jaw, and like them, were much smaller than those towards the middle. Indeed, when we view the shallowness of the groove containing the sockets of the teeth at both ends of the jaw, and the diminished size of the teeth, together with the depth of the alveolar groove and the greater size of the teeth in the middle of the jaw, there is sufficient reason to conclude, that the first formed teeth are those in the middle of the jaw, that the jaw lengthens at the symphysis and at the base, and that the new teeth formed at these places are the smallest, and that there is no absorption. Those teeth which have not cut the gum in the fore part of the jaw, are every way analogous to the uncut ones at the base. This lengthening of the jaw gives to the aged individuals a more produced snout. Some observers have stated, that the teeth of the porpess are loose in their sockets, while others assert that they are fixed, the former having examined old, the latter young individuals, or even different parts of the same jaw, the middle teeth being fixed, the others loose.

With a dorsal fin.

121. DELPHINUS. Teeth numerous in both jaws. This is an extensive genus, which CUVIER has proposed to subdivide into the beaked species (Delphinus), and the short-nosed species (Phocæna), the former represented by the Dolphin, the latter by the Porpess.

At Plate I. fig. 4. there is a representation of the Porpess, *D. phocæna.* Fig. 5. represents the teeth, *a* one in the middle of the lower jaw, and *b* one in the middle of the upper jaw. Among the British species, some have compressed and obliquely placed teeth, as the *phocæna*,—subulate, as the *delphis*,—conical, as the *orca* and *melas* *, —and truncated, as the *truncatus*. It is probable, that the incisors in the upper jaw are limited to two, as in the porpess.

122. HYPEROODON. Bottle-nose. Two small teeth in front of the lower jaw. *H. bidens.* See HUNTER, Phil. Trans. 1778, Tab. xix †.

No dorsal fin.

123. DELPHINAPTERA. Beluga. *D. albicans.* This whale is remarkable for the whiteness of its skin, all the others being black on the upper parts. (SCORESBY, Arc. Reg. Tab. xiv.

---

* The Delphinus gangeticus of ROXBURGH (Asiatic Researches, vol. vii., and Phil. Trans. 1818), belongs to this group.

† HUNTER gives the lower jaw in one place, and the upper in another, as the station of the teeth.

No teeth in the mouth.

124. MONODON. Narwal. Although there are no teeth within the mouth, there is a straight tusk, projecting anteriorly from the left upper lip, spirally twisted, sinistrally solid at the free extremity, and tubular where it is inserted in the skull. Within the skull, on the other side, appears the rudiments of a second tooth.

This second tooth, on the right side, a figure of which, belonging to a male, we have given in Plate I. fig. 6., is solid throughout, spirally twisted dextrally in a very slight manner, blunt, and somewhat puckered at the apex, bent a little towards the base, the face of which is oblique, smooth in the centre, and uneven towards the margin, and bordered by a ring of tubercular eminences. The cavity containing this kind of tusk, is, in some cases, closed with bone in front. Sir E. HOME denominates this body a *milk tusk* (Phil. Trans. 1813, p. 128.), and more recently (Phil. Trans. 1820, p. 147.), has concluded, that " as the permanent tusk in the narwhal begins to form in a direct line immediately behind the origin of the milk tusk, the great purpose of the milk tusk is evidently to open the road for, and to direct the course of the permanent tusk, till it is completely pushed out by it." This opinion appears to us to be untenable. The blunt point, the curved base, and its tuberculated margin, are ill suited to its character as pioneer. There is, besides, no evidence that these milk-tusks are ever shed, that permanent ones are formed behind them, or that the sinistral protruded tusk was ever preceded by any other. The circumstances of the case rather render it probable, that this body is a nucleus or support to the pulpy substance which secretes the tooth, that it is absorbed at an early period when a tusk is formed, but remains in its place when no tusk has been developed. An examination of the dentition of very young narwals, can alone determine the truth of this conjecture.

B. *Blow-hole single, without a bony septum. Teeth in the lower jaw.*

125. PHYSETER. Spermacete. No elevated fin on the back. The *Ph. macrocephalus* and *Catodon,* are the two species which constitute this genus.

126. TURSIO. A high dorsal fin. *T. vulgaris* and *microps,* are recognised species.

MARSUPIALIA.

Uterus destitute of a placenta.

In all the animals of this division, the uterus is complex, such as has been already described *. The young at birth are blind, and remain a long while dependent on the mother. They are either deposited in a ventral pouch containing the teats, or they attach themselves to the mother by their feet and tails. The ventral pouch and teats are supported by two bones, which are attached to the os pubis, and project along the abdomen, on the central side of the teeth. In these, the muscles belonging to the pouch and teats have their origin. These marsupial bones have likewise peculiar muscles attached to them, for the regulation of their own motion on their joint at the pubis. The scrotum of the male is placed in front of the penis.

In general habits, the marsupial animals bear a near resemblance to the glires, although, in the arrangement of their teeth, some of the genera approach the feræ.

All the toes of the hind-feet free. The uterus is double,

---

* Vol. I. p. 397.

and the penis is bifid, to correspond with the two ora tincæ. The corpora lutea are glandular.

With tusks.

In these, the tusks are long, the incisors small, and the grinders have conical points,—the usual characters of insectivorous animals. The great toe is without a nail.

127. DIDELPHIS. Opossum. Incisors, ten above, and eight below. Tail prehensile. The great toe capable of acting as a thumb. The pouch in some of the species is perfect, as *D. opossum,* while in others, its place is marked by a fold of the skin on each side, as the *D. philander.* It is probable that the discovery of other characters may induce naturalists to divide them into distinct genera. The *D. palmata,* in which the feet are webbed, constituting the genus *Cheronectis* of ILIGER, is too little known to admit of classification.

128. DASYURUS. Incisors, eight above, and six below. Two species, *D. cyanocephalus* and *D. ursinus,* are described and figured by Mr HARRIS in Lin. Trans. ix. p. 174. Tab. xix.

A. Without tusks.

129. PHASCOLOMYS. Wombat. Two long incisors in each jaw. Grinders with two transverse ridges on the summit. Herbivorous. Cœcum large, with an appendage. No tail. *Ph. ursina.*

First and second toes of the hind-feet closely united as far as the claws.

A. With tusks.

With tusks in both jaws.

130. PERAMELES. Incisors, ten in the upper jaw, and six in the lower. *P. nasutus.*

QUADRUPEDS.

Without tusks in the under jaw.

a. The great toe capable of acting like a thumb. Two long incisors in the lower, and six in the upper jaw. The rudiments of tusks sometimes visible in the lower jaw.

131. PETAURUS. Winged. The skin at the sides between the fore and hind legs expanded, fitting for partial flight. *Didelphis pelaurus* of SHAW. Probably the *D. pygmea* of the same author may warrant the reception of the genus Phalangista of HIGER.

132. BALANTIA. Without wings. *B. lemurina.*

b. Wanting the great toe.

133. HYPSYPRYMNUS. With a long tail. Incisors in the upper jaw six, in the lower two. *H. minor,* or Kangaroo-rat.

134. KOALA. No tail. Lower jaw with two long incisors, the upper with two long and some lateral short ones.

B. Without tusks. The corpora lutea are glandular; the uterus has one os tincæ, and the penis is entire.

135. MACROPUS. Kangaroo. Hind-legs remarkably long. The *M. major* from New Holland, is the only species whose history is well known.

MONOTREMATA.[*]

Quadrupeds which do not suckle their young.

The snout is produced and destitute of hair. The toes

---

[*] Μονος unus and τρημα foramen, a name given by GEOFFROY, in reference to the union of the openings of the urinary and sexual organs with the anus, to form one aperture.

are five on each foot, armed with strong claws. On the hind-feet of the male, there is a hollow, conical horny process at the setting on of the heel, which acts as a sheath to an awl-shaped bone. This bone has a narrow slit in the apex, is hollow throughout, and terminates in a flask-shaped bag, that rests on the ligaments of the bones of the foot. In the ornithorinchus, it is placed on the external and posterior sides of the leg. Sir JOHN JAMISON, by whom this structure of the spur (as it is improperly called,) was first brought into notice, observed, that the bag at the base contained a venomous fluid, which, upon the spur being pressed, squirted through the tube. His own servant suffered severely from a puncture with this venomous appendix *. The teeth are composed of vertical horny fibres. There are no external ears. The sternum is destitute of the ensiform process. The clavicles are united into a broad bone (which at first might be mistaken for the sternum), emarginate at the anterior extremity. The dorsal extremity of the scapula is imperfect, the clavicular edge is expanded, and scarcely touches the sternum. There are vestiges of marsupial bones.

It is in the reproductive organs, however, that the peculiarities of the tribe are conspicuously displayed. The penis gives passage to the semen alone, the urine being poured into the termination of the rectum †. In the female, there are no teats. The ovaria are double. Their surface is more uneven than in the mammalia. The corpora lutea

---

* Lin. Trans. xii. p. 584; and a notice of some observations by Dr BLAINVILLE, Ann. Phil. x. p. 112.

† In the ornithorinchus, the seminal urethra terminates in two openings corresponding with the two ora tincæ. In the Echidna, the primary orifices are four in number, with a circle of minute ones around each, leading us to expect (if analogy can be trusted) four uteri or oviducts in the female.

are imbedded, but they appear rather in the form of simple cavities in the substance of the ovarium, than glandular bodies like those of the mammalia *. The oviducts suffer an enlargement in their course, and terminate by separate openings in a common passage with the urine.

It does not yet appear in what manner the ovum is perfected. According to Sir John Jamison, " the female is oviparous, and lives in burrows in the ground †." According to Sir E. Home, to whom we owe nearly all the accurate information we possess, relative to the structure of the marsupial and monotrematous tribes, " the yolk-bags are formed in the ovaria; received into the oviducts, in which they acquire the albumen, and are afterwards impregnated. The fœtus is aërated by the vagina, and hatched in the oviduct, after which, the young provides for itself, the mother not giving suck ‡." All these, however, must be viewed as mere conjectures, scarcely warranted by the appearances of the organs which have been examined.

If these animals are oviparous, and we can scarcely entertain a doubt on the subject, as the eggs have been transmitted to London, it would be interesting to know the *manner of incubation*, and whether oviparous or ovoviviparous, the *kind of food* by which the young are nourished immediately after birth.

There are only two genera known, the species of which are natives of New Holland.

136. Ornithorinchus. Duck-bill. Snout produced

---

* The form of these cavities has induced Sir E. Home to term them yolk-bags. (Phil. Trans. 1819, p. 236.) They differ, however, from the organs so named in birds, in not being external and pedunculated, while they agree with the corpora lutea in their imbedded station. In the kangaroo, he assigns the office of producing the *ovum* to the corpus luteum, and the *yolk* to the oviduct, while the uterus furnishes the albumen.

† Lin. Trans. xii. p. 585.   ‡ Phil. Trans. 1819, p. 238.

depressed. Two corneous grinders in each jaw. Cheek pouches. *O. paradoxus* *.

137. ECHIDNA. Snout produced rounded, palate and tongue covered with reversed corneous processes. Without cheek-pouches. Two species have been figured by Sir E. HOME, E. *hystrix* and *setosa* †.

Quadrupeds are usually preserved in a dried state in a museum. When the specimen intended for preparation is procured alive, it ought to be killed by *pithing*, which consists in passing a sharp instrument between the skull and first vertebra, so as to divide the spinal marrow. In small animals, the puncture of a needle will be sufficient. When quite cold, the process of *flaying* may be begun. The skin is opened by a longitudinal incision along the side, and carefully separated from the subjacent parts, as far as practicable. The nearest hind-leg is now detached from the body by the separation of the joint at the pelvis, and by reversing the skin over its surface, an opportunity is given to remove the flesh from the bones, even unto the setting on of the toes. The tail is next separated at the rump, and the disengaged skin, now easily reversed, will allow the operator to remove and clean the bones of the other hind-leg. The skin is now to be pulled over the body towards the head, separating in its course the cutaneous muscles and the fore-legs, and over the head, to the nose and mouth, separating the ears by their base, and preserving the eye-lids. The carcase is disengaged from the head

---

* Its structure is described by Sir E. HOME. Phil. Trans. 1802, p. 67.
† Phil. Trans. p. 348. Tab. x. and xiii.

at the atlas. The whole flesh, (with the tongue and eyes,) is now removed from the head, and the brain scooped out by a gauge, through the foramen magnum, enlarged to a suitable size. When the whole flesh is removed, and as much of the fat as possible dried up by means of saw-dust or bran, the fleshy side of the skin, all of which, being now exposed, is to be covered with the lather of the arsenicated soap. The skin is now ready for stuffing.

A piece of softened iron wire is now fixed in the skull, and the other end extends, backwards, a little farther than the original length of the animal. This wire is a substitute for the vertebral column. A thin layer of tow or cotton is placed over the skull, the orbits of the eyes, cheeks, and throat are filled up with the same material, and artificial eyes are placed in the proper position *. The projecting vertebral wire is now to be covered with tow, to nearly the original thickness of the neck of the animal, and the skin of the head and neck returned to its natural situation. Pieces of wire are now to be fixed in each leg at the toes, extending a little beyond the soles, and continued the whole length, making them project beyond the heads of the humerus and femur. Tow is to be placed round the bones, the skin returned over them and stuffed to the natural size. The vertebral wire is now to be inserted in the tail, and the projecting ends of the leg-wires twisted round it in such a manner, as to place the four extremities in their natural relative position. Into all the vacuities, tow is now to

---

* Glass-eyes are prepared by the glass-blowers, either with a coloured iris or plain. In the last state, the iris is painted, on the glass behind, of the natural colour. Black glass beads are used for the smaller animals. A drop of black wax on a card is sometimes resorted to, or a ball of wax is painted of the requisite colour. But the shifts of the ingenious are without number.

be inserted, or pushed by a blunt wooden or whalebone probe. The animal is now to be fixed on a board, by means of the projecting wires of the soles of the feet, pressed into a suitable form and attitude, the incision sewed up, and the whole dried in a current of air, or before a fire.

Where the animals are of a large size, nearly the whole bones must be removed, a frame of wood prepared of the original size, and the skin cautiously stretched over it, filling up the intervals with tow or straw. The bats are best preserved by emptying the skin of its contents at the back or belly, as the one or the other is wished to be preserved entire for exhibition, by expanding their wings, and fixing them with a thread to a piece of card-paper.

The art of stuffing well can only be acquired by long practice. In general, the body is made to appear too long, and the legs being too much distended with the tow, look as if they were swollen.

BIRDS.

Ovarium single.

Birds are distinguished from the animals of every other division, by having their bodies covered with feathers. The structure and mode of growth of these appendices of the skin, have already been described, but we have still to consider the characters which they furnish for the purposes of classification.

The feathers receive particular names from the parts of the skin on which they grow. Those of the wings are the most remarkable, as constituting the principal organs of progressive motion in the air and in the water. These are divided into quills and coverts. The quill feathers which grow towards the extremity of the wing, on those bones which are analogous to the fingers in quadrupeds, are termed the *primaries*. They extend from the tip to the first obviously moveable joint. They differ in number, relative length, abbreviation of the web on one or both sides, and in colour. The quills on the next joint towards the base, are called the *secondaries*. These are usually shorter and broader than the former. At the joint which separates these two kinds of quill feathers, there is a tuft of three or four stiff feathers, constituting the *winglet* or bastard wing. The *tertiaries* are those which grow from the humeral joint of the wing. The *coverts*, distinguished into upper and under from their position on the wings, are placed in several rows, those feathers being the largest which cover immediately the quills. The *scapulars* cover the sides of the back, and are usually longer and more lax than those of the neighbouring parts. The *tail-feathers*, which are strong as those of the wing, have likewise their upper and under coverts. Besides feathers, many birds have hairs, particularly as whiskers or vibrissæ. The *oil-bag* situated on the rump, has usually one opening, but in the goose there are two; the summit is surrounded by a tuft of soft feathers*. Those places of the skin of birds unprotected by feathers, likewise exhibit peculiar characters. The jaws are covered with the horny bill which, in its various forms, furnishes important characters. At the base of the bill, there is sometimes a naked skin, termed *cere*, or, when it extends from the bill to the eye, *lore*. The legs and toes are covered with scales which are closely united with the skin, and appear either *reticulated* or plaited. The soles of the feet are more or less rough, with tubercles and papillæ.

---

* In many birds, as the parrot, wood-pigeon, and heron, the feathers, especially those under the wing, are covered with a soft mealy powder.

The feathers of birds are annually renewed; but the change takes place at different intervals, in compliance with the conditions of the season. We have seen a Bernacle goose (shot 6th January 1819,) with the black feathers on the neck in progress of moulting, while, on every other part of the body, the plumage was complete. The quill feathers, in particular, appear to drop off at intervals, in succession, and as there is seldom more than one of these wanting in each wing at a time, the power of flight is but little impeded. Nor are the outermost quill feathers first shed, but usually the fourth or fifth, and in some birds the innermost of the primaries.

The bones of birds, although bearing a close resemblance to those of quadrupeds, both with respect to number and position, exhibit several peculiarities by which their skeleton may be recognised. The sutures of the bones of the cranium, speedily become ossified after birth. At the union of the upper mandible with the frontal bone, there is a thin intervening osseous plate, by which, a considerable degree of motion is admitted. The orbits are separated, in some species, by a membranous, in others, by an osseous septum, descending from the frontal bone. In front of the opening of the ear, and attached to the temporal bone, the *os quadratum* is situate. It occupies the place of the zygomatic process; on the side, it is articulated with the slender cheek-bone, anteriorly with the posterior palatine-bone, and inferiorly with the lower jaw. The anterior and posterior palatine bones, are more or less connected with the inferior edge of the orbital septum. The occiput is articulated with the spine, by means of one condyle.

The spine of birds is remarkable for the number of cervical vertebræ, varying in the different species from nine to twenty-three. The lumbar vertebræ are ossified in one piece with the haunch-bones. The sternum is greatly expanded, and, in the size of its mesial crest and division of its posterior extremity, furnishes several characteristic marks of species or genera. The vertebral extremity of the ribs is bifurcated, the sternal end has an osseous appendix, and on the middle, there is a flat process projecting obliquely backwards over the succeeding rib. The clavicles are united to the anterior edge of the sternum, where they are received into an oblique groove. The scapular extremities of the clavicles are kept asunder by the merry-thought, which forms the porch as it were to the thorax. The scapular bones are long and narrow. The humerus is articulated with the scapula and clavicle in a shallow cavity. The ulna which supports the secondary quills, has, in many species, a row of tubercles on its dorsal aspect. The radius is slender. The carpal bones are small, and two in number. The metacarpal bone consists of two branches united at each extremity. On its anterior edge, near its base, the thumb bone is situated, which supports the bastard wing. This bone, in many birds, as the water-rail, land-rail, and arctic gull, supports, at its extremity, a nail or claw, more or less obvious externally. At the extremity of the metacarpus, there are two fingers, the largest of which consists of two phalanges, the smallest of one styloid phalanx.

The pelvis of birds is imperfectly developed; the ossa innominata, lumbar vertebræ, and sacrum, form only one bone, open ventrally where the symphysis pubis occurs in quadrupeds. The former is short, and does not appear externally. The tibia (usually, but improperly, termed the thigh,) is perfect; the fibula is ossified to its femoral extremity, and never reaches its whole length. The tarsus (improperly termed the leg,) is trifid at its lower extremity for the articulation of the phalanges of the three toes. The fourth toe is wanting in some species, in others, it consists of one or more joints. Whatever number of joints the fourth or hind toe possesses, the inner toe has one, the middle two, and the outer three joints more, in the greater number of birds.

Birds, in accomplishing progressive motion on land, make use of their posterior extremities, in walking, hopping, or running. In flying, the wings alone produce the motion, while the tail regulates the course. In swimming on the surface of the water, the legs are exclusively employed, but when motion is accomplished beneath the surface, the wings are then chiefly in exercise.

The third eye-lid, or membrana nictitans, is here so perfect, as (when drawn out) to cover the whole eye-ball. The external margin of the sclerotic coat is split into two laminæ, between which, a circle of osseous plates is interposed. The iris is variously coloured, according to the species, and is frequently used as a discriminating character. It is subject, however, to change with age. The optic nerve terminates in a white line, from the sides and ends of which the retina is produced. The *marsupium nigrum* or pecten, is suspended the whole length of this line, penetrates the vitreous humour, and reaches nearly to the lens. It is composed of vascular folds covered with a black pigment. Its use is unknown.

There is no external ear, although the feathers are so arranged as to supply the place of a concha. In many species these feathers are peculiar in their form and even colouring.

The nostrils exhibit many important characters. These are chiefly derived from their form, their position in the bill, and their relation to the feathers at the base.

The digestive organs exhibit many varieties of form. There is, properly speaking, no pharynx, no uvula nor epiglottis. The nasal and tracheal openings are two narrow slits, capable of being enlarged or closed at pleasure.

The gullet, in some species, is furnished with a membranaceous enlargement or crop, and in all, towards the entry to the stomach, there is a thickening of the walls, for the reception of the zone of gastric glands (ventriculus succenturiatus). The stomach is either in the form of a gizzard, the walls of which consist of powerful muscles and are lined by a thick cuticle, or a membranaceous bag. Externally, there is but little difference between the large and small intestines. At the commencement of the former, there are usually two *cœca*, varying in dimensions in different species. In a few birds, however, they appear to be wanting. In many birds, particularly among the *waders*, there is an appendix attached to the small intestine, which performs the office of a mucous gland. It is the remains of the ductus vitello-intestinalis, or communication between the yolk-bag and the small intestine of the cheek *. The rectum terminates in the *cloaca*, which is of considerably larger dimensions than the intestines, and receives the orifices of the ureters, sexual organs, and the *bursa Fabricii*. This last organ is in the form of an oval bag, filled with a mucus secreted from its glandular walls. It opens into the cloaca by a linear aperture. Its use is unknown.

The liver of birds is larger in size, and more uniform in shape, than in the quadrupeds. Its two lobes are nearly of equal size. The gall-bladder is absent in a few birds. The hepatic and cystic ducts unite in some species, and in others, open separately into the intestine.

The pancreas is situate between the folds of the duode-

---

* See " An account of an Appendix to the small Intestines of Birds," by JAMES MACARTNEY, Esq. Phil. Trans. 1811, p. 257.

num, in the form of a long narrow conglomerated gland. Its ducts are seldom united with the biliary ones.

The spleen is small, and situate between the left lobe of the liver and the stomach, in the immediate neighbourhood of the gastric zone.

The circulating system is chiefly distinguished from that of quadrupeds, by the structure of the heart. The pulmonic ventricle, instead of having a membranous valve (such as is found in both ventricles of quadrupeds, and also in the systemic ventricle of birds), is provided with a strong, tense and nearly triangular muscle, which BLUMENBACH considers as assisting in driving the blood with greater force into the lungs *.

The organs of respiration present several remarkable peculiarities. The trachea is composed of cartilaginous or bony rings, which are usually complete, especially towards the upper extremity. The larynx is in the form of a simple slit, with a cartilaginous margin. It is strengthened by several thin plates of bone, which have not been very accurately examined. At the division of the bronchiæ, there are membranous spaces, or osseous capsules, which essentially contribute to the formation of the voice. These osseous capsules are very obvious in the males of the goosanders and many ducks. Besides these peculiarities, the trachea in many birds suffers one or more enlargements in its course. In other cases, it is convoluted towards its bronchial extremity, and the convolutions are either contained in the cavity of the thorax, or received into a chamber hollowed out in the breast-bone †.

---

* Comp. Anat. p. 241.

† The reader who is desirous of further information concerning the trachea of birds, may consult " An account of some peculiar advantages in the

The lungs extend on each side much farther than in quadrupeds, are firmer in their texture, and have larger cells. On their dorsal aspect, they are applied close to the ribs, and fill up hollows between them. On their central aspect, they are covered by a delicate pleura, having numerous perforations. These apertures lead to air-cells which occupy the cavity of the thorax, and the sides of the abdomen. The cells communicate freely with one another, and with other cavities situated underneath the skin, in the middle of the bones, and extending even to the quills. These cells, wherever placed, have still a direct communication with the lungs, and receive and part with the air through the windpipe *. The walls of the cells are occasionally furnished with muscular threads.

Various conjectures have been offered by different observers, concerning the use of these air-cells. Few seem to consider that they are subservient to the aëration of the blood Some regard them as aiding the voice, while others conclude that they serve to vary the density of the animal, and enable it to accommodate itself to the different actions of flying, swimming, or diving.

The kidneys form a row of irregular lobes on each side of the lumbar vertebræ, and fill up cavities in the haunch

---

structure of the asperæ arteriæ, or windpipes of several birds, and in the land-tortoise." By Dr Parsons, Phil. Trans. 1766, p. 204.—" An Essay on the Tracheæ or Windpipes of various kinds of birds." By Dr Latham, Lin. Trans. iv. p. 90. and the Suppt. Mont. Ornith. Dict. at the end.

* The intimate connection subsisting between the lungs and the air cells, is easily exhibited by making an incision into the abdomen of a fowl, and then obstructing the windpipe by a ligature. Respiration will be carried on through the incision. We have seen, in the heron, respiration performed during a whole day through the broken humerus. See " An account of certain Receptacles of air in Birds, which communicate with the Lungs." &c., by John Hunter. Phil. Trans. 1774, p. 205.

bones. A ureter proceeds from each, and terminates in the cloaca. The urine has a white chalky appearance, and even in those species which subsist entirely on vegetable matter, it abounds in uric acid.

The male organs of generation are simple in their structure. The testicles, which are two in number, are situate close to the spine, at the commencement of the kidneys. They vary greatly in size, and are always very small at the beginning of winter, enlarging with the approach of spring and the breeding season. They are of a roundish form, or yellowish-white colour, and consist of numerous seminal tubes. Each sends out a spermatic duct, which opens into the cloaca at the summit of a conical process, or, uniting with its neighbour, forms a common duct. The two papillæ in which the ducts terminate, are distinctly visible in the common fowl. In the drake, however, the ducts terminate in a long wormshaped tube, which, when at rest, is concealed in the cloaca.

In the female organs, the ovarium is single, and the ova are inclosed in the yolk-bags, which are supported on short foot-stalks. The largest yolk-bags are placed on the outside. The oviduct is expanded at its extremity for the reception of the ovum, and, in general, pursues a lengthened tortuous course. It enlarges in size and in the thickness of its walls towards the cloaca. Its first portion nearest the ovarium, is covered internally with numerous papillæ, the orifices of the glands which secrete the glaire or white of the egg. The part which follows, is termed the uterus, and exhibits internally numerous longitudinal folds. Here the egg receives the shell. The last portion or vagina, terminates on the left side in the cloaca.

Birds exhibit very remarkable differences with regard to the size of their eggs, in proportion to the body, and their number. They likewise differ in the method of constructing their nests, the period of incubation, and the condition of the young brood. These differences never fail to arrest the attention of the practical naturalist, and their contemplation is well calculated to instruct and amuse *.

The characters by which birds are separated from other animals, are so obvious, as to be easily detected even by an inexperienced observer. It is extremely difficult, however, to distribute them into subordinate groups. All the species have so many points of resemblance, by possessing the same organs, that characters can only be obtained from the modification of these. Even when the modifications of any one character are employed, we find so many transitions from one kind of development of an organ to another, as to render it impracticable to draw a definite line of separation. If we employ the common divisions of *land* and *water* birds, we shall find that there is an extensive group called Grallæ or Waders, which hold an intermediate rank. They frequent marshy ground, and the margins of rivulets. Some of the species of this group, bear a very close resemblance to the land birds in their habits. Thus, the water-rail is intimately connected with the gallinule and coot, as waders, on the one hand; while, on the other, it

---

* Hunter conjectures, that some birds can remove their eggs from the nest when they discover danger, at least, he has reason to suspect this of the sparrow. (Phil. Trans. 1792, p. 30.) I have credible testimony, that the partridge can remove its eggs to a safe place, carrying them under its wing. The opinion that some birds will lay more than their ordinary number of eggs, by daily abstracting one from the nest, though countenanced by Pennant (Brit. Zool. i. p. 400.), on the authority of Lister, has never been established by recent observations. Montagu, indeed, repeated Lister's experiment on the swallow without success, (Ornith. Dict. i. p. xi.) nor have we been more fortunate with the magpie, sparrow, wren, and chaffinch.

bears a great resemblance to the corncrake, a bird which, in many respects, agrees with the partridge. Even among the true water birds, or such as have webbed feet, there are a few, as the bean goose, which feed on grain and grass in the fields, like the gallinaceous birds, while others, as the black-backed gull, have a propensity to feed on carrion like the eagle. These instances of dissimilar manners in birds of similar forms, and *vice versa*, might be multiplied to such an extent, as to demonstrate, that many of those groups which modern ornithologists denominate Natural Families, are, in several respects, artificial combinations, and that all our systematical arrangements must depend on characters, between which there are no absolute limits. Such we confess to be the case with the characters which are now to be employed.

FISSIPEDES.

Toes free.

In this extensive group, the feet are formed for grasping or walking. The species, consequently, reside on land. In those which frequent marshy ground, the toes are, in some cases, flattened below, or even bordered with a web, to enable them to walk on the mud, or even to swim a little. Even in such birds, which approach the Palmipedes, swimming is performed with difficulty, and diving is nearly impracticable. In many genuine land birds which frequent dry ground, the middle toe is connected, by means of a web, with the outer toe, as far as the first joint. This is probably intended to increase the sole of the foot, and render it better adapted for standing or walking on the ground.

I. Tibial joint feathered. The legs are usually short, and the feathers frequently grow upon what is termed the *knee-joint*, at the union of the tibia and tarsus, and even descend a considerable way down the tarsus. These may be considered as the genuine *terrestrial* birds; although a few species may be found among them, as the king's fisher and water crow, which, in their manners, approach the waders.

Three toes directed anteriorly.

A. Nostrils hid under an arched covering. The bill is vaulted; the toes have serrated edges, and the feathers of the wings are short. The stomach is a strong gizzard. The food consists chiefly of vegetables, occasionally mixed with insects.

a. Bill arched from the base. Eggs numerous. This group includes the GALLINÆ of LIN. The characters of the species, as connected with systematical arrangement, have been very imperfectly explored. The males, in general, have spurs.

1.] First and second toe united at the base.

(A.) Hind-toe fully developed *.

(*a.*) Front or crown of the head appendiculated.

Appendix consisting of feathers. The genera of this group have a crest of feathers, which they can elevate and depress at pleasure.

Tail coverts remarkably produced.

1. PAVO. Peacock. The size, the motions, and the colouring of the tail, distinguish this genus from all others. There are two species, the *P. cristatus*, or common peacock, and the *bicalcaratus*, or double spurred kind, which TEMMINCK has placed in a separate genus, termed Polyplectrum.

---

* It is probably among the Gallinaceous birds, and in this section, where the doubtful genus Didus of LIN. will be placed, when the species shall have been more carefully examined.

Tail coverts not produced.

(1.) Throat covered with feathers.

Cheeks feathered.

2. CRAX. Bill strong and thick, with a cere at the base. *C. alector.*

3. ORTALDIA. Bill slender, without cere. *Phasianus motmot, Gm.*

Cheeks naked.

4. LOPHURA. Tail vertical, with the middle feather arched. *Phasianus ignitus.*

5. LOPHOFERA. Tail horizontal. *Phasianus Impyanus.*

(2.) Throat naked.

6. PENELOPE. A naked space round the eyes.

With fleshy or hard processes as a crest.

Crest soft and flexible.

7. MELEAGRIS. Turkey. A conical flexible process in front. *M. Gallopavo.*

8. GALLUS. Cock. A longitudinal serrated crest. *G. domesticus.*

Crest hard.

9. NUMIDIA. Pintado. Head naked. Cheeks with wattles. *N. meleagris.*

10. OURAX. Head feathered. *Crax pauxi.*

(*b.*) Crown of the head destitute of appendices.

In many of the genera, the occipital feathers are a little produced.

Tarsus feathered.

Toes feathered.

11. LAGOPUS. Grous. Toes closely feathered above, but plain below. *L. Scoticus* and *vulgaris.*

Toes nearly naked above; furnished on each side below with a pectinated margin. *Tetraonidæ**.

Tail divided and recurved.

12. TETRAO. Black-cock. *T. tetrix.*

Tail not divided.

13. UROGALLUS. Capercailie. Tail rounded. *U. vulgaris.* Formerly resident in Scotland.

14. PTEROCLES. Tail pointed. Hind-toe small. *P. alchata* and *fasianellus.*

Tarsus naked.

Naked skin on the cheeks.

15. PHASIANUS. Pheasant. Tail roofed, with the feathers produced. Male with a spur. *P. colchicus.*

16. PERDIX. Partridge. Tail even and short. Males with only the rudiments of spurs.

Cheeks entire; clothed.

17. COTURNIX. Quail. Tail short. Males without spurs.

(B.) Hind-toe imperfect.

Tarsus feathered.

18. SYRRHAPTES. Toes downy, and connected nearly to the tips. *Tetrao paradoxus.*

19. TINAMUS. Bill long, slender, soft at the end. The true place of these two genera does not appear to be satisfactorily determined.

Tarsus naked.

20. CRYPTONYX. Head of the male with a crest of feathers. Hind-toe destitute of a nail.

---

* This family was first pointed out by FORSTER. Phil. Trans. 1772, p. 397.

2.] Toes divided to their origin.

21. OPHISTHOCOMUS. Head with a crest of feathers. Hind-toe perfect. *Phasianus cristatus.*

22. ORTYGIS. Destitute of a crest or hind-toe. *Tetrao andalusicus.*

b. Bill swollen at the base, nearly straight, and subulate towards the extremity. This includes the genus Columba of LINNÆUS. The eggs are limited to two each hatching; but many broods are produced in a season. They drink, not by sipping, and then holding up their heads, as other birds, but by a continued draught. The toes are divided to their origin. Twelve tail feathers. No spurs.

Head ornamented.

23. VERRULIA. Head covered with a warty skin, like some of the gallinaceous genera. *Columba carunculata.*

24. *Goura.* Head with a crest of feathers. *Columba coronata.*

Head plain.

25. COLUMBA. Pigeon. Bill flexible*.

26. VINAGO. Bill solid, compressed.

The numerous species of columbine birds, will probably constitute many more genera, when their forms and structure are subjected to a more minute examination.

B. Nostrils exposed, or hid only by feathers.

ACCIPITRES.

Bill and claws strong and hooked. These are rapacious animals. The feet are formed for seizing the prey, being

---

* The observations of HUNTER warrant the conclusion, that the crop of the pigeon becomes periodically glandular, at the period of hatching the eggs, and secretes a curdy matter, with which the young, at their birth, are nourished.—Ob. An. Ecp., 191.

warty underneath, and furnished with strong curved claws, especially on the hind and inner-toes. The under mandible is nearly straight, with a sloping end. The upper mandible is more or less arched from its base; and at the end it is bent down like a hook over the other, and is sharp pointed. The limbs are strong. The tongue emarginate. The females largest.

a. Bill covered with a cere, in which the nostrils are lodged. The eyes are directed laterally, as in other birds. Outer toe incapable of having its position, or its motion reversed. This includes the *diurnal* birds, constituting two families, Vulturidæ and Falconidæ, formed from the old genera Vultur and Falco.

Vulturidæ. Head naked. The eyes are on a level with the surface. Tarsus reticulated. Crop protuberant.

Nasal orifices placed transversly.

27. VULTUR. Head and neck bare. A collar of long feathers round the base of the neck.

Nasal orifices placed longitudinally.

28. SARCORAMPHUS. Condor. Neck bare of feathers. Cere carunculated.

29. PERCNOPTERUS. Neck covered with feathers.

Falconidæ. Head feathered. The young birds do not arrive at maturity of plumage until the third year. In their immature state, they have inconsiderately been described as distinct species. Indeed, the description of the species are usually so superficial, that it is difficult to found genera on essential characters.

Eyes level with the surface.

30. GYPAETOS. Nostrils covered with porrected hair. Crop protuberant. Tarsus feathered. This genus, represented by the *Vultur barbarus*, occupies a middle station

between the true vultures and eagles, and may be placed among either, with nearly equal propriety. With the latter, however, it is, perhaps, more intimately connected.

Margin of the orbit of the eye above protuberant, making the eye itself to appear as if sunk in the head.

(A.) The second feather in the wing the longest: the first, however, nearly equal. The bill is arched from the base. The birds of this division are termed Noble, and are reckoned, without very good reason, more docile than those of the following group.

Claws flat, or grooved below.
31. FALCO. Falcon. Beak with a sharp notch near the end. *F. communis.*
32. HIEROFALCO. Gerfalcon. Notch on the bill blunt. *Falco candicans* and *lagopus,*

Claws rounded beneath.
33. PANDION. Osprey. *F. Haliœtus.*

(B.) The fourth feather in the wing is the longest, and the first is very short.

Nostrils contracted.
34. CYMINDES. Tarsi short, half feathered, reticulated. Nostrils nearly closed. *F. cayennenis.*

Nostrils patent.

(*a.*) Head crested.

Tarsi feathered.
35. PLUMIPEDA. Wings shorter than the tail. Crest occipital and pendent. *Falco superbus.*

Tarsi naked.
36. SERPENTARIUS. The two middle feathers of the tail produced. A naked circle round the eye. *F. serpentarius.*

ACCIPITRES.

37. MORPHINUS. Tail feathers nearly equal. *Falco guianensis.*

(*b.*) Head plain.

Tail divided. Tarsi short, bill, toes, and claws, weak in proportion to the body. Wings exceed the tail in length.

38. ELANUS. Tarsi half feathered, reticulated. *Falco furcatus.*

39. MILVUS. Kite. Tarsi plated. *F. milvus.*

Tail entire.

(1.) Between the eye and the bill, closely covered with short feathers.

40. PERNIS. Honey-buzzard. Tarsi half feathered and reticulated. *F. apivorus.*

(2.) Between the eye and the bill naked, or covered with hair.

Bill straight at the base, and then bent at the tip. Wings reaching the length of the tail, when at rest.

41. AQUILA. Tarsi feathered to the toes. *F. chrysaëtos.*

42. HALIÆTUS. Tarsi half feathered and plated. *F. ossifragus.*

Bill bent from the base.

Wings reaching to the extremity of the tail. Tarsi plated.

43. CIRCUS. Feathers of the ears forming a collar from behind the eyes to the throat. *F. pygargus.*

44. BUTEO. Buzzard. Without a collar.

Tarsi lengthened. Wings not reaching to the end of the tail.

45. NISUS. Sparrow-hawk. Bill with a sharp notch. Tarsi plated. *F. Nisus.*

Tarsi short.
46. ASTUR. Goshawk. Tarsi plated. *F. palumbarius.*
47. CACHINNA. Tarsi reticulated. *F. cachinnans.*

b. Bill without cere. Eyes directed forward, and surrounded by a circle of radiating wiry feathers. The external toe capable of reversing its position and motion. On the top of the head, in some species, are two tufts of long feathers, termed *ears* or horns. Outer web of the first quill feather serrated. These are crepuscular feeders.

(1.) The feathers, or concha of the ear, occupying nearly the whole side of the head. Last joint of the toes plated. Wings extending beyond the tail. First feather in the wing longest.

Bill arched from the base.
48. OTUS. Horned owl. Head eared. Toes closely feathered. *O. vulgaris* and *brachyotus.*
49. ULULA. Head destitute of ears. *Strix nebulosa* of GML.

Bill straight at the base. Without horns.
50. ALUCO. Barn owl. Toes thinly feathered. *Strix flammea,* LIN.

(2.) Concha of the ear extending only to about one-half of the side of the head. The disc of feathers around the eye is less perfect than in the preceding genera.

Head eared. Bill black. Irides yellow.
51. BUBO. Eagle owl. Toes closely feathered. *Strix bubo.*
52. SCOPS. Feet naked. *Strix scops;* now *S. Aldrovandi.*

Head naked.
53. STRIX. Toes closely feathered. *S. stridula.*

PASSERES.

Bill nearly straight in the gape. In some cases, there is a slight hook at the end. No cere. Eyes lateral. The character, indeed, of this extensive group, is purely of a negative kind, comprehending all the birds which do not possess the positive characters on which the other tribes are founded.

I. The first joint of the outer and middle toes connected by a membrane. In some cases, this membrane extends to the second joint.

(A.) Gape remarkably large. *Fissirostres.* Bill short, depressed, and slightly hooked. The species are all insectivorous. This group is the best characterized of all the Passeres [*].

Upper mandible, with a notch at the extremity.

54. PROCNIAS. This genus includes the *Ampelis carunculata* of GMELIN, and the *Hirundo viridis* of TEMMINCK, two species which appear to be the types of genera, the carunculated throat of the former being remarkable.

End of the bill entire.

Nostrils open. Feathers closely set. Wings very long, and flight rapid.

55. CYPSELUS. Swift. Hind-toe directed forwards. The three others, with only three phalanges. Hirundo apus.

56. HIRUNDO. Swallow. Toes of the common form. The nests of those species which are the summer visitants

---

[*] If, by inserting the Fissirostres here, I have separated the Dentirostres from the Accipitres, with which they have several relations, I have brought them nearer to the Conirostres, with which they are equally connected, and have united the Procnias with its congeners.

of Britain, have long been admired for the materials and arrangement of their construction. But the edible nests of the *H. esculenta*, a species found in the Indian Archipelago, are still more remarkable. They form an article of trade to the China market, where those of the first quality fetch their weight in gold. They are employed to make soup, to which is ascribed powerfully restorative qualities. The substance of which these nests consists, has much the appearance of isinglass, and is disposed in irregular transverse threads, with a few feathers interposed. Neither the analytical experiments of DOBEREINER, nor those of BRANDE, demonstrate it to be of animal origin [*]. The relatively small portion of ammonia, indeed, which it yields, and its facility of incineration, rather lead to the conclusion, that it is a vegetable gum. It was once supposed to be procured from the scum of the sea. Those individuals, however, residing fifty miles from the sea, employ the same materials as those which dwell on the shore. The other species in those districts, likewise employ a portion of the same substance in the fabrication of their nests. It is much to be regretted, that the recent historians of these regions have added so little to the history of this singular substance.

Nostrils tubular.

57. CAPRIMULGUS. Goatsucker. The species of this genus give indications of several characters for subdivision.

---

[*] Sir EVERARD HOME, (Phil. Trans. 1817, p. 332.) having found the margin of the orifice of the gastric glands of the H. esculenta divided into lobes,—a form he had not observed in other birds,—concluded that the substance of the nest was secreted by these lobes. Though the use of these lobes may puzzle (to use the author's own words), " the weak intellects of human beings," and give rise to " many wild theories," we cannot admit that there is a shadow of proof, not even from analogy, to conclude that these secrete the materials of the nest. The reasoning, indeed, which is employed to support this " wild theory," derived from the supposed history of the bird, is at variance with the statements of Sir THOMAS S. RAFFLES and Mr CRAWFORD.

(B.) Gape of the ordinary size.

(*a.*) Upper mandible, with a groove or notch in the margin, on each side, near the end. This division constitutes the DENTIROSTRES of CUVIER. In reference to this mark, it may be stated, that when it is very well marked, the species are usually rapacious in their manners.

Upper mandible hooked at the end. The birds of this division are ranked by many, among the Accipitres, with which, in manners, they correspond. They differ, however, in the want of cere.

1. Bill compressed.

(*a a.*) Ridge of the upper mandible arched.

Base triangular. Upper mandible arched. *Laniusidæ.*

58. LANIUS. Shrike. Ridge of the upper mandible rounded. Nostrils surrounded with bristles. *L. excubitor.*

59. GRAUCALUS. Mandibular ridge sharp. Nostrils covered with feathers. *Corvus Papuensis.*

Bill rounded at the base. Upper mandible slightly arched, nearly straight at the base, and rounded above.

Front feathers notched by the base of the bill.

60. BARITA. Bill nearly conical, large. *Paradisæa viridis.*

Front plain.

61. OXYPTERUS. Bill sharp-pointed. Wings extending beyond the tail. Feet short. *Lanius Leucorinchus.*

62. PSARIS. Bill very large. *L. cayenus.*

(*b b.*) Ridge of the upper mandible nearly even. Beak conical. Base of the bill triangular. Wings short. *Tangaradæ.* The characters of this family have been imperfectly determined, and the genera into which it may be divided yet remain to be instituted. M. CUVIER has given indications of some of the groups.

2. Bill depressed.

Under mandible slightly arched.

63. EDOLIUS. Ridge of the upper mandible acute. Nostrils covered with feathers.

Under mandible straight.

(*a.*) Bill at the base surrounded with strong hairs. *Muscicapadæ.*

(1.) Ridge of the upper mandible straight.
Ridge indistinct.

64. TYRANNUS. The bill is long and very strong. *Lanius pelanqua.*

Ridge acute or distinct.

65. MUSCIPETA. Bill long, edges bent. *Lanius sulphuratus.*

66. MUSCICAPA. Flycatcher. Bill short, with straigh sides.

(2.) Ridge of the upper mandible bent.

67. GYMNOCEPHALUS. Face in part bare of feathers. *Corvus calvus.*

68. CEPHALOPTERUS. Feathers at the base of the bill forming an elevated tuft. *C. ornatus.*

(*b.*) Bill with slender short hairs, and shorter, in proportion, than the preceding. *Ampelisidæ.*

Neck bare of feathers.

69. GYMNODERES. Feathers on the head downy. *Corvus nudus.*

Neck clothed.

70. BOMBYCIVOCA. Bohemian chatterer. Shaft of the secondary quill-feathers enlarging at the end into a flat horny substance, resembling red sealing-wax. *Ampelis Garrulus.*

71. AMPELIS. Secondaries plain. *A. carnifex.*

Bill destitute of a hook at the end. The notch is less distinct than in the preceding division

1. Bill compressed.

Nostrils remarkably large.

72. PHILEDON. Nostrils covered with a cartilaginous scale. Bill slightly bent. Tongue ending in a pencil of hairs. *Merops molluccensis.*

73. MENURA. Nostrils without scale, and covered with feathers. *M. vulgaris.*

Nostrils common.

A naked space above the eye.

74. GRACULA. A belt of downy feathers at the base of the bill. *Paradisea tristis.*

Above the eyes feathered.

Nostrils thickly covered with feathers.

75. PYRRHOCORAX. The habit of the crow. *Corvus Pyrrhocorax.*

Nostrils naked, or covered with hair.

Tail long.

76. TURDUS. Thrush. Bill slightly arched.

77. ORIOLUS. Bill stronger, and the legs shorter, than in the thrush.

Tail short.

78. MYOTHERA. With the bill of the thrush. It has longer limbs. *Turdus cyanurus.*

79. CINCLUS. Dipper. Beak straight. Mandibles equal. *C. vulgaris.* This bird is able to sink to the bottom of the water in pools, and walk thereon, like the hippopotamus among quadrupeds.

2. Bill subulate, slender, slightly depressed at the base.

Bill a little enlarged at the base. Legs long.

80. SAXICOLA. *Motacilla rubicola.*

81. SYLVIA. *Motacilla phœnicurus.* These two genera depend on the difference in size of the enlargement at the base of the bill, which is greatest in the first.

Bill slender throughout.

Claw of the hind-toe produced.

82. ANTHUS. The species of this genus were formerly confounded with the larks. *Alauda pratensis.*

Hind claw common.

Tail and scapulars produced. The former frequently in motion.

83. MOTACILLA. Wagtail. Hind toe curved. *Motacilla alba.*

84. BERDYTES. Hind toe nearly straight. *Motacilla flava.*

Tail and scapulars common.

Bill approaching to concave at the sides.

85. REGULUS. Bill straight. *Motacilla regulus.*

86. TROGLODYTES. Bill slightly bent. *Motacilla troglodytes.*

Bill uniformly conical at the sides.

CURRUCA. Nightingale. In this genus there is a slight compression at the tip of the bill, and the upper ridge is a little arched. *Motacilla luscina.*

87. ACCENTOR. Edges of the bill inflected. *Motacilla modularis.*

(*b.*) Bill destitute of the terminal notch. This character, in point of fact, is not absolutely negative, there being few birds of this, or any of the divisions in which vestiges of a notch may not be detected. In the birds of this group, however, the notch is indistinct.

CONIROSTRES. Bill strong and conical.

Mandibles gibbous towards the end.

88. BUPHAGA. The bill towards the base is cylindrical. *B. Africana.*

Mandibles destitute of gibbosity.

Lower mandible carunculate at the base.

89. GLAUCOPSIS. The bill is excurvate or arched. *G. cinerea.*

Mandibles without caruncles.

Hind-toe pointing forwards.

90. COLIUS. Both mandibles bent. Tail long. *C. capensis.*

Hind-toe pointing backwards.

Mandibles crossing each other laterally.

91. LOXIA. Cross-bill. Both mandibles with hooked points crossing each other at the plane of the gape. *L. curvirostra.*

92. CORYTHUS. Hawfinch. Upper mandible bent over the under. *Loxia enucleator.*

Mandibles acting in opposition.

(1.) Ridge of the upper mandible nearly straight.

With a palatine tubercle.

93. EMBERIZA. Bunting. This is a well marked genus. The species, however, particularly with regard to the snow bunting, have been unnecessarily multiplied, by attending only to the colour of the plumage.

Palate plain.

Hind-toe produced and straight.

94. ALAUDA. Lark. The bill is more slender in pro-

portion to its length, than the other neighbouring genera, *A. arvensis.*

Hind-toe of ordinary size and shape.

Base of the bill with numerous hairs.

95. PARUS. Titmouse. Bill short and sharp pointed. *P. major.*

Base of the bill plain.

Commissure of the bill straight.
Bill slender, angular, and pointed.

96. SITTA. Nut-hatch. Hind-claw strong. *S. Europæa.*

Bill strong rounded.

Bill more or less inflated at the base. The following genera depend on the gradations of this character, beginning with those in which it is most obvious.

97. PLOCEUS. *Loxia Philippina.*
98. COCCOTHRAUSTES. *Loxia coccothraustes.*
99. PYRRHULA. Bullfinch. *Loxia Pyrrhula.*
100. PYRGITA. Sparrow. *Fringilla domestica.*
101. VIDUA. Tail coverts produced. *Emberiza regia.*

Bill exactly conical. The following genera exhibit a gradation in the length of the bill.

102. FRINGILLA. Chaffinch *F. cœlebs.*
103. CARDUELIS. Gold-finch. *Fringilla carduelis.*
104. LINARIA. Linnet. *Fringilla linaria.*

Commissure of the bill interrupted.

105. CASSICUS. Bill exactly conical and very large. Base of the upper mandible ascending on the front. *Oriolus cristatus.*

106. STURNUS. Stare. Bill depressed, and of the ordinary size. *S. vulgaris.*

PASSERES.

2. Ridge of the upper mandible obviously curved,

(aa.) Nostrils covered with deflected feathers.

Feathers at the base of the bill wiry.

Feathers of the front loose, and capable of being erected into a crest.

107. GARRULUS. Jay. End of the upper mandible hooked. *Corvus glandarius.*

Front feathers plain.
Tail long.
108. PICA. Magpie. *Corvus pica.*

Tail of the ordinary size.
109. CORVUS. Crow. *C. Corax.*
110. CARYCCATACTES. Nutcracker. Corvus carycatactes. The slenderness of the bill, and its want of curviture, indicates this genus to belong to a different group.

Feathers at the base of the bill downy.
111. PARADISEA. Bird of paradise. Side-feathers very long. Two tail feathers without webs. This genus contains several species.

(bb.) Nostrils naked.

Head with naked carunculated spaces.
112. EULABES. Feathers at the base of the bill downy. *Gracula religiosa.*

Head covered.
113. CORACIAS. Roller. Bill compressed. *C. garrula.*
114. COLARIS. Bill depressed. *Coracias orientalis.*

TENUIROSTRES. Bill slender, produced.

(1.) Claws long and hooked. The birds of this group are

climbers, and run up and down trees and walls with facility.

Mandibles serrated.
115. CYNNYRIS. Tongue long forked. *Certhia splendida.*

Mandibles plain.
Mandibles depressed at the face. Do not use the tail in climbing.
116. TRICHODERMA. Bill triangular, long. *Certhia muraria.*
117. DECÆUM. Bill enlarged at the base. *Certhia erythronotos.*

Mandibles subulate.
Use the tail in climbing, indicated by the feathers being worn at the points.
118. CERTHIA. Creeper. *C. familiaris.*
119. DENDROCALEPTES. Body larger, and bill broader than the preceding. *Graculus scandens.*

Tail not used in climbing.
120. VESTIARIA. Bill bent, nearly semicircularly. *Certhia vestiaria.*
121. NECTARINA. Bill slightly bent. *Certhia cyanea.*
(2.) Claws of the ordinary size.
Feet short. Tongue filiform, and divided at the end. Plumage of metallic brilliancy.
122. TROCHILUS. Humming bird. This is an extensive genus, subdivided into those species which have curved bills, and those which have straight ones.

Feet of the ordinary size.
Tail very long.
123. PROMEROPS. Tongue filiform, and divided at the end. *Mepros cafer.*

PASSERES. 247

Tail of the ordinary size.

Head with a crest of long feathers.

124. UPUPA. Hoopoe. Crest feathers in a double longitudinal row.

Head plain.

125. EPIMACHUS. Nostrils covered with downy feathers. *Upupa magna.*

126. FRIGILUS. Chough. Nostrils covered with wiry feathers. *Corvus graculus.*

II. First and second toes adhering nearly to their extremity.

The first toe is produced, and nearly of equal length with the middle one.

(A.) Hind-toe developed.

Upper mandible notched at the extremity.

127. PIPRA. Manakin. Bill compressed. Nostrils wide. Head crested. *P. rupicola.*

Upper mandible destitute of a notch.

Mandibles serrated.

128. BUCEROS. Hornbill. Tongue small and pointed. Bill remarkably large, inflated. In some of the species there is a horny protuberance, at the base of the upper mandible, of a large size.

129. PRIONITES. Tongue plumosely barbed. Feet short. Tail long. *P. Brasilienses.*

Mandibles plane.

Bill angular lengthened. Feet short.

130. MEROPS. Bee-eater. Bill slightly arched, and sharp pointed. *M. apiaster.*

131. ALCEDO. King's-fisher. Bill straght and blunt. *A. Ispida.*

Bill depressed.

132. TODUS. Tody. Bill long, obtuse, and covered at the base with bristles. *T. Viridis.*

(B.) Hind toe imperfect,

133. CEYX. Bill straight and blunt. *Alcedo tridactyla.*

Two toes only directed forwards,

The outer toe, when it exists, is directed backwards. Birds possessed of this character have been long termed Climbers, or *Scansores.* Other birds, however, in which the toes are arranged in the ordinary manner execute the same motions. The species inhabit woods, and either feed on fruits or insects.

A. Toes equally divided, two before and two behind.

(1.) Anterior toes separate.

a. Bill remarkably large and hollow. The birds of this group have been termed *Levirostres.*

Upper mandible with an acute vertical crest.

134. CROTOPHAGA. Ani. Bill compressed, without serratures on the edges. *C. major.*

Mandible destitute of the vertical crest.

Base of the bill surrounded with pencils of stiff wiry feathers.

135. BUCCO. Barbet. Feet naked. *B. tamatia.*

136. TROGON. Eurucui. Feet feathered. *T. strigillatus.*

Base of the bill destitute of wiry tufts.

Tongue plumosely barbed on the margin.

137. RHAMPHASTOS. Toucan. Bill about the size of the whole body, and serrate at the edges. *R. viridis.*

SCANSORES.

Tongue with a simple margin.

Base of the bill with a cere.

138. PSITTACUS. Parrot. The bill in this genus is more solid than in the Toucans. The tongue is fleshy, and rounded. The species are numerous, and are all the inhabitants of warm countries.

Base of the bill without a cere.

139. SCYTHROPS. Bill channelled at the side. *S. psittacus.*

140. MALCOHA. Bill plain, rounded at the base.

b. Bill of the ordinary size.

Gape wide.

141. CENTROPUS. Claw of the hind toe much produced. *Cuculus Ægyptiacus.*

142. CUCULUS. Cuckoo. *C. canorus.* The species of this genus may be subdivided by the number of feathers in the tail, the form of the nostrils, and the curvature of the beak.

Gape narrow. Tongue long.

143. PICUS. Woodpecker. Bill angular at the base, and wedge-shaped at the tip. Tongue with reversed bristles. Tail feathers rigid. *P. martius.*

144. YUNX. Wryneck. Bill rounded, pointed. Tongue smooth. *Y. torquilla.*

(2.) Anterior toes connected.

Anterior toes adhering.

145. GALBULA. Bill produced, feet short. *G. viridis.*

Toes united by a membrane. Bill short, and inflated above.

146. MUSOPHAGA. Bill ascending on the forehead. *M. violacea.*

147. CORYTHRIX. Bill not ascending. Head crested. *Cuculus persa.*

B. Two toes before, and one behind.

148. PICOIDES. Resembling the woodpeckers. *Picus tridactylus.*

II. Lower end of the tibial joint naked. Tarsi naked. The legs and neck, in general, are lengthened, in conformity with the habits of the birds. They are denominated *Waders* (*Grallæ*) because they usually frequent marshy grounds, and the margin of rivers, and wade among the mud or water. A few species, however, reside constantly on the dry land.

A. Wings developed. All the species of this group are qualified for flying, by the size of their wings, and for running, by the length of their legs.

Gape remarkably large in proportion to the size of the bill.

149. GLAREOAL. Pratincole. Bill short, arched and pointed. Nostrils linear and oblique. Wings long, pointed. Tail forked. The *G. Austriaca* is the only authentic species. From the size of the gape, and the form of the tail and wings, it has been associated, by some, with the swallows.

Gape of the ordinary size.

Bill expanded at the extremity.

150. PLATALEA. Spoonbill. The extremity of the bill is broad, flat and thin like a spoon, and the two mandibles fold upon each other, like flaps of leather. The nostrils are small and oval. The legs are reticulated. *P. leucorodia.*

Bill tapering. The remarkable form of the bill, in the two preceding genera, prevent them from being confounded with any of the other grallæ. In those that follow, the points of resemblance are so numerous, that it is difficult to select marks for classification, from among the numerous modifications which are exhibited by the same organs.

CULTRIROSTRES. Bill strong, long pointed, and sharp edged. This form of bill enables the species, according to their size, to seize fish and frogs. The second group, or *Presserostres*, have bills so feeble, that they are confined in their prey to worms, insects, and the smaller crustaceous animals.

Margin of the bill serrated.

151. PHŒNICOPTERUS. Flamingo. Bill bent, as if broken. Anterior toes webbed nearly to the extremity. Legs very long. *P. ruber*. This genus may be considered as the connecting link between the Grallæ and Palmipedes.

Margin of the bill plain.

(a.) A membranaceous space around the nostrils.

Upper mandible notched at the extremity.

152. CANCROMA. Boat-bill. Bill gibbous. Nostrils at the base with a produced groove. *C. cochlearia*.

Upper mandible destitute of a notch.

Gape extending so far back as the eyes.

153. ARDEA. Heron. Nasal groove reaching nearly to the end of the bill. Legs reticulated.

This genus may admit of a subdivision, even from the structure of the bill, but still more readily by the disposition of the feathers. It is probably in this place of the system, where the genus Microdactylus of GEOFFROY, the Decholophus of ILLIGER, should be placed, represented by the *Palamedia cristata* of GMELIN.

Gape not reaching to the eye. The nasal membrane is extended. The head thinly covered, the feet reticulated.

154. GRUS. Tongue pointed. Part of the head naked. *Grus vulgaris.*

155. PSOPHEA. Trumpeter. Tongue fringed. Head with downy feathers. *P. crepitans.*

(*b.*) Nostrils opening simply in the bill, and somewhat dorsally.

Bill slightly notched.

156. TANTALUS. Upper mandible rounded above. Part of the head naked. *T. loculator.*

Bill plain.

Mandibles receding from each other in the middle of their course.

157. ANASTOMUS. The edges of the mandibles at the open space appear as if worn.

Mandibles uniting.

Bill bending upwards.

158. MYCTERIA. Tabiru. Upper mandible triangular. Tongue minute.

Bill not bending upwards.

159. SCOPUS. Umbre. Bill slightly arched. Dorsal ridge of the upper mandible swollen at the base. *S. umbrella.*

160. CICONIA. Stork. Bill straight. *C. vulgaris.*

PRESSIROSTRES. The mandibles never possess a cutting edge throughout their whole length. In general, the edges are more or less rounded, and the whole bill exhibits a feebleness strikingly different from the strong bills of the preceding group.

GRALLÆ.

(I.) Furnished with a hind toe.

a. Toes are remarkably long, and flattened below. On their sides may be observed a margin more or less broad, but which does not unite the toes at the base. These characters enable the species, according to their development, to wade in marshy ground, and even to swim in lakes. The bill is compressed and slender. The sternum is remarkably narrow, so that the body appears to be compressed. The wings are short. They run fast, but seldom attempt to fly. M. CUVIER has divided this family, which he terms MACRODACTYLES, in such as have the winglets armed with spines, and such as have the winglets unarmed. The observations, however, which we have been able to make, induce us to conclude, that the winglets in all the genera are armed. In two genera, however, the spines are more obvious than in the others.

Spines on the wings produced. Claw of the hind toe produced.

161. PARA. Jacana. Bill tapering, obtuse. *P. jacana.*

162. PALAMEDIA. Screamer. Bill conic, upper mandible arched. *P. cornuta.*

Spines on the wings obscure.

Forehead covered. The feathers between the eyes spinous.

163. RALLUS. Rail. Bill produced, longer than the head. Under mandible even at the symphysis. *R. aquaticus.*

164. ORTYGOMETRA. Crake. Bill conical, and shorter than the head. Under mandible forming an angle at the symphysis. *O. crex*, or *corn-crake.*

Forehead with a naked stripe. Feathers between the eyes soft. A coloured band above the knee. Symphysis of the lower mandible angular. Bill conical, and about the length of the head.

165. GALLINULA. Gallinule, or Water-Hen. Toes bordered by a simple membrane. *G. chloropus.*

166. FULICA. Coot. Toes bordered by scalloped membranes. *F. atra.*

b. Toes of moderate length.

Toes bordered by a scalloped membrane. Frequent marshes and lakes. *Phalaropoda.*

167. PHALAROPUS, (Cuvier). Phalarope. Bill straight, with a blunt depressed extremity. *P. lobata.* The Grey Phalarope of British ornithologists.

168. LOBIPES, (Cuvier). Coot-foot. Bill slightly bent and acuminated. *L. hyperboreus.* The Red Phalarope of British ornithologists.

Membrane of the toes, when present, plane.

Membrane between the front toes reaching nearly to the extremity.

169. RECURVIROSTRA. Avoset. Bill much recurved. *R. avosetta.* The length of the neck and limbs, the slenderness of the bill, and the wading habits of this species, unite it with the Grallæ, while the great development of the web between the toes give it a claim to rank among the Palmipedes.

Membrane between the toes abbreviated or wanting.

(*a.*) Bill longer than the head.

1. Bill bent downwards.

Head and neck feathered.

170. NUMENIUS. Curlew. Bill round, with the nasal grooves abbreviated. *N. arcuata.*

171. PHÆOPUS. Whimbrel. Bill depressed at the end, and the nasal grooves extend nearly the whole length. *P. vulgaris.*

Parts of the head or neck bare of feathers.

172. IBIS. Bill strong, square at the base, with the na-

sal grooves produced. In some species, as *T. religiosa*, the legs are reticulated; in others, as *Tantalus ruber* of G<small>M</small>. They are plated.

2. Bill straight or very slightly bent.

(*aa.*) Bill produced, and much longer than the head.

Nasal grooves reaching nearly to the extremity of the bill.

Upper mandible with a dorsal groove near the end.

173. S<small>COLOPAX</small>. Snipe. Extremity of the upper mandible soft, when dried, punctured. Head compressed. *S. rusticola.*

Upper mandible destitute of a dorsal groove.

174. R<small>YNCHINA</small>. Bill slightly bent at the end. *R. capensis.*

175. L<small>IMOSA</small>. Godwit. Extremity of the bill depressed and soft. The limbs are longer than in the snipes, and the bill, in some of the species, has a slight bend upwards. *L. Ægocephala.*

Nasal groove reaching only half the length of the bill.

176. T<small>OTANUS</small>. Horseman. Bill slender, round, the upper mandible slightly deflected at the end.

(*bb.*) Bill not much longer than the head.

Bill depressed at the extremity. Nasal groove produced.

Outer toes united at the base by a distinct web.

177. M<small>ACHETES</small>. Ruff. Neck and ear feathers much produced in the breading season, in the males, the heads of which are in part naked, and exhibit a papillous skin. The *Tringa pugnax*, the only knowns pecies, exhibits many varieties.

Outer toes unconnected at the base.

178. CALIDRIS. Knute. Toes slightly bordered. *Tringa canutus.*

179. PELIDNA. Dunlin. Toes without a border. *Tringa alpina.*

Bill conical pointed. Nasal groove abbreviated.

180. STREPSILAS. Turnstone. Limbs short. Bill stronger than in the genera of the nearest group. *Tringa interpres.*

(*b.*) Bill shorter than the head.

181. TRINGA. Lapwing. Nasal groove extending two-thirds the length of the bill. Hind toe distinct. *T. vanellus.*

182. SQUATAROLA. Nasal groove short. Hind toe minute. Tringa squatarola.

(II.) Destitute of a hind toe.

(*a.*) Bill slender.

Bill round. Legs long.

183. CURSORIUS. Bill bent. No nasal groove. *C. Europæus.*

184. HIMANTOPUS. Bill straight, with a nasal groove. *Charadrius Himantopus.*

Bill depressed at the end. Nasal grooves produced.

185. FALCINELLUS. Bill bent. *Scolopax pygmæa.*

186. ARENARIA. Bill straight. *Charadrius Calidris.*

(*b.*) Bill strong.

(1.) Bill compressed.

Bill swollen at the end.

187. CHARADRUS. Plover. Upper mandible swollen dorsally. *C. pluvialis.*

GRALLÆ.

188. OEDICNEMUS. Both mandibles swollen at the extremity. *Charadrius œdicnemus.*

Bill wedge-shaped at the end.
189. HÆMATOPUS. Oyster-catcher. *H. ostralegus.*

(2.) Bill vaulted.
190. OTIS. Bustard. Bill slightly arched. Wings short. *O. tarda.*

B. Wings imperfect. The birds of this group usually termed *Struthiones*, are incapable of flying. The breast-bone is destitute of the mesial crest, and the pectoral muscles are weak. The muscles of the thighs, on the other hand, are uncommonly large, enabling these birds to run with astonishing swiftness. They have no hind toe.

a. Feet with three toes. All the toes with nails.

Head with a horny protuberance.
191. CASOARIS (Bontius). Cassawary or Emu. Bill compressed. Neck with caruncles. Wings, with stiff web-legs feathers. Nail of the inner toe largest. Phalanges of the toes 3, 4, 5. Feathers wiry. *Struthio Casuarius* of LIN.

Head with a horny protuberance.
192. RHEA, (Brisson). Nandou. A callous knob, in place of a hind toe. Wings capable of aiding the animal in running. Feathers of the rump produced. Phalanges of the toes, 3, 4, 5. *Struthio Rhea* of LIN. American Ostrich. CUV. Reg. An. IV. t. 4. fig. 5.

193. TACHEA *. Shanks serrate behind. Wings minute. *Casuarius Novæ Hollandiæ* of LATHAM, or New Holland Cassawary.

---

* Ταχυς celer.

b. Feet with two toes. The outer one without a nail. Phalanges, 4. 5.

194. STRUTHIO. Ostrich. Bill depressed, soft at the end. Large crop, with a smaller one between it and the gizzard. Cæcum long. Cloaca large. *Struthio Camelus.*

PALMIPEDES.

Toes webbed to the extremity.

This great division contains birds which are truly aquatic. The legs are short, and placed far behind. In walking, the body assumes nearly an erect position. Many of the species dive readily, and, when under water, employ their wings as oars to aid them in progressive motions. The skin is thickly clothed with feathers, and a close covering of down. The neck is usually longer than the legs, an arrangement which is not observed to prevail in the other groups.

1. The hind toe united with the fore toes by a continuous membrane. The outer toe is long, and the rest decrease to the hind toe, which has a mesial direction. The openings of the nostrils are indistinct, although the nasal groove reaches nearly to the extremity of the bill.

A. Base of the bill, and about the eyes, covered with a naked skin.

1. Margin of the bill sharp and entire, the extremity hooked.

Bill depressed.

195. ONOCRATULUS. Pelican. Sides of the lower mandible slender, forming the margin of a large gular pouch. *Pelicanus Onocratalus.*

Bill compressed.

196. PHALACROCORAX. Cormorant. Tail rounded. *P. carbo.* The species of this genus have a triangular bone attached to the occiput on the dorsal edge of the foramen magnum, and resting on the first cervical vertebræ. Sometimes perch on trees.

197. PELICANUS. Tail forked. *P. aquilus,* or Frigate bird.

2. Margin of the bill serrated, the extremity nearly straight.

198. SULA. Gannet. Bill strong. Nail of the middle toe serrated. No occipital osseous appendage. *S. Bassana* or Solan Goose.

199. PLOTUS. Darter. Bill slender subulated. Head small. Neck and tail long.

B. Base of the bill, and around the eyes, feathered.

200. PHAETON. Tropic bird. Bill serrated. The two middle tail feathers produced. *P. ætherius.* An inhabitant of the Torrid Zone.

Hind toe separate.

Margins of the mandibles covered with teeth. Sides of the tongue with pectinated tufts of bristles.

A. Bill broad, with transverse lamellate teeth. This includes the genus Anas of Lin. The species of this extensive group have hitherto been described in reference to the colour of the plumage, although the forms of the bill and legs yield more permanent marks of distinction.

a. Trachea of the male, with a capsular enlargement at the bronchial extremity. The limbs are short, and placed far behind.

1. Hind toe bordered with a membrane.

Base of the bill enlarged.

201. OIDEMIA*. Scoter. Bill swollen at the base, above the nostrils. *Anas nigra* and *fusca*.

202. SOMATERIA. (Leach). Eider. Base of the bill extending up the forehead, and divided by a triangular projection of feathers. *Anas mollissima* and *spectabilis*.

Base of the bill plain.

203. CLANGULA. Bill short and narrow. *A. glacialis, histrionica* and *clangula*.

204. NYROCA. Bill broad and depressed. *Anas ferina, Marila nyroca, fuligula*.

2. Hind toe without a border.

A papillous skin at the base of the bill.

205. CAIRINA. Musk-duck. Face and cheeks with a coloured warty skin. *Anas moschata*.

Base of the bill feathered.

206. TADORNA. Shielddrake. Bill broad at the end, hollow in the middle, and raised into a tubercle at the base. *A. Tadorna.*

207. ANAS. Bill plane above the nostril, and depressed. *A. boschas, clypeata, strepera, acuta, querquedula, penelope* and *crecca*.

b. Trachea of the male simple at the bronchial extremity.

This group includes those species usually denominated Swans and Geese.

208. CYGNUS. Swan. Bill of nearly the same breadth throughout. *Anas Olor* and *Cygnus*.

209. ANSER. Bill tapering. *Anser, segetum, albifrons, Erythropus, Bernicla,* and *ruficollis*.

---

* Οἴδημα tumor.

B. Bill narrow, with the margin covered with reflected teeth, and the upper mandible hooked at the end.

210. MERGUS. Goosander. Hind toe furnished with a fin. Nostrils near the middle of the bill. M. Merganser, serrator and albellus.

Margin of the bill destitute of teeth.

A. Wings short. This character renders flying difficult in all cases, and in some impracticable. The legs are placed farther back than in other birds. They constitute the *Brachypteres* of CUVIER.

Wings destitute of quill feathers. In walking, the tarsus, which is broad behind, is applied to the ground.

211. APTENODYTES. Penguin. Palate and tongue beset with reflected prickles.

Wings with quill feathers.

Bill much compressed and obliquely furrowed. No hind toe. This includes the genus Alca of LIN.

212. ALCA. Awk. Base of the bill feathered. *A. impennis* and *torda*. The A. pica is merely the razorbill in its winter dress, and also its immature state. I have had an opportunity of observing, that the black feathers, on the side of the head and throat, in the *impennis* and *torda*, during the summer season, change to white in winter.

213. FRATERCULA. Base of the bill and part of the cheeks covered with coloured skin. *Alca arctica*, LIN.

Bill more or less conical and compressed.

Membrane of the toes lobbed.

214. PODICEPS. Grebe. Bill straight, pointed. Middle claw flattened. *P. cristatus.*

Membrane of the toes entire.

With a hind toe.

215. COLYMBUS. Diver. Claws pointed. *C. gacialis.* M. CUVIER has brought together the *Immer* as a variety of *glacialis*, probably with propriety, but he errs in associating with it the *arcticus* also. This last is the male of the *septentrionalis*. *C. stellatus*, which CUVIER makes a var. of *septentrionalis*, is a true species. The bill alone is sufficient to mark the difference.

Without a hind toe.

Upper mandible notched at the extremity.

216. URIA. Guillemot. The *Colymbus Troile*, is the representative of this genus *. When in its winter dress, or immature state, it is the *C. minor* of authors. Montagu states, that this last is destitute of the notch in the bill. This, however, we have found to be a mistake.

Upper mandible destitute of a notch.

217. CEPHUS. Tyste. Bill lengthened, upper mandible slightly bent at the point. *C. grylle.*

218. MERGULUS. Bill short, margins inflected at the base. Lower mandible contracted at the top. Nasal groove wide. *Alca alle.*

B. Wings long and well adapted for flight.

Apertures of the nostrils prominent. Bill hooked at the point.

External margins of the bill with lamellate teeth, like the ducks.

219. PACHYPTILA. Nostrils tubular, separate base of the bill enlarged. *Procellaria cœrulea.*

See Plate II. fig. 1. A male. *a* the bill open to exhibit the notch.

External margins of the bill entire.

With a hind claw.

220. PROCELLARIA. Petrel. Nostrils united into a single tubular opening on the upper part of the bill. Lower mandible truncated. *P. glacialis* and *pelagica.*

221. PUFFINUS. Puffin. Nostrils opening separately. Lower mandible bent downwards. *Procellaria Puffinus.*

Without a hind claw.

222. HALODROMA. Nostrils situate dorsally. Lower mandible with a gular pouch. *Procellaria urinatrix.*

223. DIOMEDEA. Albatross. Nostrils lateral. Bill strong, sharp. *D. exulans.*

Apertures of the nostrils simple.

Bill hooked at the end.

224. CATARACTES. Skua. Nostrils covered with a corneous plate, reaching to the base of the bill. *Larus Cataractes* and *parasiticus.* The L. crepidatus is merely the young of *L. parasiticus.*

225. LARUS. Gull. Nostrils simply covered by a continuation of the bill. *L. marinus.*

Bill not hooked at the end.

226. STERNA. Tern. Bill compressed, slender, pointed. Tail forked. *T. hirundo.*

227. RYNCHOPS. Bill depressed, truncate. Upper mandible shorter than the lower. *R. nigra.*

Birds, like quadrupeds, are usually preserved in a museum in a dried state, or stuffed. The directions already given for the preservation of quadrupeds, are sufficient to guide any one in accomplishing the same object with birds. There are, however, a few peculiarities in the mode of preserving birds, which it may be worth while briefly to notice.

A little tow or cotton should always be placed in the mouth, to prevent any mucus or blood flowing out, and soiling the feathers.

The incision of the skin may be made, either from the breast to the vent along the belly, or on the side under the wing. When the opening is under the belly, the space under the wings can be afterwards exhibited in the attitude of flying. On the other hand, when the incision is made under the wing, the beauty of the belly can be better preserved, and a greater fulness given to that part in stuffing. Besides, should the feathers be any way damaged round the margin of the incision, the wing can easily be made to hide all the defects. Indeed, in skinning birds, it is indispensably necessary, for preserving the feathers from being soiled, to pin folds of paper on the margin of the incision. If feathers have once been soiled, a little soap and water applied with a sponge, and adhesion of the webs prevented, in the course of drying, by frequent stirring, will render them clean, but they usually lose much of their lustre.

As the femoral joint of birds is internal, it is never left attached to the skin, but usually separated at the tibial extremity. If a noose on the end of a cord suspended from the roof, be fixed on the tibial end of the femur, as it remains in connection with the carcase, the process of skinning will be greatly facilitated. The carcase, in this manner, may be suspended the instant the first leg is disengaged, and the remaining part of the process of flaying executed readily, without the help of an assistant.

In some cases the neck is so much narrower than the head, that the skin cannot be completely reversed, to enable the operator to take out the brain and eyes. This difficulty chiefly occurs among the grallæ and palmipedes. In such cases, an incision must be made on one side of the gape, along the cheek, to facilitate the extraction of the contents of the head, which could not be effected through the reversed skin. The more ordinary defects of stuffed birds, consist in the belly being too lean, and the neck too long.

It is sometimes a desirable object with the ornithologist, to keep alive the wounded birds which may be sent to him, for the purpose of observing their manners. To effect this, it is frequently necessary to remove a portion of the wing, to save the animal from death, by bleeding at the wounded part, or effectually to prevent its escape. The late Mr Montagu has given some valuable directions on this subject, which the reader will find at the end of the Supplement to his Ornithological Dictionary. Supposing the object in view is merely to prevent escape, and that the subject on which the operation is to be performed belongs to the duck tribe, the amputation should take place at the part of the wing corresponding with *a.* fig. 2. plate 2. where the bones of the wing of the *Corncrake* are represented. By passing the needle through the wing, close by the inside of the smaller bone at *a*, and making a ligature with the thread across the larger bone, and returning it on the outside of all, the principal bloodvessels are secured, which could not be accomplished by a ligature confined to the surface. The part of the wing may now be removed by a pair of shears or a chissel, without the loss of blood, and the wound speedily heals. In short-winged birds, such as the gallinaceous kinds, the operation is more advantageously performed at *b.*, making the ligature embrace all the vessels between the smaller bone or *ulna*, and the larger one called the *radius*. We have frequently seen amputation performed in this manner, on wounded as well as sound subjects, without the loss of a single drop of blood.

Vertebral Animals with cold blood.

The cold-blooded vertebral animals, and indeed those belonging to the remaining divisions of the system, have their temperature, in a great measure, regulated by external circumstances. The skin is either naked, or protected by scales, and never exhibits either hair or feathers. In all the tribes, the brain does not fill entirely the cavity of the skull destined for its reception. Organs of sight and hearing, more or less perfectly developed, may be detected. The circulating system is not so perfect as in the warm-blooded classes, as there is always some deficiency either in the systemic or pulmonic vessels. This condition of the circulating vessels gives rise to the following division into Reptiles and Fishes.

REPTILES,

Furnished with a systemic heart.

The skin of reptiles is either naked, or fortified with scales or plates. Among many of the tribes it is periodically renewed. The cuticular secretions are few in number; and they seldom serve to lubricate the skin.

The bones, unless in the larger kinds, scarcely ever attain the same degree of firmness as those of quadrupeds and birds. Their number and connection vary exceedingly in the different tribes. The organs of motion are fit to perform almost every kind of progression. Some are found with two or four feet, either divided or palmated, and fitted for walking, climbing, or swimming. Others move by what is peculiarly termed a serpentine motion. Many species inhabit the land, and not a few live in the water. To fit them for residing in the latter element, some of the groups, wanting feet, have their bodies compressed behind, while others are furnished with fins, destitute, however, of cartilaginous, or osseous filaments, for their support, as in fishes. Their teeth, in general, are fitted for retaining their food, rather than for masticating, and the gullet is usually dilatable. The food, in a few genera, is derived from the vegetable kingdom, but, in the greater number, animal food is exclusively employed. A considerable quantity is consumed at a time, but the intervals between the meals are remote.

The circulation of this group may be considered as imperfect, since a part only of the blood is aërated, which issues from the heart, and that portion, instead of proceeding directly to the different organs, is again mixed with the circulating fluid. The aërating organs consist, in general, of lungs situate in the common cavity of the abdomen. There is an imperfect larynx at the commencement of the trachea, incapable of producing a distinct voice. The lobes of the lungs are of unequal size, and the cells are of much larger dimensions than in the warm-blooded animals.

The kidneys are always present, either united into one mass, or variously subdivided. The ureters either pour their contents into a bladder of urine, which empties itself into the cloaca, or they terminate directly in the cloaca, the urine passing into a lateral pouch until voided. The urine itself varies much in quality and appearance in the different tribes. In some it is pure uric acid, in others very diluted urea.

Although reptiles can secrete a limited quantity of heat or cold, for their preservation in extraordinary circumstances, they usually remain nearly of the same temperature with the surrounding medium. They are remarkable, however, for the facility with which they become torpid, when the temperature sinks towards the freezing point. So necessary, indeed, does a high temperature appear to be, to the comfortable exercise of their several functions, that we find them chiefly inhabiting the warmer regions of the earth.

The reproductive system of reptiles exhibits few peculiarities. The sexes are distinct, on separate individuals. In some tribes, the male has external organs, while in others these are wanting. Impregnation is either external or internal, according to the tribes: the females are either oviparous or ovo-viviparous. The young of some of the genera undergo remarkable changes of form, before reaching maturity.

Heart with two auricles.

In addition to this well marked internal character, an external character may be given of more easy detection. The skin is never naked, being either protected by scales, knobs, or an osseous shield.

CHELONEA.

Body protected by a corneous shield. Body furnished with feet.

The reptiles of this group, denominated by the French naturalists *Cheloniens*, were included in the genus Testudo of Linnæus. The body is protected dorsally and ventrally, by a hollow shield, open at each end for the issuing of the head and fore-feet at one time, and the tail and hind-feet at another. This shield is named *back-plate* or *breast-plate*, according to its position. It is covered by numerous pieces, nearly resembling horn in texture and composition, exhibiting various forms and modes of union with one another. In some cases, however, the external covering is a continuous skin. The lateral line of junction between the two plates, is more or less obviously marked by the peculiar forms of the marginal plates.

The shield is strengthened dorsally, by its intimate connection with the vertebræ of the back, ventrally, by the sternum, and laterally by the ribs. The vertebræ of the head and tail are alone moveable. The scapular and clavicular bones are united with both plates, and form an osseous ring for the passage of the trachea and gullet. The legs vary remarkably in form. Some of the species swim, others walk, or rather crawl; and all of them are slow in their motions.

The testudinal animals are destitute of teeth; but their jaws, with few exceptions, are provided with a corneous covering, like the bill of birds, in some cases, variously notched. The upper jaw is fixed, and the under jaw has a cavity for the reception of the temporal condyle. The tongue is small. The gullet is frequently beset with hard conical processes, having a cardiac direction, and considered as destined to prevent the return of the food. Many of the species are phytivorous, others live on fish. The liver is, in general, divided into two lobes, sometimes a little removed from each other. The auricles are large in proportion, with thin walls; and, at their opening into the ventricle, are furnished with valves. Walls of the ventricle covered with fleshy eminences. The lungs are double, and nearly of equal size. The windpipe divides near the larynx; and, in some species, each branch makes a turn before reaching the lungs.

The kidneys are rather diminutive in size. The bladder of urine, however, is of extraordinary dimensions. The urine itself is transparent and watery, and, besides a little mucous and common salt, contains uric acid *.

In the reproductive system, the sexes are observed to be on separate individuals. The external male organ is cylindrical and pointed, with a groove along its whole length. The oviducts have each an enlargement, or uterus. The eggs, which are fecundated internally, resemble those of birds, in being covered with a calcareous shell. These are deposited by the females in the sand, and left to be hatched by the

---

* Dr J. Davy, Phil. Trans. 1818, p. 306.

heat of the sun. The young are perfect at birth. The testudinal reptiles furnish the navigators of the tropical seas with wholesome and refreshing food, and are held in high estimation by epicures in general.

A. Lips corneous.

Entrance to the cavity formed by the two plates closed by a lid.

1. CISTUDA. Box tortoise. Back-plate emarginate in front, with two notches behind.

The lid is formed by a plate, having a cartilaginous joint, and gives full protection to the members of the animal, when withdrawn into the cavity. CUVIER subdivides this genus into such as have two lids, one to each aperture of the shield, and such as have only one at the opening for the head.

Entrance without a lid.

a. Breast-plate continuously solid.

Head and feet capable of being withdrawn into the shield. The back-plate is rounded, and divided into compartments by large scales. Fore feet with five, and the hind feet with four, toes. Those of the first genus live on the land, those of the second frequent fresh water.

2. TESTUDO. Toes united and covered with a common scaly skin. *T. Græca.*

3. EMYS. Toes webbed. Claws long. *Testudo Europæa.*

Extremities incapable of being withdrawn into the shield.

4. CHELONURA. Tail about the length of the shield. Back-plate carinated, with sharp processes behind. *Testudo serpentina.*

b. Breast-plate interrupted by intervening cartilaginous spaces. The extremities are incapable of being withdrawn into the shield. The fore legs are remarkably produced, with the toes united, to serve as a fin. The species of this group live in the sea.

5. CHELONIA. Back-plate covered with corneous scales. *Testuda Mydas.*

6. CORIUDO. Back-plate destitute of scales. *Testudo coriacea.*

B. Lips fleshy, with a produced snout. Toes webbed.

7. CHELYS. Back-plate scaly.

There is a protuberance to the hind feet occupying the place of a web, but destitute of a claw. The toes before, and the four behind, are armed with claws. The tail terminates, as in many of the other genera, in a hard point. *Testudo fimbria,* Bruguiere, Journ. d'Hist. Nat. vol. i. p. 253. tab. xiii. f. 1. 2.

8. TRIONIX. Back-plate destitute of scales, but covered with a coriaceous skin, studded near each extremity with hard knobs.

Three toes on each foot only having claws. Inhabiting fresh water. *Testudo ferox.* Pennant, Phil. Trans. 1771, p. 268. Tab. x. fig. 1, 2, 3.

SAURIA.

Body covered with scales.

This group, including the genera Draco and Lacerta of LINNÆUS, and denominated Sauriens by the French naturalists, includes animals which have usually a lengthened body, ending in a long tail. The skin is protected by scales, which, in some of the genera, assume the form of plates. The vertebral column is complete, and capable of motion. The ribs are united with the sternum, and, by their motions, assist respiration. The feet are widely placed, so that walking is performed in an irregular manner. Some of the species swim with ease, while others are capable of flying imperfectly. The toes are armed with claws.

The jaws of lacertine animals are furnished with teeth, varying in number with the age of the individual. The tongue, in many species, is slender and capable of great extension. The animals prey chiefly on animal food. The auricles are not so large as in testudinal animals, and in some cases the ventricle is more complicated. The lungs are divided by membranaceous plates, into numerous polygonal cells. The voice is restricted to a roaring and hissing. The liver, in some genera, is divided into two lobes; in others it is single.

The urinary organs differ in this group, chiefly in the presence or absence of a receptacle or bladder of urine. Where present, the ureters terminate directly in a receptacle. The urine itself contains a large quantity of uric acid, together with carbonat and phosphat of lime *.

The female is either oviparous or viviparous. The male is furnished with an exsertile penis. Impregnation takes place internally.

A. Furnished with four feet.

a. Tongue very long, and capable of being pushed out to a considerable distance. Five toes to each foot.

Skin shagreened. Toes enveloped by the skin. Tongue clavate, hollow above.

9. CHAMELEON. Toes divided into two bundles of two and three in each.

The skin is remarkable for the changeability of its colours, according to the states of the animal. The eyes are covered with a thick membrane, leaving only a narrow slit opposite the iris; and each eye is independent in its motions. This genus contains several species, chiefly distinguished from one another by the appendices of the head.

---

* Phil. Trans. 1818, p. 306.

REPTILES.

Skin with imbricated scales.

The tongue is divided, at the extremity, into two threads. The toes are separate and armed with claws. Eyelids formed by a longitudinal slit. The rudiment of a third eyelid may be detected. The vent is transverse. Penis bifid.

Teeth on the palate as well as the jaws.

10. LACERTA. Lizard. Head depressed and covered with large plates. Scales on the belly and around the tail, disposed in transverse bands.

There is a collar on the under side of the neck, formed by a transverse row of large scales. The *L. agilis*, a native of this country, is usually considered as oviparous. M. de Sept-Fontaines, communicated to the Count de la Cepede, a proof that it is sometimes ovoviviparous, he having opened a female, and found within her twelve young ones perfectly formed, from eleven to thirteen lines long *. In the year 1803 I kept a female of this species for two months, until it died in September, after giving birth to four young ones perfectly formed, and measuring an inch and a half in length.

Destitute of palatine teeth.

(A.) Head depressed and covered with large plates.

The genera of this group may readily be distinguished from the lizards, by the absence of the collar, and the scales under the neck being small. The scales on the belly, and around the tail, however, are large and square.

Tail compressed.

11. DRACÆNA. Scales on the back large and carinated, tail furnished with a crest. *La Dragonne* of La CEPEDE.

---

* The Natural History of Oviparous Quadrupeds or Serpents, by the Count de la CEPEDE. Trans. 4. vols. 8vo. Edin. 1802., vol. i. p. 384.

12. CUSTA. Dorsal scales small and smooth.

Some of the species, as *Lacerta bicarinata*, have the tail with a crest, while, in others, as *L. teguexin*, the tail is smooth. The teeth are notched. Under each thigh there is a row of pores.

Tail rounded.

13. AMEIVA. Head sub-pyramidal. *Lacerta Ameiva.*

(B.) Scales on the head, belly and tail, small and imbricated.

14. MONITOR. This genus was formerly termed Tupinambis, by a mistake of Seba. There appear to be indications of three divisions. 1. Tail compressed and carinated, as *M. elegans.* 2. Tail round and carinated, as *Lacerta Nilotica.* 3. Tail round and plain as the *monitor* of Egypt.

a. Tongue short, and limited in its motion. Number of toes indeterminate.

(A.) Feet furnished with suckers.

The suckers occur on the under side of the toes, the surfaces of which are broad. They consist of transverse pouches, with fringed margins. These enable the animal to climb perpendicular walls, like the common house fly *. The tail has circular folds.

Teeth serrated.

15. ANOLIUS. The antepenult joint of the toe enlarged to form the sucker.

The body is shaped like the lizards. The throat, in some of the species, is capable of inflation, and of changing colour. There are palatine teeth. Several species are natives of America, some of which have rounded tails, and others where a crest may be observed.

---

* See an interesting " Account of the feet of those animals, whose progressive motion can be carried on in opposition to gravity," by Sir EVERARD HOME. Phil. Trans. 1816, p. 149.

Teeth minute, and close set.

The animals of this group differ from those of the preceding, in having a depressed form and nocturnal habits. They are known by their eastern name, Gecko.

16. ASCALABOTES. M. CUVIER has subdivided this genus into four groups, all of which appear to be susceptible of farther division, from marked characters. In some, the disk of the sucker extends over the whole under-surface of the toes, and is either grooved, as the *thecadactyles*, or plain as the *platydactyles*. In others, the sucker is towards the base of the toes, as the *hemidactyles*, or near the extremity, as the *ptyodactyles*. The *S. fimbriatus*, in the last of these groups, has the toes half webbed, and an expanse of membrane at the side of the tail *.

(B.) Feet destitute of suckers.

1. Toes, in part, united by a common integument.

This group includes the Crocodiles, the largest of the reptiles. The scales are large, and many of them are longitudinally carinated. The tail is compressed, with a crest above, which is double at the base. The toes are five before, and four behind, the three inner ones, only, on each foot armed with claws. They have no clavicular bones. The external ear is capable of being closed by two fleshy lips. The jaws are each furnished with a row of strong teeth, some of which are produced beyond the others in the lower jaw. The vent is longitudinal, while in the other lacertine tribes it is transverse. The external organ of the male is

---

* M. CUVIER has added to this group the Stellio phyllurus from New Holland, which is destitute of suckers. Its true place in the system is not satisfactorily determined.

single. The females are oviparous. The species are inhabitants of fresh waters.

Some of the teeth in the lower jaw produced, and received into a cavity in the upper. Feet half webbed, and not denticulated.

17. ALLIGATOR. Snout broad and blunt. Teeth unequal, the fourth on each side of the lower jaw produced. *Crocodilus sclerops.*

The produced teeth not received into a cavity in the upper jaw, but passing along the sides. Feet webbed and denticulated.

18. CROCODILUS. Crocodile. Snout oblong and flat. Teeth unequal. *Lacerta crocodilus*, LIN.

19. GAVIALA. Gavial. Snout produced and rounded. Teeth nearly equal. *Lacerta Gangetica.* Phil. Trans. 1756, p. 639, tab. xix.

2. All the toes free.

(*a.*) Tail tapering, distinct from the body. Legs developed.

With lateral expansions for flight.

20. DRACO. Dragon. On each side of the body there is a membranaceous wing, scarcely connected with the legs. It is supported by the first six false ribs, which, instead of being bent round towards the belly for the protection of the viscera, proceed laterally from the body. With this substitute for a wing, the animal is assisted in its leaps from one tree to another. There is a long throat-pouch and lateral short ones, supported by productions of the os hyoides. The teeth are of three kinds in each jaw,—four incisors, two tusks, and twelve grinders. Three species have been determined.

Destitute of lateral expansions.

Furnished with palatine teeth. Head plated. Throat with a dilatable pouch. Body and tail with imbricated scales. Thighs with glandular pores.

21. IGUANA. Body and tail furnished dorsally with a longitudinal crest, of vertical scales. *Lacerta Iguana.*

22. POLYCHRUS. Destitute of a dorsal crest. *Lacerta marmorata.*

Destitute of palatine teeth.

(AA.) Tail covered with rings of large scales.

Head enlarged behind by muscles of the jaws.

23. STELLIO. Ear surrounded with spines. Thighs destitute of glandular pores. Toes long and pointed. *Lacerta stellio.*

Head simple. Glandular pores on the thighs.

24. CORDYLUS. Both the belly and back covered with rings of scales, many of which are spinous. *Lacerta cordylus.*

25. MASTIGURA. Scales of the body small, smooth, and uniform. *M. spinipes.*

(BB.) Scales of the tail not annulated.

Scales on the back forming a crest.

Crest covered with scales. It is supported by spinous processes of the vertebræ.

26. BASILICUS. Basilisk. Teeth strong and compressed. *Lacerta basilicus.*

Crest formed of single scales. Scales on many parts of the body carinated and pointed.

27. CALOTES. Crest confined to the back. *Lacerta calotes.*

28. LOPHURUS. Crest continued along the tail, giving it a compressed form.

Back destitute of a crest. The head is enlarged behind by the maxillary muscles.

29. AGAMA. Scales on different parts of the body spinous. *Lacerta muricata.*

30. TRAPELUS. Scales smooth and without spines. *T. Ægyptiacus.*

(*b.*) Body and tail nearly of the same thickness throughout. Feet imperfectly developed, and scarcely fitted for walking or aiding the animal in progressive motion. In general appearance, the species resemble serpents. *Scincusidæ.*

31. SCINCUS. Scales regularly imbricated. The genus Seps of DAUDIN differs chiefly in the superior length of the body, and the hind legs being more remote from those in front.

32. CHALCIDES. Scales square and formed into rings. There are several species easily distinguished by the number of the toes on the feet.

B. Furnished with two feet.

The animals of this group resemble the Scincusidæ in their lengthened form and imperfect feet, and, by wanting one pair, make a still nearer approach to the serpents.

33. BIPES. Fore-feet wanting. Underneath the skin the rudiments of scapular and clavicular bones may be detected. The species of this genus may admit of still farther division from the condition of the toes.

34. CHIROTES. Hind-feet wanting. *Lacerta lumbricoidis.*

II. Body destitute of feet.

OPHIDIA.

In this division of Reptiles, which includes the SERPENTS, the skin consists chiefly of a cuticle and corium. The cuticle exhibits considerable differences in thickness, and, as happens to the lizards, is periodically cast off and renewed. The corium is destitute of a villous surface, and intimately connected with the muscular web underneath. The appendices of the skin consist exclusively of scales. These are divided by naturalists into two kinds. The true *scales* or *squamæ*, are usually produced longitudinally, and pointed or rounded posteriorly. They have, in general, some asperities on the surface. The *shields* or *scutæ* are of much larger dimensions, and produced transversely. In some cases they embrace the whole of the body, forming rings; in other cases, a half or a quarter of one side only. They usually present a smooth surface, and occur chiefly on the belly, between the throat and vent. They differ considerably in number in different species, and even in individuals of the same genus, so that the character furnished by their number should be employed with caution in the discrimination of species. There does not appear to be any well marked secretion from the skin, nor are there glands from which it could proceed. Between the nose and the eye, indeed, on each side, there is a bag or crumen in some species, not unlike the bags which occupy the same position in some of the ruminating animals. Here, however, there is no appearance of glandular structure yet discovered, and the cuticle with which they are lined, appears continuous, and falls off along with that which covers the rest of the body. Dr TYSON first observed these crumens in the rattlesnake, and Dr RUSSELL and Sir E. HOME have since attended to their structure more particularly *.

The osseous system of serpents is chiefly characterised by the articulation of the vertebræ, the number of ribs, and the absence of the extremities. The head is connected with the atlas by a single tubercle, having three articular surfaces. The posterior articular surface of each vertebra is in the form of a rounded eminence, which is received into a corresponding depression in the anterior end of the one which follows. There is thus throughout the vertebral column a series of ball and socket joints, admitting a great extent of flexure, in some cases restrained by the spinous or articular processes. The ribs in some species amount to nearly three hundred pairs. The dorsal extremities of the ribs are enlarged, with a terminal depression, which receives a rounded protuberance of the vertebra. Each vertebra, therefore, has a rib on each side, limited by a modified ball and socket-joint. These ribs are not continued beyond the anus. They embrace a great part of the sides, and

---

* The last of these observers, when considering their use, offers the following remarks : " As amphibious animals, in general, have no glands to supply the skin with moisture from within, but receive it by coming in contact with moist substances, it is possible the bags, in the snake, may be supplied in that manner, and the more so, as the cuticular lining appears perfect. Another peculiarity is remarkable in snakes furnished with the bags described above, namely, an oval cavity, situated between the bag and the eye, the opening into which is within the inner angle of the eyelid, and directed towards the cornea. In this opening there are two rows of projections, which appear to form an orifice capable of dilation and contraction. From the situation of these oval cavities, they must be considered as reservoirs for a fluid, which is occasionally to be spread over the cornea; and they may be filled by the falling of the dew, or the moisture shaken from the grass, through which the snake passes."—Phil. Trans. 1804, p. 74.

terminate in cartilaginous processes united to the scales of the belly *.

The progressive motion of serpents is executed by the assistance of the ventral scales, which serve as so many feet. The body is raised up into one or more arches, and then suddenly unbent, is pushed forwards, the posterior edges of the scales, opposed to the ground at the hinder extremity of the circle, offering sufficient resistance. By thus alternately elevating and unbending the body, serpents move in their characteristic manner backwards or forwards, with an astonishing degree of rapidity. When at their utmost speed they appear scarcely to touch the ground. In their more ordinary movements, the ventral plates or scales serve as feet. These scales slide under each other by a kind of inclusion, so as to permit the ventral surface to shorten or lengthen at the will of the animal. When some of the foremost scales are pressed on the ground, those behind are brought forward, and in their turn supporting the body, enable the fore part to advance. To qualify the scales to do this with greater advantage, they are connected with one another, by means of muscular threads and a longitudinal band, and are likewise aided by the peculiar mechanism of the ribs. These last we have seen are connected with the ventral scales by a flexible cartilage. They are capable of moving on their vertebral joint, either ventrally or dorsally, anteriorly or posteriorly. These motions are aided by the muscles which are inserted into them, and originate in ribs contiguous to those to be

---

* M. Lacepede has remarked, in the viper, boa, and rattlesnake, that there are uniformly two ribs, and one vertebra, corresponding to each shield of the belly. In one small group, some of the ribs are united by cartilaginous productions on the belly, and there is even the rudiment of a sternum.

moved, and in the vertebræ. The body, in general, is of a rounded form, but, when preparing for progressive motion, the ribs are drawn somewhat dorsally, so as to flatten the scales of the belly, and by moving anteriorly or posteriorly, give to the scales, with which they are connected, a corresponding degree of motion. The ribs in this case act as limbs to the scales, which may be compared to feet. This singular use of the ribs of snakes, in assisting progressive motion, was detected by the acute TYSON, in his admirable anatomical examination of the rattlesnake *. The same subject was still farther illustrated by Sir EVERARD HOME, in his dissection of the Boa constrictor †.

Besides the capabilities of executing progressive motion, many serpents can twist their bodies round the branches of trees, or suffer a considerable portion to hang down. In this attitude, the larger kinds are ready to fall down upon their prey passing beneath, such as deers and antelopes. Such animals are not only retarded by their weight, but incommoded by the foe twisting itself in wreaths round their body, and by contractile efforts crushing it to death.

This method of seizing their prey is confined to the larger kinds. The smaller sorts are able, by their mouth and teeth, to seize and retain their victims. There is no mastication, the food being swallowed entire. To facilitate the deglutition, the under jaw consists of two bones, as in birds, and, like these animals, they are joined to the cranium by the intervention of a bone, similar to the os quadratum. The upper jaw is also loosely connected with the head, and, in some species, admits of considerable motion at the point of junction. The mouth can thus be opened very wide, and larger animals admitted, than, from the ordinary size of those devoured, one would be led to suppose.

---

* Phil. Trans. No. 144, copied by RAY into his Synopsis Animalium, p. 291.   † Ibid. 1812, p. 163.

The armature of the mouth is of two kinds,—common teeth, and poison fangs. The common teeth form a single row on each side in the lower jaw, and usually a double row in the upper. The external row, above, is connected with the maxillary bone; the inner row is supported, as in many fishes, on the palatine bones. These teeth are subulate and recurved.

The *poison fangs*, vulgarly termed the *sting*, are confined to the upper jaw, and occur on each side towards the extremity. In some, the fang is the largest tooth, the foremost in the row, and is followed by several common teeth in the maxillary bone. In others, the maxillary bone is abbreviated, and the deficiency at its proximal end is supplied by a long moveable peduncle destitute of teeth. The large fang is placed in this short maxillary bone, and is followed by a few other teeth of similar form. These fangs are to be viewed as the osseous openings of the ducts from the poison-bags, which are situate at the base. They contain a tubular cavity from their base, passing through the tooth on its convex side, to the apex, where it ends in a narrow slit. When the serpent bites an animal, the poison flows from the bag through this slit, into the bottom of the wound, where, to most advantage, it can produce its deleterious effects. Whether this dreadful apparatus of venom is to be considered as connected with the digestive system, as an instrument for conquering the prey, and obtaining food, or for defensive purposes, has not been satisfactorily determined. The latter conjecture is the most plausible.

The true structure and mode of formation of the poison fangs, appear to have been first investigated with care by Mr SMITH. He found that the poison tube was formed by a fold on the surface, the edges of which met in the middle, but continued separate at the point and base of the tooth. The hollow in the middle of the tooth in which the secreting pulp was situate, can easily be traced as unconnected with this tubular external fold. The same observer found several of the common teeth having the rudiments of this external groove*. It is usually supposed, that upon the first fangs being destroyed, others are developed to supply their place. There is, however, some difficulty in accounting for the manner in which the poison duct, in such cases, can change its course. It is probable that the grooves in the adjoining common teeth have given rise to the opinion. More accurate observation, however, can alone determine the point.

The poison itself has much the appearance of oil, but, in its general properties, it resembles gum. Its noxious properties continue even after it has been dried. If it is instilled into the wound, in any quantity, and enters any of the larger vessels, death speedily follows. In other cases, there is previously great pain produced, the part swells and becomes discoloured, and exhibits marked indications of violent local action. The virulence of the poison depends not only on the species of serpent, but on its condition at the time, and the habit of body of the animal which has received the bite.

The remedies which have been suggested to destroy the influence of the venom, after the bite of the animal, are numerous, but few of them can be safely relied on. Some recommend making an incision into the part wounded by the fangs, and sucking out the venom and contaminated blood. Others advise the unfortunate patient to compress the part by li-

---

* Mr SMITH on the Structure of the Poisonous Fangs of Serpents, Phil. Trans. 1818, p. 471,

gature, and thus, by retarding the circulation, give time for the poison to become diluted and neutralised by the local fluids. These remedies are obviously such, as, in ordinary cases, should be resorted to. On the other hand, some recommend the internal and external use of ammonia, the external application of caustic potash, or of heated turpentine. Many vegetables have likewise been employed as antidotes. The subject, however, though one of great interest, is in some measure unexplored, and can only receive useful illustration from the researches of those who reside in the districts where the larger and more venomous species so frequently prove destructive to human life.

The tongue is usually slender, divided at the extremity, and sheathed at the base. The alimentary canal is short in proportion to the size of the body. The gullet and stomach are capable of great dilatation, to receive the large animals which are swallowed. Digestion takes place slowly. The liver is, in general, uniform in its appearance, and nearly entire. There is a gall-bladder. The spleen is of a lengthened form, and placed at the commencement of the intestines.

The heart possesses two auricles and one ventricle. The portion of blood transmitted for aeration, is returned by a systemic vein, to the systemic auricle, from this it passes into the common ventricle, which receives, through the other auricle, the blood returned from the body.

All serpents breathe air. Though many species can dive, they are compelled to return, at intervals, to the surface of the water, in order to respire. The larynx is too simple and membranaceous to admit of the existence of voice. A hissing sound, however, can be produced. The trachea is a membranaceous tube, terminating directly in the lungs, without the usual bronchial divisions. The lungs are single, in the form of an elongated bag, with large cells on the walls. This bag lies in the cavity of the abdomen.

The kidneys are lengthened and lobulated. The ureters terminate usually by a single orifice in the cloaca, in the form of a papilla directed towards the bag, which may be considered as the bladder of urine. This bladder is formed by a fold of the coats of the intestine; it receives the contents of the ureters, unmixed with the fæces.

In this receptacle the urine frequently becomes inspissated, and is voided in lumps at distant intervals, which have frequently been mistaken for fæces. In its ordinary state, Dr John Davy, to whom the public is indebted for several valuable observations on the urinary organs of the amphibia *, describes it as of a butyraceous consistence, becoming hard, like chalk, by exposure to the air, and consisting of pure uric acid.

The reproductive organs of serpents present few peculiarities. The external organs of the male are double, each short, and surrounded with bristles. These, in the rattlesnake and viper are bifid, and beset with bristles. They appear in a pouch near the anus. In the female, the external openings are double, corresponding with the condition of the male organs. Impregnation takes place internally. Some species are oviparous, others are ovoviviparous. The young do not undergo any remarkable metamorphosis.

Serpents are found in the greatest numbers, both in reference to species and individuals, in tropical countries. In such regions, likewise, they attain the largest size. Few species are found in the temperate and colder districts. In all cases they seem greatly invigorated by heat, and in its absence, speedily sink into a torpid state.

---

* Phil. Trans. 1818, p. 304.

Wherever man takes up his abode, there the serpents have to engage in an unequal contest, which ends in their extirpation. Man not only attacks them in open combat, with the lance, the sword, or the musket, but he falls upon them, when gorged with food, and prepared only for slumber, or when in a lethargic state in consequence of cold. He likewise drains the marshes where they procured their food, and cuts down the forest to which they were accustomed to retire.

But while man is their personal foe, wages against them a war of extermination, and thus powerfully influences their geographical distribution, he receives powerful support from many of the domestic animals which accompany him in his dispersion over the globe. The hog is not afraid to give battle even to the most venomous; and, in general, comes off victorious. The goat, likewise, readily devours the smaller kinds of serpents, and hence the Gaelic proverb, " Cleas na gaoi ther githeadh nathraeh," like the goat eating the serpent,—importing a querulous temper in the midst of plenty *.

When unrestrained by opposition from man, and the physical conditions of their life, they are well qualified for extending their geographical limits. Neither forest, mountain, marsh nor rivulet, can retard their progress. Almost all the species can swim, and many of them with great ease. Indeed it is probable, that many of those stories which have been propagated, regarding *vast sea snakes*, have originated, in the appearance of some of the larger serpents at sea, where they have been driven by accident. Some of the Asiatic species reside almost constantly in the waters, either fresh or salt.

Independent of the claim which serpents have to our notice as constituting an extensive division of the animal king-

---

* Statistical Account of Scotland, vol. xii. p. 449.

dom, they furnish an interesting subject of inquiry, from the superstitious opinions which have been entertained concerning them, and the strange properties with which credulity has invested them. It is not our province to give a history of the errors of the human mind, when untutored by that philosophy which gives precedence to accurate observation, but to unfold those characters which the different groups exhibit, and by which they may be recognised. In this inquiry, it is indeed painful to consider, that human ingenuity has hitherto failed to convert serpents into any thing that is useful,—for it is not worth while to regard them in this light, when occasionally furnishing a repast to a few naked savages, or serving to amuse, when dancing to the signals of a juggler, before a few indolent Asiatics. It is still more painful to consider the destruction of human life by their venomous fangs, or the quantity of misery which they have occasioned to those who have survived the noxious bite.

Serpents admit of a very natural distribution into two sections, according as the scales on the belly are similar or dissimilar to those on the back. Other characters correlative with these, intimate the propriety of the arrangement.

A. *Serpents with the Ventral and Dorsal Scales similar.*

The serpents of this division, which are destitute of ventral shields, have the lower jaws intimately united, and supported at the base by two ossa quadrata, which are articulated immediately to the cranium. The upper maxillary bones are united to the cranium and to the intermaxillaries. This arrangement limits the power of gaping, or of swallowing objects larger than the ordinary aperture of the mouth. The body is nearly of equal thickness at both extremities. Progressive motion is chiefly accomplished by unbending the arches into which the body is successively thrown. The body can advance or retreat by either extremity, with nearly equal readiness. The eyes are so small, that, with many, they have passed unobserved.

a. Serpents with a third eye-lid.

In the animals of this group, the teeth are small, and nearly of equal size. The tongue is notched in a crescent form. The ribs are more or less united, to supply the place of a sternum. When irritated, the body is thrown by the muscles into a very rigid state, in which condition it breaks into fragments by the slightest stroke. None of the species are considered venomous. The condition of the rudiments of some of those organs possessed by the Saurian reptiles, which may here be detected, justify the subdivision of the three genera here contemplated.

(1.) Traces of scapular and clavicular bones occur under the skin. There is likewise a minute sternum and an imperfect pelvis. Though the rudiments of the extremities can be detected within the skin, there are no vestiges of legs or feet on the outside.

35. OPHISAURUS. The tympanum is externally visible. The maxillary teeth are conical; and, besides these, there are two groups on the palate. *O. ventralis.*

36. ANGUIS. Snake. The tympanum is concealed. No palatine teeth. The maxillary teeth are compressed and recurved. *A. fragilis.*

(2.) No vestiges internally of pelvis, scapula, clavicle, or sternum. The anterior ribs, however, are united by intervening cartilaginous productions, which serve instead of a sternum.

37. ACONTIAS. The teeth are conical. There is the rudiment of a second lobe of the lungs. *A. meleagris.*

b. Serpents destitute of a third eye-lid.

The serpents of this division are likewise destitute of all vestiges of sternum, pelvis, or shoulder-bones; neither do the ribs unite to complete the circle.

(*a.*) Body covered with scales.

Scales in rings.

38. AMPHISBÆNA. Scales quadrangular, and disposed in circular bands round the body. Before the vent, there is a row of perforated tubercles, the openings of excretory ducts. Oviparous. *A. fuliginosa.*

Scales imbricated.

39. TYPHLOPS. Scales small and placed in an imbricated order. Snout depressed and produced. Tongue long and forked. *T. lumbricalis.*

40. TORTRIX. Ventral scales slightly enlarged. Tongue thick and short. *T. scytale.*

(*b.*) Body uniformly covered with tubercles.

41. ACROCHORDUS. The tubercles may be viewed as modifications of scales. They have each three ridges. The top of the head is fat, and covered by small imbricated scales. *A. Javenensis.*

B. *Serpents with the Ventral and Dorsal Scales Dissimilar.*

The ventral scales of this division constitute those shields or scutæ which have been already described. The lower jaw-bones are loosely connected, and the ossa quadrata have a free motion on the skull, or are supported on cartilaginous or osseous peduncles. The upper maxillaries are likewise loosely connected with the skull and the intermaxillary bones. This condition of the jaws increases greatly the power of gaping, and enables the serpents to take into the mouth animals much larger than the ordinary thickness of the body. The palatine bones are likewise loosely connected, and armed with recurved pointed teeth.

A. Destitute of poison-fangs.

The serpents of this division, like those of the preceding, are not venomous. There are four regular rows of entire teeth above, and two below.

Shields under the tail simple, like those on the belly.

42. ERPETON. This genus, instituted by LACEPEDE, is characterized by two soft eminences, covered with scales, seated on the extremity of the snout. The shields under the tail are small. Head with large plates. *E. tentaculatus.*

43. BOA. Snout destitute of soft eminences. Vent furnished with a hook on each side. *Boa constrictor.*

Shields under the tail divided, or forming a double row.

44. COLUBER. (Python of DAUDIN.) Vent furnished with lateral hooks, like the Boa. The ventral shields are narrow. *C. Javenensis.*

45. NATRIX. Destitute of anal hooks. Oviparous. *T. torquata.* Ringed snake.

B. Furnished with poison-fangs.

1. Poison-fangs single, followed by a row of common maxillary teeth. These last are not so numerous as in the harmless kinds of the preceding section.

Hinder part of the body and tail rounded.

46. PSEUDOBOA. Shields of the belly and lower side of the tail single. Head covered with large plates. A swelling at the occiput. Back carinated, and covered with a row of enlarged scales. *P. fasciata.*

47. TRIMERESURA. Shields under the tail, near the vent, single; but, toward the extremity, double. Dorsal scales not enlarged. *T. viridis.*

Hinder part of the body and tail much compressed. This form enables the species to swim with ease. They frequent arms of the sea, lakes, and rivers. Head covered with large plates.

48. PELAMIS. Hind head swollen. Scales rather reticulated than imbricated. Ventral shields minute. *P. bicolor.*

49. HYDROPHUS. Head not swollen. The shields are more distinct than in Pelamis. *H. ayspisurus.*

2. Upper maxillary bones destitute of common teeth, supporting only the poison fangs on each side.

The shields entire on the belly, and under the tail beyond the vent.

a. The entire shields under the tail continued to the extremity.

Tail furnished with a rattle, (crepitaculum.)

This very remarkable organ occurs as an appendage to the tail, the last vertebra of which is enlarged to serve as a mould for its production. This vertebra is somewhat conical, rounded at the extremity, subquadrangular, with three rounded circular ridges, the largest of which is next the body. The first formed cup of the rattle, which, in substance, is similar to the scales, exhibits the form of the vertebra, from which, however, it is separated by a thin membrane. It is connected directly with the scales on the back, and, by the intervention of a row of small scales, with the shields on the belly. Thus formed, the cup has three ridges externally, with corresponding grooves internally.

When this first cup has arrived at maturity, a second begins to form underneath. The terminal and middle ridges of the second grow within the grooves of the middle and basilar ridges of the first. As these ridges of the second increase in size, they push from the vertebra the first formed joint, and the second cup appears with its basilar ridge, only visible externally, its middle and terminal ones covered by the basilar and middle ridges of the first. In this manner are the rattles formed to the number of twelve, or, if certain relations are to be credited, to the number of twenty or even thirty cups.

As the middle and terminal ridges of the last formed cup are rather less in diameter than the grooves of the basilar and middle ridges of the preceding one, in which they are contained, a kind of ball and socket-joint is formed at each division, admitting an imperfect degree of motion when the body moves.

Each cup, when covering the vertebra, appears to be nourished in the same manner as the skin. When detached from the vertebra, however, though still connected with the last formed one, it dries and becomes brittle and elastic. When the serpent moves its body, the cups of the rattle, likewise moving upon one another, make a noise not unlike the folding of dried parchment. This noise is said to be audible at the distance of twenty yards, and is thus useful in giving warning of the approach of the destructive reptile to which it is attached. As the cups of the rattle consist merely of dried matter which, in the dry season, is brought into a condition to make a noise when the animal moves, so, in like manner, under the influence of external circumstances, the rattle, in the wet season, is soft and mute.

It does not appear to be as yet determined, whether a new cup is formed with each renewal of the skin, or whether the succession of the cups takes place in a different order. The number of cups found in a rattle, can scarcely be considered as the number which have actually been formed, but merely as the number of those which have outlived accidents. There are two nasal bags or crumens, and two scent bags at the vent.

50. CAUDISONA. Rattle-snake. The head in this genus (the title of which is derived from the old trivial name of this most remarkable species), is covered with scales similar to those on the back. *C. horrida.* Boquira.

51. CROTALUS. Millet. Head covered with large plates. *C. miliaris.*

Tail destitute of a rattle.

52. SCYTALUS. Destitute of nasal bags. The species of this genus, natives of India, are but imperfectly known.

b. The entire shields under the tail not continued to the extremity.

53. ACANTHOPHIS. Shields double towards the extremity of the tail, which terminates in a spinous process. There are two species, *A. palpebrosus*, and *A. Brownii.*

54. LANGAIA. Shields behind the vent, forming rings which surround the body. The under side of the tail towards the extremity, covered with small scales. Snout produced. *L. nasuta.*

Shields behind the vent divided.

A. Tail round.

a. Head behind larger than the neck.

Head furnished with crumens.

55. TRIGONOCEPHALUS. Head much enlarged behind. Tail frequently armed with a spinous process. *T. lanceolatus.*

Destitute of crumens.

(*a.*) Hind-head furnished with a hood.

56. NAJA. Head covered with large plates. *N. lutescens* *.

---

* When the animals of this group are irritated, the skin on the neck is expanded and drawn forwards, and appears behind the head as a kind of

b. Hind-head destitute of a hood.

aa. Furnished with cuticular appendices on the forehead.

57. CERASTES. Scales on the head similar to those on the back. *C. vulgaris.*

bb. Fore-head destitute of appendices.
Head covered with small plates.

58. CHERSEA. Aesping. Head covered in the middle with three plates. *C. vulgaris.*

59. HEMACHATUS. Head covered with four rows of large plates. *H. vulgaris.*

Head covered with common scales.

60. COBRA. Scales on the head carinated. The head itself, from the eyes to the snout, subtriangular. *C. atropos.*

61. VIPERA. Viper. Scales on the head rough. *V. Berus.* Common viper or adder.

b. Head, behind, of the same size as the neck.

This uncommon narrowness of the head arises from the shortness of the os quadratum, and the osseous peduncle by which it is united with the skull. The power of swallowing large objects is consequently limited, as the jaws have but little motion.

62. ELAPS. Head covered with large plates. *E. lacteus.* Milky viper.

B. Tail compressed.

63. PLATURUS. Head covered with large plates. *P. laticaudatus.*

---

hood. This motion is produced by the cuticular muscles of the neck aided by the moveable ribs. This mechanism, as displayed in the common cobra di capello, is exhibited by Sir EVERARD HOME, in his " Lectures on Comparative Anatomy," vol. II. tab. ii. iii. iv.

HEART WITH ONE AURICLE.

BATRACHIA.

The skin is naked, and usually lubricated by a mucous secretion. The vertebræ, in some genera, are hollowed into cups on each end like those of fishes. The ventricle is destitute of fleshy columns, and sends out the blood by one opening. There is no external organ of generation in the male. Fecundation is generally external, and the eggs are deposited in the water. The young are hatched in that element, and, at the first, possess external branchiæ, which, in one genus, are persistent, and continue to supply the place of aërating organs, while in the other genera, they are absorbed when the lungs have acquired their proper degree of development.

A. Furnished with feet.

a. Furnished with a tail.

(1.) In the adult state, furnished with lungs.

The lungs are in the form of two bags, with the walls imperfectly cellular or honeycombed. Free air is respired in the immature state, when recently excluded from the egg, the young are furnished with fimbriated processes on each side of the neck, with apertures at the base, which serve as gills. These are supported by cartilaginous arches. At this period, the nose, which, in the adult, is the aperture through which the animal breathes, is unconnected with the mouth. As the animal reaches maturity, the connection between the nose and throat is established; the lungs become developed, while the branchial cartilages coalesce, and the gills are absorbed. The animal, if it resides in the water, is obliged, at intervals, to come to the surface to respire.

Furnished with four feet. Toes destitute of claws.

64. SALAMANDRA. Salamander. The tail round, in the

adult state. *Lacerta Salamandra.* The species, the type of the genus, is ovoviviparous, and the sexual impregnation is internal.

65. TRITON. Eft. Tail compressed in all stages of their growth. *Lacerta palustris.*

The males in spring are furnished with a divided membranaceous dorsal crest, which disappears after the business of procreation. The species are imperfectly determined.

Feet only two in number. Toes with claws.

66. SIRENA. Toes four on each foot. *Sirena lacertina* *.

That this animal, in its adult state, is destitute of gills, may be safely inferred, by a comparison of the observations on its internal structure, made by HUNTER †, with the excellent account of the structure of the Apneumona or Proteus, by CONFIGLIACHI and RUSCONI. The concluding observations of these observers are so remarkably interesting, that we consider no apology due for inserting them in this place, as they are given in the Edinburgh Philosophical Journal, vol. v. p. 104.

" Having thus terminated the anatomical description of the Proteus Anguinus, we proceed to examine the two following questions; *first,* Whether it be true, as many believe, that this reptile can respire, at the same time, by gills and by lungs? *Secondly,* If the *Sirena lacertina* is to be regarded as a larva or a perfect animal? To determine these questions, it will be necessary to compare the branchial structure, the organs of circulation, and the supposed lungs of the proteus, with the corresponding parts in the *sirena* and in the *larvæ* of the salamander and of frogs.

---

* Phil. Trans. 1766, Tab. ix. p. 189.

† Phil. Trans. 1766, Tab. ix. p. 307.

"With respect to the branchial structure, there is a remarkable difference, not only as to form, but to texture, between the arches of the proteus, and those of the siren and larvæ above mentioned. In the siren and larvæ, the branchial arches are four on each side, and their margins are furnished with small points,—in the proteus, there are but three on each side, and these are smooth. The arches of the proteus have an osseous structure,—those of the siren and larvæ are cartilaginous. This difference did not escape M. CUVIER, who, speaking of the proteus, says, ' l'appareil osseux qui porte les branchies, est beaucoup plus dur que ne l'avons trouvé dans la *sirene*, et dans *l'axolotl :*' and in his anatomical description of the latter animal, he farther says, ' *l'appareil qui supporte les branchies à de grands rapports avec celui de la sirene*, et je crois que, lors de la metamorphose, il en reste une partie pour former l'os hyoïde de la salamandre.' Now, if the branchial arches of the siren be, as M. CUVIER asserts, entirely cartilaginous, although the cranium, the lower jaw, and the vertebræ, be perfectly ossified; and if these arches, both in form and number, be similar to those of the *axolotl*, which M. CUVIER himself regards as a larva,—may it not be presumed that the former animal is a larva also? If, farther, the branchial arches of the proteus, which is a *perfect* animal, be osseous, and entirely different from those of the siren and all the larvæ hitherto known, have we not in these facts the strongest reasons for regarding the siren as an *imperfect* animal, and, therefore, essentially different from the proteus?

"With regard, next, to the organs of circulation, there are, in the larvæ of the frog and salamander, as many arteries given off on each side by the trunk that springs from the heart, as there are branchial arches, viz. four. In the siren and axolotl (which have also eight branchial arches),

M. Cuvier speaks only of six arteries, three on each side, going to the gills; but as, by the aid of injections, we have found, say the authors, that, in the larvæ above named, there are eight vessels, and that the artery which runs along the interior arch of each side, and which M. Cuvier has not seen, is that which in process of time becomes the pulmonary artery, so, guided by analogy, we hold it for certain, that, as the siren is furnished with eight branchial arches entirely similar to those of the other larvæ, there are also eight arteries, four on each side, corresponding to them. And, proceeding on this opinion, we may remark a striking difference in the circulating system of the siren and proteus, since the artery, properly called Pulmonary, which is found in the siren and larvæ above mentioned, does not exist in the proteus. Doubtless in the proteus, the air-bladder, like every other part of the body, is duly supplied with blood; but the blood sent to it is furnished by an artery coming off, on each side, from one of the aortic trunks, and which artery, descending along the canal of the bladder, gives to it a branch, and is then continued to the ovary or testicle in each sex respectively.

" Besides these differences in the arterial, there are others in the venous system; for the vessel which returns the blood from the air-bladder of the proteus, does not empty itself directly into the cava or the auricle, as is observed in other reptiles; but into the vein which carries back the blood from the organs of generation, which itself enters the cava above the middle of the kidney; hence in the proteus, not only the true pulmonary artery, but the vein also, is wanting. This anatomical fact, ascertained by repeated injections, might alone be sufficient to demonstrate, that the two air-bladders with which the proteus is furnished are not true lungs: but as some, perhaps, may not yield to the force of these arguments, we shall continue the comparison, especially as applied to the organs of respiration.

"In the larvæ of the frog and salamander, the *trachea* opens directly into the lungs. These organs have the form of two sacs, and, from being longer than the trunk, cannot be extended in a straight line through it, but at the lower end are folded a little from one side of the abdomen to the other. So, in the siren, we see the trachea to open directly into the lungs, which, as in the above-mentioned larvæ, says M. Cuvier, ' sont deux longs sacs cylindriques, que s'etendent jusqu' à l'extrémité posterieure de l'abdomen, et se replient même alors en avant.' But, in the proteus, neither do the supposed lungs reach to the pelvis, nor does the supposed glottis open into the air-bladders, but issue in a cavity which communicates with the air-bladders by two long conduits. Thus, then, the structure of the branchial arches, the distribution of the bloodvessels, and the form and size of the lungs in the proteus, differ entirely from the corresponding organs in the siren and larvæ of the salamander.

"If, farther, we consider the mode in which frogs and salamanders respire air, and compare it with that of the proteus, we shall obtain still further evidence of the differences subsisting between them. All zoologists, including M. Cuvier, now admit that frogs first receive air into the mouth through the nostrils only, and from thence force it into the lungs by an action resembling deglutition. But neither the proteus nor the siren are able to respire in this manner; for the nostrils in the former do not open into the mouth, but beneath the upper lip; and in the siren, ' *les narines, simplement creusées sur les côtés du museau, ne pénètrent point dans la bouche*,' says M. Cuvier. Neither do these animals respire air in the manner of serpents, for they are both destitute of ribs. When also the proteus takes air into the mouth, it escapes rapidly through the branchial apertures: nor is there any ground for believing that any portion of it enters the very narrow clink of the glottis to pass into this cavity, and from thence through the two membranous canals to the air-bladders. No muscular structure suited to produce such effects exists, and the fine membranous canals, subject to compression every instant from the stomach, altogether unfit them for performing the office of air-tubes or bronchi. In all reptiles that respire air, the structure of the organs is such as to permit free inspiration and expiration, however different the form may be; but in the proteus, the want of ribs and diaphragm, the fact that the nostrils do not open into the mouth, the extreme narrowness of the aperture termed *glottis*, and the narrowness, length, and compressibility of the air-tubes, all shew, that in this animal none of those arrangements exist, which nature has instituted with such great solicitude and skill in other reptiles, to carry on with ease and certainty the respiratory function. But it is needless to multiply arguments, to prove that the air-bladders of these animals in nowise perform the office of lungs, since it has been already shewn that, when taken out of the water, they die just as fishes do.

" M. CUVIER justly observes, that those animals can alone be deemed truly amphibious, ' qui respirent, a la fois, l'air èlastique en nature, et celui qui contient l'eau :' and he then goes on to state, that the *sirena lacertina* respires through its whole life by lungs and by gills, and is therefore a permanently amphibious animal; but that the larvæ of other reptiles make use of these two different organs only for a short period, and are therefore only temporarily amphibious. With all due respect, however, to so great a zoologist, we, say the authors, are of opinion, that before pronouncing the siren to be permanently amphibious, it would have been proper to have made upon it, or upon animals which resemble it, experiments similar to those we have made on the proteus. If, in his researches with regard to ambiguous reptiles, he had not contented himself with examining only their skeletons, but had examined also the *larvæ* of the salamander, while yet alive, we are certain that his investigations would have conducted him to opinions entirely opposite to those which he has been led to form.

" In our investigations on this point, we have directed our attention to the above mentioned larvæ, to observe particularly the changes which occur in their intimate structure, when they are transformed into *perfect animals*. Between the siren and these larvæ there is the greatest resemblance, not only in regard to the structure of the branchial arches, but also to the nostrils; for, in the siren, as well as in these larvæ, the nostrils do not open into the mouth. This circumstance prompted us to examine the condition of the bones of the face in these larvæ, and we have thereby satisfied ourselves, that the larvæ of the salamander is unable to breathe by lungs, until the maxillary bones, the zygomatic arches, and the palatine bones are sufficiently developed to form the canal of the nostrils, in such a manner that its posterior extremity may open into the mouth. Before this canal is so formed, these larvæ are unable to respire atmospheric air, and, if taken out of the water, they then soon die; and, therefore, guided by analogy, we incline to believe, that, to the siren, whose nostrils ' ne pénètrent point dans la bouche,' the same things ought to happen. Moreover, as its lungs are similar in all respects to those of the salamander, and are furnished with a true glottis, we are farther of opinion, that the siren is the larva of some reptile, the genus of which is as yet unknown, and which will differ from its larva in not possessing gills, and in having a trunk somewhat longer.

" We consider that the porteus is not an amphibious animal, having a double circulation, as some have maintained, but a *perfect reptile*, different entirely from all others. It is a reptile, in respect to its having a single circulation, and a fish, in regard to its mode of respiration,—in other words, it is a reptile which respires air mixed with water, while others respire atmospheric air: so that, were it allowable to revive the old idea of a chain of beings, the proteus might be regarded as the link which would connect reptiles with fishes."

(2.) Gills permanent.

There are no lungs, but in the situation which they occupy in the preceding group, there are two bags with simple membranaceous walls.

67. APNEUMONA. Feet four, with three toes before and two behind.

The only well characterised species, is the *A. anguina*, which inhabits the waters of subterranean caverns in Carniola in Germany, where it lives excluded from the light. The genus to which it belongs, has been usually denominated Proteus, which we have ventured to change for another title expressive of the want of lungs. The term Proteus had been long preoccupied, as the name of a genus of infusory animals, remarkable for the mutability of their forms; two species of which are delineated by MULLER, in his Animalcula Infusoria, Tab. II. Fig. 1.–12. and 13.–18. Besides, the reptile under consideration is nowise remarkable for mutability of form.

b. Destitute of a tail.

This section includes the different kinds of frogs and toads. The eggs are fecundated as they are deposited. They are surrounded by a glaire, which has properties intermediate between gelatine and albumen, and which increases greatly in bulk when placed in water. The substance termed *star-shot-jelly*, is considered as this glaire brought into that state by a frog having been swallowed by a bird, and the warmth and moisture of the stomach, making the jelly in the oviducts expand so much, that the bird is obliged to reject it by vomiting *. The young or *tadpoles*, as they are termed, lose the tail, gills and beaks, and acquire four legs, when they reach a certain period, and exchange a dwelling in the water for a residence on land. At the period of this change, the intestines are loaded with fat, which Sir E. HOME considers as destined to furnish materials for the development of the parts which takes place †. The thumb of the male during the season of sexual intercourse, is furnished with an enlarged knob. The urine is watery, and contains urea ‡.

Jaws and palate furnished with teeth. The skin is smooth, and the hind-legs are longest.

68. HYLA. Tree-frog. Extremities of the toes expanded, with suckers beneath.

The species are numerous, and climb trees in search of insects. The male has a throat-bag, which is filled with air during croaking.

69. RANA. Frog. Toes simple. The male has cheek-bags, which are filled with air during croaking.

The common frog is the type of the genus. The spawn is deposited in masses. The adult animals leap easily. According to old WALTON, " the mouth of the frog may be opened from the middle of April till August, and then the frog's mouth grows up, and he continues so for at least six months without eating ∥."

---

* See Phil. Trans. 1810, p. 212. and 217.
† Phil. Trans. 1816, p. 301.   ‡ Phil. Trans. 1821, p. 98.
∥ The complete Angler. BAGSTER's 2d Edition, London, 1815, p. 237. See likewise PENNANT, Brit. Zool. vol. iii. p. 11.

Jaws destitute of teeth. Body warty.

70. BUFO. Toad. Tongue short and thick.

The common toad (which the ignorant and the prejudiced persecute, though harmless), is the type of the genus. Eggs deposited in chains.

71. PIPA. Mouth destitute of tongue.

The Rana Pipa is the type of this genus. The toes on the fore-feet are divided into lobes at the extremity. The eggs are fecundated externally, when they are collected by the male and spread over the back of the female, to which they adhere. The skin of the latter, by degrees, forms a cell round each egg, in which it remains until hatched; and even the young do not quit the cells, until they pass from the tadpole state. The analogy of this arrangement of the reproductive system, to the condition of the marsupial quadrupeds, is very remarkable.

Destitute of feet.

72. CECILIA. Eyes obscure. The absence of scales, the cup-shaped vertebræ, the shortness of the ribs, the absence of the os quadratum, and the simplicity of the heart, intimate, that this genus should not, as heretofore, be included among the Serpents. Several species have been determined.

FISHES.

Destitute of a systemic heart.

As fishes are destined to reside constantly in the water, their whole organization is suited to a residence in that element, and exhibits, in the greatest perfection, those combinations which are suitable to aquatic animals.

The Skin of fishes consists, as in the other vertebral animals, of a true skin, a rete mucosum, and a cuticle. The *corium* is remarkably thick in those species which have small scales; while in those which have large scales, it frequently assumes the appearance of a thin membrane. It is much more closely attached to the muscles in this tribe than in any of the other vertebral animals. This organ, in the cod, for example, consists almost entirely of gelatine, and is much esteemed as an article of food, and is used also in fining, as a substitute for isinglass. Eel-skins are likewise used in the manufacture of size, in consequence of the gelatine which they contain. There is no appearance of the corpus papillare or villous surface. The mucous web is remarkable for the brilliant tints which it exhibits in many species, communicating to the incumbent scales their peculiar lustre.

The *cuticle* appears in fishes in a soft state, and, in many instances, is a simple mucous substance enveloping the body. It is detached at certain seasons of the year in large pieces. The *scales* are implanted in this layer, and, in their position, the facility with which they are reproduced, and the purposes which they serve, resemble the hairs on the skins of quadrupeds. They cover the body of fishes like tiles on the roof of a house, pointing backwards. The posterior edge, which in general is free, is usually crescent-shaped, fringed in some species, and smooth in others. By means of a lens, longitudinal ribs may be perceived finely decussated by transverse striæ. These ribs sometimes radiate from the centre, and the crossing striæ are concentric. When macerated in weak acids, they are found to consist of alternate layers of membrane and phosphate of lime, and hence are supposed to increase in every direction by the addition of new layers.

Instead of imbricated scales, some fishes are protected by *osseous plates*, covered, like the scales, by the cuticle, and presenting an even surface. Among some of the sharks, as the *Squalus acanthias*, instead of scales there are flat, bent, bristly laminæ; and in the remora there are hard, rough tubercles. These osseous plates in the sturgeon, resemble in shape the shell of a limpet.

The naturalist employs the appearances exhibited by the form, surface and size of the scales, as a character in the discrimination of nearly allied species, although the disposition of the longitudinal and the transverse rays, together with the condition of the margin, would furnish more permanent marks. The scales, in the description of a fish, are likewise considered in regard to their adhesion to the skin. Thus some scales, which adhere but slightly, are said to be *deciduous;* while others, which cannot be rubbed off but with difficulty, are termed tenacious or *adhesive*.

Besides the scales many fishes are furnished with *spinous processes*. These sometimes accompany the fins; while in other instances they appear as the armature of the head and cheeks. They appear to be of the same consistence and composition as horn. Those found on the head are in general fixed; but those connected with the fins are moved by peculiar muscles. These organs may be considered as defensive weapons, and act, in some instances, not merely by their form and consistence, but by some *venomous* secretion by which they are covered *.

---

* Thus the common weever (*Trachinus draco*) inflicts a wound with the spines of the first dorsal fin, often followed by violent burning pains, inflammation, and swellings; so that the fishermen are in the practice of cutting off the offensive organ before they bring the fish to market. The spines of the *Squalus acanthias*, or piked dog-fish, and *Doras carinatus*, are likewise considered by fishermen as capable of inflicting a dangerous wound.

The surface of the skin of fishes is almost always covered with a slimy fluid, to protect them from the penetrating influence of the surrounding element. This mucus is poured out from small pores, situated under the scales in every part of the body of some fishes, while in others, these excretory ducts are arranged in a determinate order. The openings of the ducts, in some species, have corresponding apertures in the incumbent scales. These ducts were first observed and described by Steno, in his works, " *De Musculis et Glandulis,*" p. 42. and " *Elementorum Myologiæ Specimen,*" 1669, 8vo. p. 72. The subject was afterwards investigated by PERRAULT, LORENZINI, and RIVINUS, and more recently by MONRO. To this last author we are indebted for many excellent observations and sketches *.

---

* " In the skate, numerous orifices, placed pretty regularly over the surface, have been observed by STENO to discharge this slimy matter. With respect to these last, I have remarked some memorable circumstances. First, I have discovered one very elegant serpentine canal between the skin and muscles, at the sides of the five apertures into the gills. Farther forwards it surrounds the nostrils; then it passes from the under to the upper part of the upper jaw, where it runs backwards as far as the eyes. From the principal part of this duct, in the under side or belly of the fish, there are not above six or eight outlets; but from the upper part near the eyes there are upwards of thirty small ducts sent off, which open upon the surface of the skin. The liquor discharged from these has nearly the same degree of viscidity as the synovia in man. But besides the very picturesque duct I have been describing, I have remarked on each side of the fish, a little farther forwards than the five breathing holes, a central part, from which a prodigious number of ducts issues, to terminate on almost the whole surface of the skin, excepting only the snout or upper jaw. At these centres all the ducts are shut; and in their course they have no communication with each other. In these two central parts, or on the beginning of the mucous ducts, a pair of nerves, nearly as large as the optic, terminate; and, which is a curious circumstance with respect to them, they are white and opake in their course, between the brain and their ducts; but when they divide, they become suddenly so pellucid, that it is impossible to trace them farther, or to distinguish them from the coats of the ducts." *Struct. and Phys.* p. 21.

In the osseous fishes, the openings of the mucous ducts are chiefly observable in the fore part of the head, and in the *lateral line*. This line extends from the head to the tail, along each side of the fish, and exhibits several striking peculiarities. It is not observable in the lampry; in general it is single, but in the sand-eel it has the appearance of being double. It is usually of a different colour from that of the sides, and varies according to the species in position and direction. After death it sometimes disappears, and hence some difficulties have arisen with regard to the discriminating marks which it furnishes.

The mucus which is poured out upon the skin by these ducts, in some cases appears to be the liquid known by chemists under that name, while in other instances it appears to be of the nature of albumen. In the eel, for example, the skin turns white from the coagulation of the albumen, when plunged into boiling water. Chemists, however, have not turned their attention to the subject.

The BONES of fishes vary in form, proportion, and number, according to the species. The skeleton is more complicated than that of man, and is difficult to prepare and preserve Hence the osteology of fishes is a subject but little attended to by naturalists. Avoiding all minute details, we propose to consider the skeleton, as consisting of a head, spine, and ribs.

As the head of fishes is covered with a skin only, its form is easily ascertained, and it exhibits remarkable differences in shape according to the species. In all the species it is large in proportion to the size of the body, and consists of a great number of separate pieces. These amount to eighty in the perch. But as these bones are soon ossified together, it becomes very difficult to trace the original lines of separation in aged individuals. The occiput appears like a vertical truncation of the cranium, and is united to the spine by a single tubercle placed below the foramen. The motion of the head is very limited in every direction. In some of the cartilaginous fishes, the head is joined to the vertebral column by two condyles; but this articulation is equally incapable of extensive motion as the former.

The *vertebral column* is either cylindrical, angular, or compressed. The vertebræ may readily be distinguished from those belonging to the higher classes, by the peculiar form which they exhibit. The body of each is of a cylindrical figure, with a funnel-shaped depression at each end. It consists of concentric rings, which are supposed by some to increase in number with the age of the animal. The vertebræ are destitute of articular processes, and, when in union, form, throughout the whole column, cavities composed of two cones, joined at the base. These cones contain a cartilaginous substance formed of concentric fibres, of which those next the centre are the softest. By means of this cartilage the vertebræ are united, and upon it they perform all their movements. In the cartilaginous fishes all the vertebræ are consolidated together, so that the spinous processes can only be distinguished.

The vertebræ may be divided into the cervical, dorsal, and caudal. In osseous fishes, the *cervical vertebræ* are in general wanting, although in some cases they exist, as in the herring, to the number of four. In the cartilaginous kinds, they are ossified into one piece. The *dorsal vertebræ* are easily recognised, by wanting processes on the inferior part. These have generally on the sides transverse processes, to which the ribs are attached. The *caudal vertebræ* are possessed of spinous processes, both on the superior and inferior surfaces. In those fish which are flat these are very long, as in the flounders. The first caudal vertebra is in general of a peculiar shape. The cavity of the trunk is terminated by its central process. In the flounders it is large, round in the fore part, and terminated below by a sort of spine. The last caudal vertebra is however, more remarkable than the first. It is almost always of a triangular form, flat, and placed vertically. Upon its posterior extremity it bears articular impressions, which correspond to the small and delicate bones of the fin of the tail.

The number of the bones of the vertebral column in different species, being exceedingly various, suggested to ARTEDI the use of this character in the separation of nearly allied species. Among the species of the genus Cyprinus, for example, a difference in the number of vertebræ has been observed to the amount of fourteen. In ascertaining this character, ARTEDI recommends the greatest circumspection. The fish should be boiled, the fleshy parts separated, and the vertebræ detached from one another, and these counted two or three times in succession, to prevent mistakes. This character is of great use, as it is not liable to variation, individuals of the same species exhibiting the same number of vertebræ in all the stages of their growth.

The number and size of the *ribs* are likewise extremely various. The cartilaginous fishes may be considered as destitute of true ribs. Where they exist, as in the osseous fishes, they are articulated to the body of the vertebræ, or to the transverse processes. They are forked in some fishes, and in others double; that is, two ribs proceed from each side of every vertebra. In the genus Cyprinus they are of a compressed shape; in the cod they are round; and in the herring, like bristles.

The number of the ribs, likewise, furnishes a character in the discrimination of species, which may be safely relied on in the absence of more obvious characters.

Besides these bones which we have enumerated, there are many more osseous spiculæ, which serve to support the fins, and to strengthen the muscles.

The composition of the bones of fishes has never been investigated with sufficient care. It is well known, that they never acquire so great a degree of hardness and rigidity (with the exception, perhaps, of the bones of the ear), as the bones of the mammalia or birds: hence we may safely conclude from the facts connected with the process of ossification in other animals, that the bones of fishes abound in gelatinous and cartilaginous matter, while the portion of earthy or saline matter is small. The earthy salts are phosphate and carbonate of lime, and the phosphate of magnesia, the former predominating in quantity *. In one division of fishes, termed the *cartilaginous*, the proportion of earthy matter is so small, that the bones never become indurated, but continue in all the periods of the life of the fish soft and flexible. These animals are, therefore, supposed to grow during the whole course of their existence. In such fishes, the bones are not *fibrous* as in the *osseous* kinds, but *cellular*, and the walls of the cells formed of ossified membrane.

When the bones of some fishes are boiled in water, they undergo a change of colour. This circumstance is well illustrated in the case of the gar-fish, or sea-pike (*Esox belone*), the bones of which, by boiling, become of a grass green colour; and in the bones of the viviparous blenny, which experience a similar change. This alteration of colour has fostered some of the prejudices of the vulgar, but has failed to arrest the attention of the chemist.

The bones of fishes, when reduced to powder, are mixed up with farinaceous substances, and used instead of bread,

---

* In a species of Chætodon, described by Mr BELL as the *Ecan bonna* of the Malays, many of the bones, as the ribs and spinous processes, appear as if diseased. They are enlarged at particular places, like tumors, which, when cut through, are spongy and full of oil. Phil. Trans. 1793, p. 7. Tab..-vi.

by some of the northern nations. In Norway, and even in some of the remote districts of our own country, fish bones are given as food for cows, and are greedily devoured by them.

The ORGANS OF MOTION present many striking peculiarities.

If we attend to the vast variety of forms, exhibited by different kinds of fish, we shall be disposed to conclude, that *shape* exercises but little influence on their movements. While some are cylindrical and lengthened, others are nearly globular: some are depressed, while others are compressed. The general form, however, approaches to ovate, the body being thickest at the thorax, and tapering a little towards the head and tail.

The *fins* of fishes correspond with the wings of birds, the former being calculated to give the motion to the body in the water, the latter in the air. These organs vary in number, size, situation, and structure, in different species.

The *number* of fins varies according to the genera, and even according to the species. It is difficult to fix on those fins which exercise the greatest influence on the habits of the animal, as there is not any one fin common to all fishes, although all fishes have at least one of these organs more or less developed. The *size* of the fins is equally various in the different species, as it bears no constant proportion to the figure or magnitude of the fish, nor to its habits or instincts.

The *situation* of the fins furnishes the ichthyologist with some of the most obvious and useful characters. Those fins which are situate on the back are termed *dorsal*, and vary greatly in number and shape. The fin which surrounds the extremity of the tail is termed the *caudal* fin, and is always placed perpendicularly. It is forked in some, even, or rounded in others. Between the caudal fin and the anus, are situated the *anal* fins, which vary in number and shape according to the species. Between the anus and throat are placed the *ventral* fins. When they do exist, they never exceed two in number, and are parallel to each other. The *pectoral fins* are usually two in number, and are placed on each side, a short way behind the gill opening. By LINNÆUS and others, the ventral fins are considered as analogous to the feet of quadrupeds, and the characters furnished by their position are employed as the basis of his classification. Those fishes which are destitute of ventral fins, are termed, in his system, *apodal*; those which have the ventral fins placed nearer to the anterior extremity than the pectoral fins, are termed *jugular*; those having the ventral fins on the belly, immediately below the pectoral, he calls *thoracic*, and when the ventral fins are placed behind the pectoral fins, they are termed *abdominal*. These distinctions are of great importance in an artificial system, and may be employed with success in the inferior divisions of a natural one.

The *structure* of the fins of fishes has long occupied the attention of naturalists. In general, these organs consist of numerous jointed rays, which are subdivided at their extremities. These are covered on each side by the common integuments, which form, in some instances, soft fibres projecting beyond the rays. These fins, with articulated rays, were considered, by the older ichthyologists, as furnishing characters for arrangement, of great importance, and are still highly valued by many naturalists. Fishes possessing these roft rays, are termed *malacopterygii*. Besides these articulated rays, there exist in the fins of some fishes one or more rays, made up of a single bony piece, enveloped like the former, by a common membrane. Some fishes have one or more fins consisting entirely of these bony rays. Fishes with such rays are termed *acanthopterygii*. In a few genera, the posterior dorsal fin is destitute of rays, and has obtained the name of *pinna adiposa* or flesh-fin.

As these rays serve to support the fins, and are capable of approaching or separating like the sticks of a fan, we may conclude that they move upon some more solid body as a fulcrum. Accordingly, we find in the sharks, for example, that the rays of the pectoral fins are connected by a cartilage to the spine. In the osseous fishes, the pectoral fins are attached to an osseous girdle which surrounds the body behind the branchiæ, and which supports the posterior edge of their aperture. This osseous girdle is formed of one bone from each side, articulated at the posterior superior angle of the cranium, and descending under the neck, where it unites with the corresponding bone. Between the rays of the fin and this bone, which resembles the scapula, there is a range of small flat bones, separated by cartilaginous intervals, which may be compared to the bones of the carpus. The rays of the ventral fins are articulated to bones which correspond to the pelvis in the higher classes of animals. The pelvis is never articulated with the spine, nor does it ever form an osseous girdle round the abdomen. In the jugular and thoracic fishes, it is articulated to the base of the osseous girdle which supports the pectoral fins. In the abdominal fishes, the bones of the pelvis are never articulated to the osseous girdle, and are seldom connected with each other. They are preserved in their situation by means of certain ligaments. The rays of the caudal fin are articulated with the last of the caudal vertebræ, which is, in general, of a triangular form, and flat. The rays of the dorsal fin are supported by little bones, which have the same direction as the spinous processes, and to which they are attached by ligaments.

As connected with the fins, we may here notice those organs which are termed *cirri* or *tentacula*, according as they are placed about the mouth, or on the upper part of the head. They are in general soft, but often contain one jointed ray. They do not differ in structure from the fins, and are so closely connected with them, that it is difficult to point out their use. It is not probable that they are organs of touch, as many have imagined, but rather peculiar modifications of fins.

The muscles which move the fins, and all the other organs of the body, are of a paler colour than in the animals of a higher order. They are also more uniform in their substance, being in general destitute of tendinous fibres. In the greater number of fishes, there are no muscles peculiar to the head. The sides are furnished with the most powerful ones, to execute the lateral movements of the animal. These muscles are disposed in layers or arches, with the convexity towards the head. The different muscles are strengthened by small detached spines, imbedded among the fibres of the muscle, and giving them additional strength. Between the layers there is in general a quantity of viscid albuminous matter interposed. After death, this fluid speedily undergoes a change, and can seldom be observed in fishes which have been kept a few days. But in recent fish, when boiled, the albumen appears coagulated in the form of white curd, between the layers of the lateral muscles.

The motions of fish are performed by means of its fins. The caudal fin is the principal organ of progressive motion. By means of its various flexures and extensions, it strikes the water in different directions, but all having a tendency to push the fish forward; the action resembling, in its manner and effects, the well known operation of the sailor termed *skulling*. The ventral and pectoral fins assist the fish in correcting the errors of its progressive motions, and in maintaining the body steady in its position. BORELLI cut off, with a pair of scissars, both the pectoral and ventral fins of fishes, and found, in consequence, that all the motions were unsteady, that they reeled from right to left, and up and down, in a very irregular manner. The dorsal and anal fins serve to maintain the body in its vertical position. But from the circumstance of some of these fins being wanting, and others evidently too small to produce the desired effects, those fins which are present, appear to be capable of executing all the movements for which the others, when present, are designed.

The medium in which fishes reside, prevent us from making any accurate observations on the velocity of their motion. Mackrel, and some other marine fishes, will seize a bait moving at the rate of six or eight miles an hour; and some of the voracious sharks will keep up with a vessel in her voyage across the Atlantic. The darting of a salmon or trout in the water, resembles the rapidity of an arrow, but such motion cannot be kept up for any length of time. This the angler is well aware of, who, with his hook fixed on very slender gut, will kill, by fatigue, the strongest salmon in the course of an hour or two, and a large trout in the course of two or three minutes. These facts seem to indicate, that however numerous and powerful the muscles of a fish may be, they are incapable of supporting a continued exertion.

Besides the action of swimming, fishes are likewise capable of *leaping*. They accomplish this by a violent effort of the caudal fin, or, according to some, by bending the body strongly, and afterwards unbending it with an elastic spring.

A few species are capable of sustaining themselves in the air for a short interval, and are termed *Flying-fish*. Such fish have the air-bag, an organ to be noticed hereafter, of uncommonly large dimensions; hence the body has great buoyancy. The pectoral fins are likewise of an extraordinary size. Having by a leap raised themselves above the surface of the water, they continue in the air and move forwards, seldom farther than a hundred yards, by the action of their pectoral fins. The continuance of their flight is interrupted by the drying of the membrane of these fins, when they again fall into the water.

There is one species of fish (Perca scandens of LIN. Trans. vol. iii. p. 62.), which appears capable of *climbing*. By this motion, according to Lieutenant DALDORFF of Tranquebar, it sometimes raises itself five feet above the surface of the water, mounting up the crevices of trees. The spines of its gill-cover retain it in its position; and when the body is bent to one side, the spines of the anal fin fix themselves in the bark; and when the body is then brought back to its ordinary shape, the head has reached a higher elevation. The spines of the expanded gill-covers again keep a firm hold, and a similar twisting of the body takes place in another direction. The spines of the dorsal fin contribute likewise to this extraordinary progression. The flying-fish leave the water to escape from other fishes which prey upon thém; but the object to be gained by these movements of this fish has not been ascertained, nor has even a conjecture been offered on the subject.

A few fishes possess an organ of adhesion, which is generally termed a *Sucker*. In some of these it is situate on the upper part of the head, while in others it is placed on the thorax. In the celebrated fish called the Remora, it is *coronal*, of an oval form, and consisting of transverse rows of cartilaginous plates, connected by one edge to the surface of the head, while the other edge is free, and finely pectinated. A longitudinal partition divides the plates in the middle of the head. In the spaces between the plates, and on each side of the partition, a row of fleshy tubercles may be observed. In the cyclopteri this organ is thoracic, of a circular form, and consists of numerous soft papillæ. In the lampry, the mouth contracts, and the lips act as a sucker.

The use of this organ to the fish, it is difficult to ascertain. When, by its means, the fish attaches itself to the sides of other fishes, or to the bottom of ships, it is carried forward without any exertion of its own; and, during storms, adhesion to rocks may save a weak fish from being tossed about by the fury of the waves; but there may, perhaps, be other purposes to which it is subservient, which still remain to be discovered.

The sucker furnishes to the ichthyologist, characters for the discrimination of the species which are obvious and permanent; but these have seldom been described with accuracy or minuteness.

The organs of motion are extensively employed by the systematic ichthyologist, in the formation of his divisions. It does not appear, however, that naturalists have determined the exact value of the characters which they furnish, either for generic or specific distinctions. LA CEPEDE, in some instances, has formed genera from a difference in the number of the dorsal fins; while, into the genus Gadus, species with one, two, and even three fins are admitted. As the number of the fins is invariably the same, in the same species, and as these organs may be supposed to exercise considerable influence on the habits of fishes, the character thus exhibited may be safely employed in generic distinctions. The characters furnished by the structure of the fins have not been overlooked, especially the rays. The circumstance of being bony or jointed, is often noticed in specific distinctions, although well entitled to regulate divisions of a higher kind, as the character furnished is permanent. Those characters furnished by the fins, which are employed exclusively in the construction of species, are derived from their form, and the number of their rays. But as these characters are liable to vary in different individuals of the same species, they should be employed with great caution. In many fishes there are numerous rays on each side the different fins, so concealed under the skin that it is impossible to count them, while others do not reach the extremity of the organ. Hence the number of rays must vary with the mode of enumerating, and perhaps with the age of the animal. The extent of variation occasioned by the last cause, has not been satisfactorily determined.

In expressing the number of rays in the different fins, it is the practice of many naturalists to employ abbreviations. Thus 1. D. 10. intimates that the first dorsal fin has ten rays; A. 10. that the anal fin has rays to the same amount. When the fin consists of both soft and spinous rays, the following symbol is employed, $V \frac{3}{6}$. Here the ventral fin has three spinous, and six soft rays. When there are spinous rays on both sides of the soft rays of a fin, the circumstance is thus expressed, $C \frac{3-3}{20}$ Here there are three spinous rays on each side of twenty soft rays in the caudal fin. These contractions are extremely useful in shortening description, and are in common use.

There are other abbreviations occasionally employed to express the relative distance of the different fins. Supposing the length of a fish, from the extremity of the snout to the origin of the caudal fin, to be twenty lines, and to its termination to be twenty-five lines, the result is thus contracted, LC : A : : 20 : 25; the standard of comparison being always the length from the snout to the beginning of the tail. LC : DI : : 20 : 6, LC : DF : : 20 : 8, will intimate that, while the length of the body is twenty, the length from the snout to the beginning of the dorsal fin is six, and to its end, or posterior base, eight. Dr BROUSSONET, who proposed these abbreviations of relative position, likewise proposed to divide the body into as many regions as there are measures equal to the length from the snout to the base of the pectoral fin. Thus D 3 will indicate the dorsal fin to be in the third region, while V. 3. 4. will intimate, that the ventral fin is situate in the third and fourth regions *, This method presupposes a proportional increase in the dimensions of the different parts, during the growth of the animal.

The AIR-BAG, as intimately connected with the organs of motion, here merits some notice. This organ is called by some the Swimming-bladder; by others the Air-bladder. It is the *vesica natatoria* of WILLOUGHBY, and the *vesica aërea* of ARTEDI. In this country it is called the *sound*. When present, it is situate in the anterior part of the abdominal cavity, and adheres to the spine. It is wanting in the chondropterygii, and even in some of the osseous fishes, as the flounder and mackrel.

It is very different in shape, according to the species. In the herring, and some other fishes, it is oblong, and pointed at both ends. In the salmon it is obtuse at both ends. In the burbot it is obtuse in the posterior end, and bifid at its anterior extremity. In the carp it is divided transversely, and in the silurus, longitudinally, into two lobes.

In general there is a duct *(ductus pneumaticus)*, by means of which this air-bag communicates with the œsophagus, or the stomach. In the sturgeon, there is a round hole, nearly one inch in diameter, in the upper and back part of the stomach, communicating with the air-bag. The hole is surrounded by thin muscular fibres placed between the membranes of the stomach and air-bag, which decussate

---

* Phil. Trans. 1781, p. 442.

at opposite sides of the hole. These are considered by MONRO, as having the effect of a sphincter muscle. In the salmon, the last quoted author found a hole so large as to admit readily the largest sized goose-quill, leading directly through the coats of the œsophagus into the air-bag. The œsophagus in this fish has a thick muscular coat, but the fibres of that coat do not seem to form a distinct sphincter around the hole. In other fishes, the duct of communication is of considerable length. In the common herring, the under part of the stomach has the shape of a funnel; and, from the bottom of the funnel, a small duct is produced, which runs between the two milts, or the two roes, to its termination in the middle of the air-bag. In some fishes, as the cod and haddock, MONRO could not perceive any ductus pneumaticus, or opening into any of the abdominal viscera. The air-bag was not enlarged by blowing into the alimentary canal, nor could he empty the air-bag without bursting it.

In the air-bag of the cod and haddock, the same acute observer examined the red-coloured organ noticed by WILLOUGHBY, and considered by him as a muscle, the surface of which is very extensive, as it is composed of a vast number of leaves or membranes doubled. In those fishes, however, in which the air-bag communicates with the alimentary canal, this red body is either very small and simple in its structure, as in the conger eel, or entirely wanting, as in the sturgeon, salmon, herring, and carp.

Naturalists, in general, are disposed to regard the air-bag as accessory to the organs of motion. Having observed that flat fish, which reside always at the bottom, are in general destitute of this organ, they have assigned to it the office of accommodating the specific gravity of fishes to the density of the surrounding element, and thus enabling them to suspend themselves at any depth. A very simple experiment has likewise countenanced the opinion. When the air-bag of a fish is punctured, the animal immediately falls to the bottom, nor is it able, by any exertion of its fins, to elevate itself again. When in a sound state, the external skin of the air-bag (regarded as possessing strong muscular power *) is supposed capable of contraction, so as to condense the air, and enable the animal to sink, or of extension, so as to allow the air to expand, and aid the animal in rising in the water.

The above theory fails in explaining all the phenomena. The eel, which resides always at the bottom, is yet possessed of an air-bag; while the sharks, which roam about in all depths, and the mackrel, which pursues its prey at the surface, are destitute of this reputed organ of equilibrium.

The air-bag of some fishes soon loses its muscular power, in consequence of the air being expanded by the action of the sun, when the fish has remained too long at the surface. In this situation the fish continues at the surface. When some fish are suddenly brought up from deep water, the diminished pressure occasions the expansion of the air contained in the bag. The organ sometimes bursts in such cases, and the contents, rushing into the abdomen, push the gullet out of the mouth of the fish. This effect is frequently produced in the cod-fish.

Various opinions have been advanced with regard to the manner in which this air-bag is filled. By some it has been supposed, that a portion of the air, which fishes are capable of abstracting from the water, is transmitted through the gullet and stomach into the air-bag, when necessary, and expelled and renewed at the pleasure of the animal. NEEDHAM long ago considered that the air, or, as he termed it, a vaporous exhalation contained in the air-bag, was

* Phil. Trans. 1804, p. 19.

generated in the blood, secreted into this organ, to be afterwards thrown into the stomach or intestines, to promote the digestion of the food.

The nature of the air contained in the air-bag, was never investigated until pneumatic chemistry had opened up new fields of discovery. In 1774, Dr PRIESTLEY turned his attention for a short time to the subject; and in the air-bag of the roach he found azote in one instance unmixed, and in another, in union with oxygen. FOURCROY afterwards examined the gaseous contents of the air-bag of the carp, and found them to consist, in a great measure, of pure azote.

The most accurate and extended experiments on this subject, are those of M. BIOT, published in the *Mem. d'Arcueil*, i. 252. and ii. 8. He found the proportion between the oxygen and the azote (for he was unable to detect the presence of hydrogen, or any sensible quantity of carbonic acid), to vary according to the species *.

* The following Table exhibits the results obtained.

Names of the Fish.	Proportion of Oxygen.
Mugil cephalus,	quantity insensible.
Ditto,	ditto.
Murænophis helena,	very little.
Sparus annularis, female,	0.09
——————— male,	0.08
Sparus sargus, female	0.09
Ditto, male,	0.20
Holocentrus marinus,	0.12
Labrus turdus,	0.16
Sparus melanurus,	0.20
Labrus turdus var.,	0.24
Sciæna nigra, female,	0.27
——— male	0.25
Labrus turdus, female,	0.24
——————— male,	0.28
Sparus dentex, female,	0.40
Sphyræna spet (Esox sphyræna, LIN.)	0.44

The depth at which the fishes had been caught, increases from the beginning to the end of the table, and the proportion of oxygen observes the same rule. This last circumstance induced Biot, and his friend De Laroche, to endeavour to ascertain the proportion of oxygen contained in sea-water at different depths. They were unable to perceive any difference. M. Configliachi has more lately repeated these experiments, and found that the proportion of oxygen in sea-water, bore no relation to the depth from which the water had been obtained *.

These experiments lead to the conclusion, that the air contained in this sac is a secretion of the organ; that in fishes which live near the surface azote is separated; but in fishes which live at great depths, the quantity of oxygen is proportionally increased. The purposes accomplished by this arrangement have never been explained in a satisfactory manner. The red organ which we have already taken notice of, as existing in some fishes, is now generally considered as the part which separates this air from the blood. But, as this organ is not always present when there is an air-bag, we are still left in doubt on the subject.

To the systematic ichthyologist, the characters furnished by the air-bag are of considerable importance, although seldom sufficiently attended to. They are easily traced, and they are not subject to variation.

To the economist, the air-bag or sound is considered as an article of value. This organ in the cod or ling, when salted, forms a nourishing and palatable article of food,

Names of the Fish.	*Proportion of Oxygen.*
Sparus argenteus,	0.50
Holocentrus gigas,	0.69
Gadus merluccius,	0.79
Trigla lyra,	0.87

* Annals of Phil. vol. v. p. 10.

held in high estimation in the northern islands of this country. But it is chiefly in the manufacture of the substance called Isinglass, that the sounds of fishes are extensively employed. The sounds of various kinds of sturgeon are chiefly made use of for this purpose. The external membrane is removed, and the remaining part is cut lengthwise, and formed into rolls, and then dried in the open air. The sounds of cod and ling are frequently employed as a substitute for those of the sturgeon. They require some dexterity to separate them from the back-bone. But when the membranes are well scraped on both sides, steeped for a few minutes in lime-water to absorb the oil, and then washed in clean water, and dried, they form an isinglass of considerable value *.

Isinglass consists almost entirely of gelatine, and is used either as food, or for the purpose of fining liquors. 500 grains of it yielded to Hatchet, by incineration, 1.5 grains of phosphat of soda, mixed with a little phosphat of lime. An inferior kind is manufactured from the bones, fins, and useless parts of fishes. These materials are boiled in water, the fluid skimmed and filtered, and afterwards concentrated, until it readily gelatinizes on cooling. This kind is much used in various manufactures, and might be procured in considerable quantity at all the fishing stations in this country, where the materials abound, but which are at present left as a nuisance on the adjoining beach †.

The BRAIN of fishes is of a less compact texture than the corresponding organ in the higher classes. In some

---

* Phil. Trans. 1773, p. 1.

† The value of these materials to the agriculturist is properly appreciated in some places, as at Wick in Caithness, where, by means of the herring-guts formed into a compost with earth, and then applied as manure, the ground in the neighbourhood has been fertilized, on which, a few years before, the very heath was of dwarfish growth. It would be fortunate if the natives of Orkney and Zetland were to imitate the example of those at Wick.

species, indeed, it is nearly fluid. In structure, however, it is nearly the same, although characterized by a few constant marks. The subdivisions of the brain and cerebellum, or their tubercles and lobes, are more numerous than in the mammalia and birds. In one genus of fishes, the Gadus, Dr MONRO * found spheroidal bodies between the dura and pia mater, and covering the greater part of their nerves, like a coat of mail, in their course towards the organs to which they are destined. He was unable to ascertain their use.

The *Spinal Marrow* in fishes can be easily seen through the large intermediate spaces of the vertical spinous processes. Like the other nerves of fishes, the size of the spinal marrow is in proportion to the size of the body, not to the brain from which it proceeds.

The external *Organs of Smell* present several remarkable differences, according to the species, varying in number, shape, and position. In many fishes the nostrils are single, while in others they are divided at the surface by a transverse membrane, and thus exhibit the appearance of being double on each side. They likewise vary in shape, being round in the cod-fish, oval in the conger-eel, and oblong in others. They are placed in the snout in many fishes, near the eyes in some, and between the eyes and the snout in others. Where the openings are double on each side, these are either placed contiguous to each other, as in the carp; at a little distance, as in the perch; or remote, as in the eel. The nostrils, in some instances, appear like short tubes

The nasal openings are furnished with a few muscular fibres, which are capable of executing a limited contractile

---

* Structure and Physiology of Fishes, Edin. 1785, p. 44.

motion. This motion, however, in living fishes, can seldom be perceived.

Proceeding to the examination of the inside of the nostril, we may observe, that, in the sharks, skates, and eels, the nasal laminæ are placed parallel to each other, on both sides of a large lamina, which extends from one end of the fossa to the other, and consists of folds of the pituitary membrane. In general, in the other fishes, whether cartilaginous or osseous, the laminæ proceed like radii from an elevated and round tubercle. The pituitary membrane, in some fishes, as in the pike, is furnished with reticular ramifications of black vessels, but in the greater number of fishes these vessels are red. Between these, are situated some small papillæ, which pour out a thick mucilage.

The *Olfactory Nerves*, at their origin, form swellings or knots, so large as frequently to have been mistaken for the real brain. These tubercles in skates and sharks are united into one homogeneous medullary mass, from each of the lateral parts of which the olfactory nerves arise. In the species of the genera Pleuronectes, Clupea, Esox, Perca, and Salmo, there are two pair of tubercles, the anterior of which is smaller than the other.

In the cartilaginous fishes, as the skate and shark, the olfactory nerve is very soft. It is, in them, a bulb which passes obliquely forward toward the nares, which are at a greater or less distance from the brain, according to the species. The spinous fishes have the olfactory nerve very long and slender. In those which have the snout elongated, this nerve is received into a cartilaginous tube. In those with short snouts, the nerve is surrounded by a fine membrane only, which appears to be the same as that which contains the fat or oily humour that covers the brain. In the haddock, and some other fishes, the olfactory nerve, in its course from the brain to the nose, passes through a cineritious ball, which resembles the cineritious matter connected, in our body, with the olfactory nerve within the cranium.

When the olfactory nerve arrives behind the folded membrane which we have described, it is dilated to be applied to the whole of its internal and convex surface. In some fishes, no previous enlargement takes place, while in others the nerve swells into a real ganglion. When expanded, it has been compared to the retina, but the filaments of which it is composed are more distinct.

The sense of smell in fishes is supposed by many to furnish them with the most delicate tests, for searching after, and distinguishing their food *.

We may observe, however, that the well known voraciousness of fishes, the eagerness with which they will seize a metal button, or any glittering object, the whole art of artificial bait and fly fishing, all seem to point out the organ of sight as the principal instrument by which they discover their food. Besides, the organs of smelling are by no means favourably situated for receiving quickly the impressions new objects are calculated to produce. In the chondropterygii the nares communicate by a groove with the angles of the mouth, but in general the organs of smell

---

* Dr Munro *primus* states, that " if you throw a fresh worm into the water, a fish shall distinguish it at a considerable distance; and that this is not done by the eye is plain from observing, that, after the same worm has been a considerable time in the water, and lost its smell, no fishes will come near it; but if you take out the bait, and make several little incisions into it, so as to let out more of the odoriferous effluvia, it shall have the same effect as formerly. Now, it is certain, that had the creatures discovered this bait with their eyes, they would have come equally to it in both cases. In consequence of their smell being the principal means they have of discovering their food, we may frequently observe them allowing themselves to be carried down with the stream, that they may ascend again leisurely against the current of the water, thus the odoriferous particles swimming in that medium, being applied more forcibly to their organs of smell, produce a stronger sensation "—Comp. Anat. p. 127, Edin. 1783.

have no communication with those of mastication or respiration; and, as the external openings are narrow, and but ill supplied with muscles, we are at a loss to conceive in what manner the water impregnated with odoriferous particles can be thus rapidly applied to the extremities of the olfactory nerve. The same water must pass through the mouth, and be spread over the extended surface of the gills; so that we may presume, until farther light be thrown on the subject, that these latter organs may likewise contribute to warn the fish of the presence or absence of salutary or noxious impregnations.

The organs of smell furnish the ichthyologist with some important characters in the description of the species. These have hitherto been too much neglected, although they have the advantage of being permanent.

The EYES of fishes, like all other red-blooded animals, are two in number. They vary greatly in position, both being, in some species, on the same side of the head, as in flounders; while in others they are nearly vertical. In general, however, they are placed one on each side of the head. The eyes of fishes are larger in proportion to the size of their bodies, than in quadrupeds, as we find the eye of the cod-fish equal in size to that of an ox.

Fishes, in general, are destitute of *eye-lids,* and are seldom even furnished with projections in place of eye-brows. In the moon-fish (Tetraodon mola), however, the eye may be entirely covered with an eye-lid, perforated circularly. In the greater number of fishes, the skin passes directly over the eye, without forming any fold; and in some cases, it does not adhere very closely to the eye. Thus the common eel and lampry may be skinned without producing any hole in the situation of the eye, the skin only exhibiting at that place a round transparent spot. In the trunk-fish (Ostracion), the conjunctiva, or external covering of the cye, is so similar to the rest of the skin, that we observe lines upon it, which form the same compartments as on the body of the fish. Some fishes may be considered as blind, as the Myxine cœca, in consequence of the uniform opacity of the skin in passing over the eye.

The form of the eye in this tribe of animals is nearly that of a hemisphere, the plane part of which is directed outwards, and the convex inwards. In the Ray, the superior part is also flattened, so that the vertical diameter is to the transverse as 1 to 2. This flatness of the exterior part of the eye is compensated by the spherical form of the *crystalline lens*. This body is more dense in fishes than in land animals *.

---

* MONRO found the crystalline lens of an ox to be 1104, while that of a cod was 1165, water being reckoned at 1000. The crystalline lens projects through the pupil, and leaves scarce any space for the aqueous humour. The *vitreous humour* is proportionally small. The portion of the axis occupied by each of the three humours of the eye, in the herring, for instance, may be expressed in fractions, as follows: aqueous humour $\frac{1}{7}$, crystalline lens $\frac{5}{7}$, and the vitreous humour $\frac{1}{7}$. The spherical form of the crystalline lens has been already stated; but the following Table, from the observations of PETIT and CUVIER, will exhibit more clearly the proportion between the axis and the diameter in a few species.

The axis is to the diameter in the

Salmon as	9 to 10
Sword-fish,	25 : 26
Shad,	10 : 11
Pike,	14 : 15
Barbel,	11 : 12
Carp,	14 : 15
Mackrel,	12 : 13
Whiting,	14 : 15
Shark,	21 : 22
Ray,	21 : 22
Herring,	10 : 11
Tench,	7 : 8
Eel,	11 : 12
Conger,	9 : 10

The *sclerotic coat* of the eye of fishes is more firm and dense than in the higher animals. It is here cartilaginous, semitransparent, and elastic, and sufficiently solid to preserve its form of itself. In the salmon it is of the thickness of a line posteriorly, and of an almost bony hardness before. This is frequently the case in other fishes, especially near its junction with the cornea, where it sometimes appears like an osseous ring. The outer layer of the *choroid coat* is either white, silvery, or gold-coloured, and is very thin and little vascular. The inner coat, to which the term *membrana Ruyschiana* has been applied, is in general black, and covered everywhere by mucous substance. In the ray, however, it is transparent. Between these two membranes of the choroid coat there is a body of a brilliant red colour. Its form is usually that of a thin cylinder, formed like a ring round the optic nerve: the ring, however, is not complete, a segment of a certain length being always wanting. Sometimes, as in the *Perca labrax*, it consists of two pieces, one on each side of the optic nerve. It is considered by some as muscular, and enabling the eye to accommodate its figure to the distance of the objects; while others regard it as glandular, and destined to secrete some of the humours of the eye. This gland, we may add, does not exist in the *Chondropterygii*, as the rays and sharks.

The *iris* is in general distinguished by its golden and silvery brilliancy. This arises from its transparency, allowing the natural colour of the choroid coat to be discerned. The pupil is different in form in the different species, but, in general, it approaches to circular or oval; in some genera, as the salmon, it projects into an acute angle at the anterior part. In the *Gobitis anableps* of Linnæus, the cornea is divided into two portions, and there is a double pupil with a single lens. In the ray, the superior edge of its pupil is prolonged into several narrow stripes disposed in radii, gilded externally, and black internally. In their ordinary state they are folded between the superior edge of the pupil and the vitreous humours; but when we press the superior part of the eye with the finger, they unfold themselves, and cover the pupil like a window-blind. In the torpedo, the pupil can be completely closed by means of this veil. No other fishes possess any thing similar to this conformation, although in most osseous fishes there is at each corner of the orbit a vertical veil, which covers a small part of the eye.

In general, the eyes of fishes are placed in a conical cup, and repose on a mass of gelatinous matter contained in a loose cellular substance. This trembling elastic mass affords the eye a point of support in all its motions. In the *Chondropterygii*, however, the eye is joined to the extremity of a cartilaginous stalk, which is itself articulated in the bottom of the orbit. In this manner the muscles act on a long lever, and have therefore great power in moving the eye.

The *optic nerves* arise under the cerebrum, and are very large. They are composed either of distinct filaments, or of a single flat band, which is sometimes folded longitudinally on itself, and contracted into the figure of a cord. They cross each other without being confounded, and we plainly see that the nerve of the left side proceeds to the right eye, and that of the right side to the left eye. This crossing is less apparent in the cartilaginous fishes, although in the ray the right nerve passes through an opening in the left. These nerves pass directly through the membranes of the eye by a round hole. Internally they form a tubercle, which is papillated in the ray, sharks, and carps. The radiating fibres, which arise from the edges of these tubercles to form the retina, are very obvious. In other genera the retina is formed from the edges of two long white caudæ, in the same manner as it arises in birds from the single white line.

The eye is one of the most important organs which fishes are known to possess. It enables them to perceive the approach of their foes, and it is the principal instrument by which they obtain their food. The amateur in artificial fly-fishing often tempts the fish with one kind of fly, but in vain; and, upon substituting another in its place, of a different form or colour, succeeds in the capture. These motions of the fish are all regulated by the eye; hence some fish will bite as readily at a bit of red cloth as at a piece of flesh.

As this organ exercises a very powerful influence on the habits of fishes, it should be carefully attended to by the ichthyologist. The characters which are furnished by its form and position are not liable to variations, and they are sufficiently obvious. Those furnished by the colours of the different parts hold a secondary rank. They are not very liable to vary, but they experience great changes after death, and should be used with very great caution.

It was long known to naturalists, that fishes possessed some means of distinguishing the vibrations of sonorous bodies, or possessed the SENSE of HEARING. Trouts and carp have been taught to come to a particular place of the pond for food upon a bell being rung; and a drum has sometimes been employed to drive fishes into a net. In general, however, it was supposed that the vibrations communicated to the water, became sensible to the fish, through the medium of the organs of touch.

The Abbé NOLLET (in the *Hist. de l'Acad. R. des Sciences*, 1743, p. 26.) ascertained, by conclusive experiments, that the human ear was susceptible to the impressions of sound, even when immersed in water. This discovery directed the attention of anatomists to the structure of the organs of hearing, and CAMPER, GEOFFROY, HUNTER[*], and VICQ D'AZYR, succeeded in pointing out the nature of the different parts. Our illustrious countryman Dr MONRO, in his work on the structure and physiology of fishes, contributed to enlarge our knowledge of the organs of hearing, in this tribe, by numerous accurate dissections.

In the osseous fishes, no external ear has hitherto been detected, and the same remark is applicable to those cartilaginous fishes which have free branchiæ. But in the cartilaginous fishes with fixed branchiæ, small apertures have been discovered leading to auditory organs. These were first observed by MONRO, in the skate and the angel shark. In the former fish they occur in the back part of the occiput, near the joining of the head with the spine. They are two in number, not larger than to admit the head of a small pin; and in large fish are found at the distance of an inch from each other.

In fishes that have free branchiæ, the internal organs of hearing are situated in the sides of the cavity of the cranium, and fixed there by a cellular tissue, consisting of vessels, and osseous or cartilaginous fræna. In the fishes with fixed branchiæ, these organs are inclosed in a particular cavity, formed in the substance of the cranium. This cavity is situate on the side and posterior part of that which contains the brain, with which it does not communicate, except by the holes that afford passages for the nerves. This sac exhibits many differences as to size and form, in the different species. Besides the ordinary viscid fluid, there are some small cretaceous bodies suspended by a beautiful plexus of nerves. These, in the osseous fishes, are three

---

[*] Phil. Trans. 1782, p. 373.

in number, and are hard and white like porcelain. In the cartilaginous fishes with three branchiæ, these bodies are in general fewer in number, and of a softer consistence, being seldom harder than moistened starch. It is supposed that these bodies assist in communicating to the nerves the vibrations produced in the water by sound. With the sac are connected three semicircular canals, filled with a viscid fluid, similar to that in the large sac. The *auditory nerves* arise so near to the origin of the fifth pair, that they have been considered as the same. In the genus Raja these pass into the cavity of the ear, by a particular foramen; in the osseous fishes, they are distributed directly into that organ.

As the ear of fishes is much less complicated in its structure, than in the higher orders of animals, we may conclude, that the sense of hearing is weak in proportion. Indeed, the difficulty of detecting any natural movements of fishes, occasioned by sound, led the ancients to conclude that they did not enjoy this sense. In ichthyology, the characters of the organs of hearing are too minute and difficult of detection, ever to be employed. They vary in different species, it is true, and may be resorted to in cases of difficulty; but for their investigation they require a dexterous hand and an experienced eye.

As the tongue of fishes (the organ in which the *sense of taste* resides in the higher orders of animals) is but imperfectly developed, naturalists are in general disposed to conclude, that the sense of taste can scarcely be said to belong to this class of beings. It presents no visible distinct papillæ, and its skin is analogous to the common integuments of the mouth. The nerves which supply it, are branches of the same nerves which proceed to the branchiæ. In the present state of our knowledge it is impossible to assign the precise influence which the sense of taste exercises on the economy of fishes. If noxious ingredients exist in the water, it appears probable, that some warning will be given to the animal of their presence, either by the nerves of the mouth, during the passage of the water to the gills, or by the latter organs. It does not appear, however, that this sense is ever used in the discrimination of food, and does not furnish any characters, for classification, to the ichthyologist.

We have already observed, that the skin of fishes is destitute of the corpus papillare, and hence anatomists have concluded, that the animals of this class possess the SENSE OF TOUCH in a very limited degree. Besides, few nerves have hitherto been traced to the skin; and as its surface is in general coated with scales, it appears but ill adapted for receiving very delicate impressions. In some species, however, such as the common trout, the sense of touch is easily displayed, if, when the fish is resting under a stone or bank, the hand be moved gently towards it, and its sides titillated. It will then exhibit the pleasure it derives, by leaning on the hand, and if the operation be performed with care, every part of the body may be gently stroked, and the fearless fish in part raised above the water.

Fishes possess no voice by which they can communicate their sensations to others. Some species utter sounds when raised above the water, by expelling the air through the gill opening when the flap is nearly closed: while others, even under water, as the salmon, utter certain sounds while in the act of depositing their spawn; but for what purpose these sounds are uttered, or by what organs they are produced, we are still ignorant.

In reference to the ORGANS OF NUTRITION, Fishes exhibit innumerable modifications of form and structure. They are all destitute of organs of prehension, and few of them can be considered as furnished with flexible lips. The maxillary structure, however, supplies these defects.

The mouth in many fishes is terminal, while, in others, it has a ventral aspect, being placed beneath a projecting snout. The opening in all cases is transverse. The bones of the upper jaw are chiefly the maxillary and intermaxillary. The latter frequently occupy a large portion of the gape, while the former extend beyond the gape like a moveable process. These bones are loosely connected with the cranium by means of palatine and cheek bones, which form a sort of inner jaw, to which the lower jaw is articulated. The articulations admit, in many cases, of extensive motion, and enable the animal to form with them a sort of tubular proboscis.

The tongue is supported by a largely divided os hyoides, which gives support to the arches of the gills, the bones of the gill-lid, and those of the gill-flap.

The teeth are situate, according to the species, not only on the lower jaw, the maxillary and intermaxillary, but on the vomer, the palate, the gill-arches, the tongue, and the walls of the pharynx. In some species, the teeth are formed in sockets, as among the Mammifera; in others, they seem to be continuations of the bones, while in a few they may be viewed as epiphyses. The last kind appear to be readily reproduced [*]. In nearly all fishes the teeth are bent inwards, and thus serve to retain the food. The pharynx appears to be a direct continuation of the mouth.

The *gullet*, on account of the absence of a neck, is remarkably short in fishes. In some, indeed, the stomach seems to open directly into the mouth. Where the gullet is obvious, it exhibits few peculiarities of structure. In some of the branchiostegi it is beset with tufts of hair resembling a fine net-work. It is in general capable of great

---

[*] Phil. Trans. 1784, p. 279.

dilation, and when the stomach is unable to hold the whole of the prey which has been seized, a part remains in the gullet until the inferior portion gives way. The zone of gastric glands is in general well marked.

The *stomach* of fishes is in general thin and membranaceous, differing little in its structure and appearance from the gullet. It frequently contains the remains of crustaceous animals, still retaining their form, but greatly altered in consistency. Hence naturalists have concluded, that the food is reduced by solution, and not by trituration. But in some fishes, particularly those which subsist principally on shell-fish, the stomach has thick muscular coats. Its shape varies in the different species, but the characters furnished by this organ are seldom regarded.

The *intestines* exhibit many remarkable peculiarities. Sometimes they proceed directly from the stomach to the anus in nearly a straight line. In other instances, they form in their course one or more flexures. In some instances, the gut is widest towards the stomach, and gradually becomes smaller as it approaches the anus, while in others the reverse of this is the case. It is furnished internally, in some species, with spiral valves, in others with lozen-shaped hollows, while in a few it has numerous fringed laminæ. Between the great and small intestines, in the chondropterygii, there is a kind of cœcum or appendix vermiformis; but in osseous fishes there is no appearance of any such organ. In the last division, however, there are bodies which have been termed *Appendices*, or Intestinula-cœca. These are situated at the origin of the gut, in a double or single row. They vary in number, shape, or size, according to the species; but continue the same in all the individuals of the same species. In place of these in the chondropterygii, there is a glandular body, which has been compared to the pancreas of warm-blooded animals.

In the sturgeon, an organ is found, in its internal structure, similar to those intestinula; but in its outward form resembling the pancreas of the skate. It is inclosed in a muscle, evidently intended to express its contents. It opens into the intestine by three large orifices, and has internally a singular reticular appearance, as exhibited by Monro, in his work on fishes, page 84. tab. ix. The character for the discrimination of the species furnished by the appendices, is of importance, as being easily investigated and permanent.

These intestines, and the rest of the viscera situated in the cavity of the abdomen, are contained in a membranaceous sac or peritoneum. This is silvery in some fishes, black or spotted in others. Willoughby observed, that this sac opens externally near the anus, by means of two small holes. These openings were afterwards examined by Monro, who found in each of these passages a semilunar membrane or valve, so placed as to allow liquors to get out from the abdomen readily, but to resist somewhat their entry into it. They serve for the exit of the eggs.

The anus in fishes occupies many different positions, according to the species. This circumstance was seized upon by Scopoli, in the construction of his system of classification, and was raised to the dignity of a primary character. This orifice is not merely the opening, whence issues the fæces, but frequently the spawn also.

The *liver* in fishes is remarkable on account of its size, in proportion to the rest of the body. It commonly lies almost wholly on the left side. Its colour exhibits various shades of brown, frequently mixed with yellow. It abounds in oil. It is entire in some fishes, as the lamprey, flounder, and salmon; or divided into two or more lobes, as in the perch and carp. These varieties of form are constant in all the individuals of the same species, but frequently differ somewhat in the species of the same genus.

The *gall-bladder* is present in the greater number of fishes; but in a few species, as the lamprey, its presence has not been détected. In the Squalus maximus, the duct is dilated at its termination in the intestine. The *bile* varies greatly in colour according to the species. In the thornback and salmon it is yellowish white, and, when evaporated, leaves a matter which has a very sweet and slightly acrid taste, containing no resin. The bile of the carp and eel is very green and very bitter, contains little or no albumen, but yields soda, resin, and a sweet acrid matter, similar to that which may be obtained from salmon bile. The biliary ducts open separately into the intestine. The liver of the cod, cut into small pieces, boiled in the stomach of the same animal, and eaten with vinegar and pepper, is a favourite dish in the northern islands of Scotland, and termed *Liver Muggie*.

The vessels of the *Absorbent System* of fishes are analogous to those of quadrupeds. They are, however, destitute of valves, unless at their termination in the red veins, and do not appear to possess conglobate glands. Dr Monro, to whom we are indebted for the first illustration of this class of vessels, gives an interesting view of their arrangement in the cod and the salmon *

---

* " The chief branches," he says, " of the lacteal vessels of the great and small intestines, and which are smaller in proportion to the blood-vessels, than in the Nantes pinnati of Linnæus, run upwards in the mesentery, almost parallel to each other, and near the mesenteric arteries. In their whole course, they communicate by a vast number of small transverse canals. At the top of the abdomen, near the gall-bladder, the lacteals of the stomach, and lymphatics of the spleen, liver, and intestinula cœca are added. The chyle, mixed with the lymph of the assistant chylopoietic viscera, passes upwards, and towards the right side, into a large receptacle contiguous to the gall-bladder, and between it and the right side and back part of the lower end of the œsophagus. From the receptacle of the chyle, large

The termination of the lymphatic veins in the skin, may be readily ascertained in this class. Coloured liquors injected into them, are discharged by numerous pores, chiefly situate on the upper parts of the body. These orifices are placed at regular intervals. As Dr Monro did not observe any appearance of extravasation in the cellular substance, he considered that these orifices were the natural beginnings of the lymphatic veins.

---

canals pass upwards to the right and left, receiving in this course the lymph from the organs of urine and generation. Those on the left side are chiefly behind the œsophagus.

" The chyle, mixed with the abdominal lymph, having ascended above the bones, which resemble our clavicles, is poured into large cellular receptacles, situate chiefly between the clavicles and the undermost of the gills, and which also receive the lymph from all the other parts of the body.

" Four lymphatic vessels, which terminate in these receptacles, and which have their extremities contracted by a doubling of their internal membranes, chiefly merit attention. The first conveys the lymph from the middle of the belly, from the ventral and pectoral fins, and from the heart. The second runs up the side of the fish, parallel to the great mucous duct, and brings the lymph from the principal muscles of the tail and body. The third is deep seated, and conveys the lymph from the spine, spinal marrow, and upper part of the head. The fourth lymphatic vessel, or rather plexus of vessels, brings the lymph from the brain and organs of the senses, and from the mouth, jaws, and gills.

" These receptacles may therefore be called the common receptacles of the chyle and lymph. The right receptacle communicates freely with the left by large canals, which pass chiefly behind the heart and œsophagus.

" From each of these receptacles in the salmon, a canal runs downwards and inwards, and opens into the upper end of its corresponding vena cava inferior, contiguous to, and on the fore and outer side of the internal jugular vein. The termination of these canals are contracted, and their internal membranes are doubled, so as to serve the purpose of valves, in preventing the passage of the blood from the venæ cavæ into the receptacles. In the cod kind, the receptacles are proportionally larger than in the salmon; and, besides transmitting the muscles of the gills, and their several nerves, contain the upper cornua of the air-bladder."—(Monro, *Struct. and Phys. of Fishes*, p. 31.)

The *spleen* varies greatly in its form and position in the animals of this class. In some it is nearly triangular, while in others it approaches to a spherical figure. It is in general entire; in some instances, however, it is divided into lobes, which adhere by very slender filaments. In the sturgeon, these lobes are seven in number, while in the angel shark they are limited to two. It is placed in some species on the stomach, or to the first part of the intestines; in others between the stomach and liver; and in a great number it is under the air-bag, and above the other bowels. It is always of a darker colour than the liver.

While in reptiles the whole blood does not pass through the aërating organs in succession, in the course of circulation, it is otherwise with fishes. The heart which they possess is pulmonic, receiving the blood from the veins, and transmitting the whole of it to the aërating organs. It is situate in the fore part of the body, in a cavity between the gills, and a little behind. The pericardium or membrane, which lines this cavity, is similar to the covering of the cavity of the abdomen, and, like it, is often spotted or silvery. In the skate, Dr Monro found the bottom of the pericardium lengthened into the shape of a funnel, which divides into two branches. These are tied closely to the lower part of the œsophagus, and open into the cavity of the abdomen. Into this cavity there is secreted a saltish liquor. The heart itself is small in proportion to the body of the animal, and varies greatly in figure in the different species. It is quadrilateral in some, and semicircular in others. It consists, as we have already mentioned, of a single auricle and a single ventricle, corresponding to the right side of the heart of warm-blooded animals. The auricle is in general larger than the ventricle, and of a thinner texture in its coats. This last division of the heart has walls of considerable thickness. It sends forth an artery, which, at its separation from the heart, forms a bulb varying in shape according to the species. This artery subdivides and proceeds directly to the gills, over whose leaves it is spread in the most minute ramifications.

The organs of respiration in fishes consist of four parts, a gill-lid, a gill-flap, the gill-opening, and the gills themselves. The two last are always present, but one, and sometimes both, of the two first are wanting. We propose to examine these parts in succession, beginning with those which are exterior.

1. *Gill-lid.* The gill-lid, or, as it is also termed, *operculum*, is situated behind the eye on each side. The bones of which it consists vary in form and number. The first, termed by CUVIER the *pre-opercle*, is united with the cheekbones, and extends, in the form of a thin plate, from the occiput to the articulation of the lower jaw. To its distal margin are attached the *opercle*, next the occiput; then the *sub-opercle*, and, last of all, the *inter-opercle*, next the tongue. The surface in some is smooth, in others rough, or tuberculated, or striated, or spinous. Its use is to give support to the gill-flap, and act as a cover to the opening of the gills. It is absent in fishes which have fixed branchiæ, and in a few with free branchiæ. When it does exist, the characters which it exhibits in its structure are subject to little variation.

The *gill-flap* is the *membrana branchiostega* of LINNÆUS, and was considered by him as a true fin. It consists of a definite number of curved bones or cartilages, with a connecting membrane. Its posterior edge is generally free, and its anterior edge or base is united with the gill-lid. It is capable of extension and contraction, and, when at rest, it is generally folded up under the gill-lid. It is wanting in the chondropterygii, and likewise in a few genera of osseous fishes. When present, it appears to assist the mouth in promoting the current of water through the gills, or perhaps forms a current over the gills, when the mouth is occupied in seizing prey.

The gill-flap furnishes to the systematic ichthyologist some of his most useful characters. He seldom pays attention to its form, but its rays are eagerly counted, as he finds that they are not subject to vary. Species of the same genus have, in general, the same number of rays, and many of the Linnæan genera depend on this circumstance for their character. But in counting their number, care must be taken to examine the structure of the gill-lid at the same time, as the student sometimes enumerates, among the rays of the gill-flap, the posterior divisions of that organ, when present, and hence finds his observations at variance with the descriptions of authors.

The gill-opening in the osseous fishes, and among the branchiostegi, is a simple aperture behind the gills on each. It is sometimes round, or semi-lunar, and in relative position it differs according to the species or genera. In the cartilaginous fishes, the opening on each side is subdivided into as many apertures as there are gills, the gills in this tribe being fixed to the membranes which act as partitions in the opening. In such fishes, these openings are on the summit, at the sides, or underneath, according to the genera.

In the fishes with free gills, these organs are in general eight in number, four on each side. Each gill consists of three parts, a cartilaginous or bony support, and its convex and concave sides. The support of each gill consists of a crooked bone or cartilage, in general furnished with a joint. At its base, it is united with the bones of the tongue, and above with those of the head. At both extremities it is moveable, and throughout is flexible like a rib. Its position is nearly vertical. From its exterior or convex side, issue a multitude of fleshy leaves, or fringed vascular fibrils, resembling plumes, and closely connected at the base. These are of a red colour in almost all fishes in a healthy state. The internal or convex side of the support next the mouth exhibits many singular differences. It is always more or less furnished with tubercles. These, in the genus Cyprinus, are smooth,—in the Cottus rough. They are lengthened into slender spines in the herring and smelt, but in the former these are serrated, while in the latter they are smooth. This concave part of the gill is of a white colour, and forms a striking contrast with the colour of the convex side.

In some osseous fishes, the gills exceed four in number on each. In the herring, for example, there is a small imperfect gill on each side, attached to the inner side of the gill-lid, on which all its motions depend. It has no bony arch nor concave side. At the entrance to the gullet there is a cartilage on each side, studded with tubercles, resembling, in appearance, the concave side of the last gill, but connected with deglutition. In the *plaise*, a similar gill may be observed on the inside of the gill-lid, but no distinct appearance of a sixth gill at the entrance to the œsophagus.

In the chondropterygii, the gills are far from being so perfect. They are fixed to partitions which serve the purposes of the bony arches in the osseous fishes. These partitions extend from the mouth to the gill-opening, and vary in number according to the genera. They are destitute of the inner or concave white side, but the fleshy leaflets are of the same structure with those on the convex part of the gills in osseous fishes *.

---

* We are indebted to that distinguished anatomist, Sir EVERARD HOME, for some important observations on the respiratory organs of the lamprey and

FISHES.

The extent of surface presented by the gills of a fish, to enable the blood to come in contact with the air in the water, is much greater than one would, without attentive consideration, be led to suppose. Dr Monro calculated,

---

myxine, the apodal chondropterygious fishes, and the least perfect in the system. " In the lamprey (he says) the organs of respiration have seven external openings on each side of the animal; these lead into the same number of separate oval bags, placed horizontally, the inner membrane of which is constructed like that of the gills in fishes. There is an equal number of internal openings leading into a tube, the lower end of which is closed, and the upper terminates by a fringed edge in the œsophagus. These bags are contained in separate cavities, and enclosed in a thorax resembling that of land animals, only composed of cartilages instead of ribs, and the pericardium, which is also cartilaginous, is fitted to its lower extremity like a diaphragm." Phil. Trans. 1815, p. 257. In the myxine, the external openings are two in number, but there are six lateral bags on each side, placed perpendicularly, to which there are six tubes from each of the openings, and close to the left external opening there is one which passes directly into the œsophagus.

" In the lamprey, the water is received by the lateral openings of the animal into the bags which perform the office of gills, and passes out by the same opening; the form of the cavities being fitted to allow the water to go in at one side, pass round the projecting parts, and out at the other. A part of the water escapes into the middle tube, and from thence, either passes into the other bags, or out of the upper end into the œsophagus. There is a common opinion, that the water is thrown out at the nostril; this, however, is unfounded, as the nostril has no communication with the mouth."—" In the myxine, the elasticity of the two tubes, and the bags into which they open, admits of the water being received; and the pressure produced by the action of the external muscles, forces it into the œsophagus, from whence it is thrown out by the opening at the lower end of that tube." Ibid.

The means here stated, as employed to bring the water to the surface of the gills, is probably, in the case of lamprey, only used during the action of the sucker, as the gill cavities seem but ill fitted for the continued absorption, circulation, and ejection of the water by the same orifice, such as our author supposes. Analogy, too, forbids us to conjecture, that the gills of the lamprey shall be watered by motions the reverse of those which are executed by other fishes.

that the whole gills of a large skate presented a surface equal to 2250 square inches, or equal to the whole external surface of the human body.

Any injury received by the gills of fishes, is attended with much pain, and a considerable effusion of blood. Some fishermen seem to be well aware of this last circumstance, and cut the gills with a knife as soon as the fish is taken. A copious bleeding takes place; and they find that a fish so killed, will keep much longer in a fresh state, than one on which this operation of bleeding has not been performed.

The number of respirations in a minute is seldom above thirty, or below twenty. In the same individual it is liable to considerable variation, depending on the will of the animal.

The blood, after being renovated in the gills, is reabsorbed by a multitude of minute vessels, which unite together; but, instead of returning the blood to a systemic heart, to be afterwards distributed through the body, this aorta exercises that function, and descends along the inferior side of the spine, in a canal fitted for its reception, giving off arteries, during its course, to the adjacent parts. The blood is absorbed again by veins, which have extremely thin coats. These are much larger in their course than in their termination; and form, in different parts of their course, considerable receptacles for blood.

It was the opinion of RONDELETIUS, that fishes were destitute of KIDNEYS and the bladder of urine; but the observations of WILLOUGHBY and others have demonstrated their existence. The kidneys of fishes are uniform in their substance, and of a reddish brown colour. They are, in general, long and narrow, and apparently united into one mass. The peritoneum covers their under surface, and they are placed longitudinally under the spine. The ureters begin by numerous roots, and run along the under surface of the kidney. They terminate either in a vesica urinaria, or a cloaca; or unite together to form a dilatation, which supplies the place of a bladder of urine.

In the chondropterygii, the ureters terminate in the cloaca, but in the other cartilaginous fishes the bladder of urine is present, although very small and thin in its coats. The urethra in most fishes is short, and commonly opens behind the anus by an orifice, which also gives issue to the sexual evacuations. Renal glands are wanting in this class.

The REPRODUCTION of fishes is a subject involved in great obscurity. The element in which they reside conceals from us the actions which they perform, so that we are unable to point out, with certainty, the uses of the different organs, or the functions which they exercise. Even in the days of ARISTOTLE, the difference in the mode of reproduction between the cartilaginous and the osseous fishes had been observed; and although many accurate observations have been made by modern zootomists, much still remains to be done, both in the field of observation and dissection.

In reference to the reproductive system, fishes may be divided into two classes. Thus, some have the sexes distinct, while in others they are united. Those with the sexes distinct, may be subdivided into such as have the impregnation External or Internal.

Fishes with the sexes distinct, and with external impregnation.—The fishes included under this division, are by far the most numerous. They have all free branchiæ. Some of them possess a cartilaginous skeleton, while others belong to the division termed Osseous. In all of them, the egg is impregnated externally, and arrives at maturity without the aid of the mother.

In the *males* of this division, the testes, known by the name of *milts*, are two in number, of a white colour, and lengthened form. The surface is usually irregularly tuberculated. They are situate on each side of the abdomen, and consist of glandulous sacs destined for the preparation of the impregnating fluid. Through the middle of each milt there passes a ductus deferens, uniting with each other at the posterior part of the abdomen, and forming a kind of vesicula seminalis. This duct either terminates in the cloaca, or by a small orifice near its caudal margin.

We possess few accurate experiments on the chemical composition of the seminal fluid of fishes. FOURCROY published[*] some experiments on the milt of the carp. He found that it was neither acid nor alkaline. It appears to consist of albumen, gelatine, phosphorus, phosphat of lime, phosphat of magnesia, and muriat of ammonia. More recently, Dr JOHN subjected the milt of a tench to a chemical analysis, and obtained the following ingredients: water, insoluble albumen, gelatine, phosphat of ammonia, phosphat of lime, phosphat of magnesia, and alkaline phosphat. He could not detect the presence of any phosphorus, which had been given as a constituent by FOURCROY and VAUQUELIN. In all these examples, however, the seminal fluid was mixed with the substance of the testes.

In the *females* of this division, the ovary, usually termed the *roe*, is double in the greater number of fishes, but in a few it appears to be single. It occupies nearly the same position as the milt in the males. It consists of a thin delicate folded membrane inclosing the ova, disposed in transverse layers, and connected by means of bloodvessels. There is no distinct oviduct. The external openings are similar to those in the male.

---

[*] Annales de Chim. vol. lxiv. p. 3.

Previous to the deposition or ejection of the roe or eggs by the female, a social union has been formed with a male. But this connection is merely temporary, and is dissolved immediately after the impregnation of the egg has taken place.

The ova are first deposited by the female, and then the male pours upon them the impregnating fluid. In many instances, they mutually form a hole in the sand, and place therein the roe; in other instances, the roe is deposited in the crevices of rocks, or on sea weeds or aquatic plants. But it would be endless to detail the various ways (even were we better acquainted with them than we profess to be) in which fishes perform this curious function of their nature.

The eggs of fishes are very various with respect to colour, but agree in being of a spherical form. The integument is more or less firm, according to the species. The yolk, which contains but little oil, instead of occupying the centre, as in the eggs of birds, is placed laterally, and is surrounded by the glaire or albuminous matter. Between the yolk and the glaire is situated the germ or embyro. The germ becomes ready for exclusion at very different periods, according to the species. Thus, the egg of the carp is said to be perfected in the course of three weeks, while that of the salmon requires as many months. But in the eggs of the same species, a great deal depends on the temperature to which they are exposed; as, in the same pond, those eggs are soonest hatched which have been deposited in the shallowest water.

As the embryo is developed, the heart first appears, afterwards the spine, eyes, and tail. The organs of motion are evolved in the following order. The pectoral fins first make their appearance, and afterwards those of the tail; the dorsal fins follow, and then the ventral and anal fins.

Fishes with the sexes distinct, and with internal impregnation.—In this division are included the chondropterygii, and likewise a few osseous fishes. In the *males* of this division, at least in those of the chondropterygii, the testes are two in number, flat, and of great extent. Each is divided into two portions; the first resembles the soft milt of oviparous fishes, and the second consists of small spherical glandular bodies. From these an epididymis is produced, chiefly composed of convoluted tubes, which terminate in a vas deferens; the under part of which is greatly dilated, and forms, as in birds, a considerable receptacle, or vesicula seminalis. Contiguous to the outer side of the dilated end of the vas deferens, there is a bag of considerable size, filled with a green liquor, which is discharged into the same funnel with the semen, and probably at the same time with it. By some, this is considered as a vesicula seminalis, while, by others, it is regarded as supplying the place of a prostate gland. The funnel through which the seminal fluid is poured, opens near the cloaca, and, in some cases, is a little produced externally.

In the males of the chondropterygii, there are certain organs situated near the anus, consisting of bone, cartilage, and muscles. These were long regarded as the external organs of reproduction. But RONDELETIUS was of opinion that these were only accessory organs, and enabled the males to retain the females more closely during coition. The celebrated ichthyologist BLOCH, from dissections, arrived at the same conclusion. They are termed *claspers.*

The male organs of the sharks and rays, are such as we have now described; but few accurate observations have been made on the male organs of these ovoviviparous fishes, which belong to the branchiostegous and osseous tribes, such as the syngathus, blennius, and muræna.

In the *females* of the sharks and rays, the ovaria, two in number, are situate at the sides of the spine, and contain ova of different sizes. From each of these proceeds an oviduct, the anterior extremities of which are united to the diaphragm and spine. Internally, these ducts are covered with glandular papillæ, and pass through a large glandular body, after which, they dilate into a large sac, which is the uterus.

When the ova pass into the oviduct, they are carried to this glandular body, which is supposed to secrete the glaire or albuminous part, and afterwards conveyed to the uterus, where they receive the shell. It is not determined at what period the egg becomes impregnated, or in what manner the operation is performed.

Some of the species of this group are ovoviviparous. The eggs, while in the uterus, are enveloped in a membranaceous bag, having the vacant spaces filled with a transparent jelly. When the young are sufficiently developed, the yolk-bag is taken into the belly, and parturition takes place.

In other species of this group, which are oviparous, the eggs are of a depressed quadrangular form, pointed or produced at the corners, and protected by a coriaceous shell. The fœtus is aërated in these, by means of two slits in the shell, for the admission and escape of the water *.

Oviparous fishes which are hermaphrodite.—Instances of hermaphroditism among fishes, were, for a long period, considered rare, and always as accidental. BASTER detected such an arrangement in the whiting, and DUHAMEL observed the same in the carp. But it was reserved for that able anatomist, Sir EVERARD HOME, to point out a parti-

---

* Sir E. HOME's paper " *On the mode of breeding of the ovoviviparous Shark.*" Phil. Trans. 1810, p. 205.

cular tribe of fishes, in which the organs of both sexes are always present in the same individual.

Having been unsuccessful in obtaining any male lampries, although he got what were considered as females in abundance, Sir EVERARD began to suspect, that the individuals of the species were hermaphrodites, and his observations on these fish, at different periods, justified his conjectures *.

---

* " I found upon examination, that the two glandular bodies projecting into the belly, one on each side of the ovarium, which have always been supposed to be the kidneys, varied very much in size and appearance at the beginning and end of the season. When the ova are so small that the animal is reputed to be a male, these glandular bodies, and the black substance upon which they lie, appear to form one mass, and the duct upon the anterior part is thin and almost transparent, containing a fluid equally so; but, in the end of May, when the ova increase in size, these glandular bodies become larger, more turgid, and have a distinct line of separation between them and the black substance behind; their structure is more developed, being evidently composed of tubuli running in a transverse direction, and the ducts leading from them are thicker in their coats, and larger in size.

" On the 5th of June, the ova were found to be of the full size; and a small transparent speck, not before to be observed, was seen in each; at this time, the tubular structure had an increased breadth, and the duct going from it contained a ropy fluid, which, when examined in the field of the microscope, was found to be composed of small globules in a transparent fluid. On the 9th of June, neither the ova nor the tubular structure had undergone any change. On the 11th of June, the ova were of the same size, but the slightest force detached them from the ovarium; the tubular structure had increased still more in size, the fluid in the ducts was thicker, more ropy, and when water was added to it in the field of the microscope, it coagulated, and what was before made up of globules, had now the appearance of flakes. The ova do not pass out at an excretory duct as in fishes, but drop from the cells in the ovarium in which they were formed, into the cavity of the abdomen, and escape by two small apertures at the lower part of that cavity, into a tube common to them and to the semen in which they are impregnated." " In the lampern or pride, and in the gastrobranchus cœcus, a similar structure is observable." Phil. Trans. A. D. 1815, Part II. p. 266.

His observations leave little room to doubt that the animals in question are hermaphrodites; without, however, determining at what precise period, or in what position, the eggs are impregnated.

Although the sexual organs of fishes had been long known, it was not until the middle of the 18th century that any experiments were performed to ascertain the effect of their abstraction. TULLY appears to have been the first person who performed the operation, and an account of his experiments has been published in the Gent. Mag. vol. xxv. p. 416, and in Phil. Trans. vol. xlviii. p. 870. When the abdomen of the fish is laid open, the milt or roe carefully removed, and the wound sewed up again, the fish appears to experience but little pain, and the wound heals in a few weeks. These experiments have frequently been performed on the carp, and they are attended with little risk. The fish grows to a large size, and its flesh is said to have a more delicate flavour. But castration has never come into general use among the proprietors of fish ponds, being seldom performed but from motives of curiosity or science.

We have already stated, that the impregnation of the egg takes place, in many species, after exclusion, and the experiments which have been conducted to establish this point, have likewise made us acquainted with the existence of *hybrid fishes*. Even in a common fish pond, where carp and trout are permitted to live in company, the carp sometimes impregnates the eggs of the trout, or the trout those of the carp. The limits, however, within which this irregularity is confined, have never been investigated with care.

Fishes exhibit very remarkable differences in regard to the *number of eggs* which they produce. The rays and sharks seem to prepare but a very limited number. RON-

DELETIUS states the number in the squalus acanthias at six; other observers have found in other species 26 and even 30. But the number of eggs in other kinds of oviparous fish exceeds almost our powers of reckoning *.

There is no regular proportion between the weight of the fish and the weight or number of eggs produced. Nor is there any estimated proportion between the number of eggs deposited, and the number of fish which arrive at maturity. The eggs are eagerly sought after by other fishes, by aquatic birds and reptiles. In the young state, they are pursued by their own species, as well as by beings belonging to other classes.

The season in which fish deposit their eggs varies according to the species, and even the habit of the individual. It is well known, that among salmon, even in the same river, a difference of some months is observable, and we believe that the same remark is applicable to all other kinds of fish. In general, before spawning, fish forsake the deep water, and approach the shore, that the roe, being placed

---

* The following Table may convey to the general reader some idea of their prolific powers, as extracted from HARMER's Remarks on the Fecundity of Fish, Phil. Trans. 1767, p. 280.

Fish.	Weight.		Weight of spawn.	Number of eggs.	Season.
	Oz.	Dr.	Grains.		
Carp, - -	25	5	2571	203109	April 4.
Cod-fish, -	0	0	12540	3686760	December 23.
Flounder, -	24	4	2200	1357400	March 14.
Herring, -	5	10	480	36960	October 25.
Mackrel, -	18	0	$1223\frac{1}{2}$	546681	June 18.
Perch, -	8	9	$765\frac{1}{2}$	28323	April 5.
Pike, - -	56	4	$5100\frac{1}{2}$	49304	—— 25.
Roach, -	10	$6\frac{1}{2}$	361	81586	May 2.
Smelt, - -	2	0	$149\frac{1}{2}$	38278	March 21.
Sole, - -	14	8	$542\frac{1}{2}$	100362	June 13.
Tench, -	40	0	—	383252	May 28.

in shallow water, may be vivified by the influence of the solar rays *. At that season, some fish forsake the salt water, ascend rivers, and, after spawning, retreat again to the ocean. Such were formerly called *Anadromi*.

The eggs of various species of fish belonging to the oviparous order, with distinct sexes, are used as articles of food. Where circumstances permit, they are consumed while in a recent state. In other situations they are salted, and form the well known article of trade called *caviar*.

The characters which the organs of reproduction furnish, in the discrimination of species, have been hitherto much neglected. Connected as they are with the existence of the animal, and exercising a powerful controul over its habits, they ought to be examined with care, and their appearances recorded in detail.

Few accurate observations have been made to determine the *age* of fishes. The element in which they reside, is supposed to preserve them from the pernicious influence of sudden changes of temperature; the slowness of the process of ossification, the coldness of their blood, and the tardiness of all their primary movements, are considered as indicating a lengthened existence. Accordingly, we find the age of the carp has been known to reach to 200 years, and of the pike to 260. The marks, however, by which the age of fishes may be determined, remain yet to be discovered.

Fishes are greatly tormented with intestinal worms. The common stickeback may be quoted as a remarkable instance. Its death is often occasioned by the increase of the *Tænia solida* of GMELIN; and it is even supposed, that this fatal

---

* The practice of cutting the sea-weeds for the manufacture of kelp, has deprived many species of a suitable situation for the deposition of the spawn, and the fry of the requisite protection. Hence the universal complaint, along the shores of Scotland, that fish are now less abundant than formerly.

issue follows the completion of the function of reproduction with regularity *.

Fishes exhibit remarkable differences with respect to their *vivaciousness*. Thus, some fishes expire almost the instant they are taken out of the water, as the herring and smelt; others are capable of surviving hours, and even days, when removed from their native element, as the eel, carp, and some others. It sometimes happens, that vivacious fishes are conveyed to a distance by birds, and left, without being killed, on rocks or fields. This has given rise to many of the absurd stories which have been told of showers of fishes. RONDELETIUS observes with propriety, that those fishes in which the gill openings are but imperfectly covered, expire soonest when taken from the water; and those fish having branchiæ, protected by a gill-lid which shuts close, or by a narrow opening, are most vivacious. The air soon dries the fine plumes of the gills, and obstructs the process of respiration and of circulation.

It is seldom that a fish is permitted to die a natural death from old age. During every period of its existence, it is surrounded by foes; and when no longer able to exercise its wonted watchfulness, or exert its powers of defence, it falls an easy prey to its more powerful adversaries. In a domesticated state, previous to death, the dorsal fins lose the power of maintaining the body in a vertical position, the levity of the belly, and the extraordinary distension of the air bag, reverse the natural position, so that the back becomes undermost, and the body floats on the surface. Similar appearances present themselves, when the waters are contaminated by noxious mineral or vegetable impregnations.

We have already stated, that fishes naturally reside in the water; but as this element is found to differ in its con-

---

* Annals of Philosophy, February 1816.

stitution and temperature according to its situation, we may expect to find the finny tribes that dwell in it, influenced by these circumstances. At a very early period, the diversity in the distribution of fishes attracted the attention of observers. RONDELETIUS at last attempted a division of this class of animals, from the different situations in which they are found, into marine, fluviatile, lake, and pond fish. It will, however, be more suitable to our present purpose, to consider them as inhabitants of the sea or of fresh-water.

The salt-water fishes are much more numerous than those which reside in fresh-water. They cannot be distinguished from fresh-water fishes by any peculiarity of structure, or external form. They are always found in the greatest numbers in tideways, and on those banks which are formed at the junction of opposite currents. They in general resort to a certain kind of bottom, in which we may suppose they find a plentiful supply of food. Some are always found near rocky shores, while others prefer the sandy bays. Some are found only in the open ocean, and are termed *pelagic;* others keep within a short distance from the coast, and are termed *littoral*.

The fresh-water fishes are not so important, in an economical point of view, as those which inhabit the ocean. Some species frequent rivers, and seem to require, for the preservation of their health, a continued current of water. Others live in lakes, and seem contented to spend their days where the water is still. Like salt-water fishes, they appear to prefer particular altitudes; and in ascending mountains, we may observe that the fish in the lakes and rivers have their boundaries, as well as the vegetables which cover their surface. Thus, WAHLENBERG found, that the pike and perch disappeared from the rivers of the Lapland Alps along with spruce fir, and when 3200 feet below the line of perpetual snow. Ascending 200 feet higher, the gwiniad and the grayling were no longer to be found in the lakes. Higher up still, or about 2000 feet below the line of perpetual snow, the char had disappeared; and beyond this boundary all fishing ceased *.

When a salt-water fish is put into fresh-water, its motions speedily become irregular, its respiration appears to be affected, and unless released it soon dies. The same consequences follow when a fresh-water fish is suddenly immersed in salt-water. But in what manner they are influenced by the change, has never been satisfactorily determined.

There are not a few fish which may be said to be *amphibious*, or capable of living either in fresh or salt water at pleasure. Such fish, in an economical point of view, are extremely valuable, as they furnish to the inhabitants of this and other countries an immense supply of food. The salmon may be given as an instance in this country, where, from one river (Tay), 50,000 head of full sized fish have been procured in one season. To the Greenlanders, their Angmarsæt, or *Salmo arcticus*, is perhaps more valuable, as it is dried hard, then broken and pounded, and formed into bread, as well as consumed in a fresh or salted state.

All these fishes seem to reside chiefly in the sea. There they grow and fatten; but when the time of spawning approaches, they forsake the salt-water, and return to rivers and lakes. But this desertion of the ocean is only temporary, and regulated by the circumstances connected with reproduction. The instant the spawning is finished, they repair again with equal rapidity to the ocean, to repair their exhausted strength, and fit them for obeying the laws of their existence. Some of these fishes appear to be capable

---

* See appendix to the 2d vol. of the " Lachesis Lapponica" of LINNÆUS, by Sir J. E. SMITH. London, 1811.

of living exclusively in fresh water, when confined in a lake or river *.

The circumstance of some fish being capable of living either in fresh or salt water, has suggested the idea of attempting to modify the constitution of salt-water fishes, so as to enable them to subsist in fresh water. If the change is attempted to be produced in young fish by degrees, and with caution, the experiment may prove successful, especially with those fish that reside chiefly near the sea-shore. But in the case of fishes which live in deep water, a change not only in the respiratory organs must be produced, but likewise in those of digestion, as they must subsist on a new kind of food. We regard such experiments as curious, but can scarcely bring ourselves to believe that they will be productive of advantage to society.

We possess but few accurate observations on the distribution of fishes, with respect to *temperature*. Living in an element subject to little variation from the change of the seasons, fishes, like sea-weeds, have an extensive range of latitude as well as longitude through which they roam. But they appear to abound in the greatest variety of species in the equatorial regions, and to diminish in numbers with regard to species as we approach the poles. In this country we may observe a certain arrangement of some of the species with respect to latitude. Thus, the fresh-water fishes

---

* The Honourable DAINES BARRINGTON, gives evidence of this in the case of the smelt and the grey mullet.—Phil. Trans. p. 312. We are likewise informed in the Statistical Account of the parish of Lismore (Stat. Acc. vol. i. p. 485.), " That about 50 or 60 years ago, there were some sea-trouts carried to these lakes, the breed of which preserve their distinction perfectly clear to this day. They retain their shining silver-scales, though they have no communication with the sea; their flesh is as red as any salmon, and their taste is totally different from that of the yellow-trout.

of England are much more numerous than those in Scotland. In the sea at the south of England, the pilchard is found in abundance, while it is rare in Scotland. In the seas of the north of Scotland, the tusk (*Gadus brosme*) abounds, in the south of Scotland it is very rare, and in England it is unknown.

The investigation of those revolutions which fishes have experienced since the formation of the globe, is attended with peculiar difficulty. The external form, on which in general the specific distinction is founded, is destroyed by pressure. All distinct traces of the softer parts have disappeared, and the geognost is left to draw his conclusions from the form of the teeth, or the outline and structure of the skeleton. Hence, the conclusions which have been drawn respecting the particular species should be received with caution. In the newer rock-formations, which have been termed *local*, such as the strata at Eningen, the remains of fishes have been observed, belonging to existing races, and still natives of the neighbouring lakes. But, in the rocks of those formations which are called *universal*, the skeletons of fishes which have been found, in all probability, belong to species now extinct. In examining the organic remains which we consider of this sort, it would appear, that the teeth of unknown sharks are more numerous than those of any other description of fish. They are found in all the floetz limestones of this country, in company with the ancient camerated shells. Vertebræ of osseous fishes are chiefly found in the strata connected with the chalk-formation, seldom in those of an older date.

Migration of Fishes.—Those fishes which enter rivers for the purpose of spawning, perform their migrations annually, but do not appear at any very precise period. Their motions appear to be regulated by the condition of their generative organs, and these are in their turn controlled by the temperature of the water in which the fishes remain, or the supply of food. In rivers where salmon spawn, it is observed that they continue entering the river for the space of seven or eight months. Those marine fishes, such as the herring, pilchard, and many others, which leave the deep-water, and approach the shores for the purposes of spawning, are equally irregular, with respect to their periods of appearing and disappearing.

Besides these movements, which depend on the generative impulse, many marine fishes appear to migrate from one shore to another, influenced by laws which have never been satisfactorily explained. Thus, haddocks have been known to visit a coast for many years in succession, and then suddenly to disappear, leading off at the same time all those predacious fish which fed upon them. Perhaps these movements may depend upon the supply of food, and be regulated by circumstances over which we can exercise no controul. Accurate observations, however, would probably ascertain the limits of these migrations, and enable us to derive advantage from motions which at present we regard as calamitous.

In tracing the history of those attempts which have been made to subject this portion of the creation to our controul, we trace at the same time the progress of civilization and luxury. In Egypt, the inhabitants had their sluices and their fish-ponds in the days of Isaiah (chap. xix. 10.), and from this early seat of the arts and sciences, the Romans probably acquired the knowledge of rearing and feeding fish. During the more prosperous days of that refined people, almost every wealthy citizen had his fish-ponds. In modern times, the Chinese bestow more attention on the cultivation of fish, than perhaps any other nation. And in Europe, the importance of the subject has been duly appreciated by the Swedes, Prussians, and Germans In the latter countries.

a considerable part of the revenue from property is derived from the carp-ponds.

In general, the rearing of fresh-water fish in artificial ponds has hitherto been chiefly attempted; few trials having been made to rear the salt-water fish in confinement. In the construction of a pond for fresh-water fish, care should be taken to have a regular supply of water, free from mineral impregnations, to cover the deepest part of the pond at least six feet. The more extensive the shallow ground at the sides is, especially if it be covered with marsh plants, so much more abundant is the supply of those minute animals, on which many fish chiefly subsist. Care should likewise be taken to introduce those small fish, which, by multiplying, may furnish a constant supply of food.

When fish-ponds are formed, it is, in general, the wish of the proprietor to have a certain number of his stock in good condition, that he may have a regular supply for his table. For the accomplishment of this object, there is usually one pond set apart for the purpose, into which are introduced those full grown fish which he wishes to feed. During the winter season little food is required; but along with the heat of spring, fishes acquire a keen appetite, and at that period a constant supply of food should be given them. They should be fed morning and evening at a stated time, and always at the same place in the pond. The food should consist of any kind of corn, boiled or steeped in water for some time until it swells. Malt is esteemed a very fattening food, and the crumbs of bread steeped in ale; but peas are considered as little inferior to either. Pikes must have an abundant supply of eels, otherwise they require a long time to fatten. Some recommend the laying of dead carrion upon stakes over the surface, that it may breed maggots, which, falling into the water, furnish an abundant supply of very acceptable food.

In the construction and management of fish-ponds, there are many circumstances of a local nature which it is impossible here to specify *. The methods employed to stock these ponds are at present more deserving of our attention. The first, and certainly the most obvious method, is to obtain living fish from similar situations. In catching these, the utmost care should be taken not to bruise them, or to rub off their scales, and to keep them as short a time out of the water as possible. The vessels in which they are to be carried should be full of water, as when the barrel is not entirely full, the fish are liable to be driven by the currents against the lid or sides. This transportation should take place only in cold weather, and in the winter season (as fishes can bear cold better than heat), and should be performed with as much expedition as circumstances will permit.

The second method of stocking fish-ponds, is, in some respects, preferable to the preceding, especially when the waters are at a distance from which the supply is to be derived. This consists in ascertaining those places in which the spawn of the wished-for species is deposited, and conveying the impregnated eggs to a similar situation in the new ponds. In this manner a vast number of individuals may be obtained at once, and with great certainty of success, provided they are supplied during the journey with fresh water, and but little agitated. The impregnated eggs may be known by a small aperture, which may be detected on one side by means of a good microscope, and which is scarcely perceptible previous to impregnation. By means of this method, however, a much longer period must elapse before fish are procured for the table than by the former,

---

* See Phil. Trans. 1771, p. 310.; and North's History of Esculent Fish, Lond. 4to. 1794.

although this objection is in a great measure obviated, by securing, from the eggs, a race of fish with constitutions accommodated to your waters.

The last method, which hás been rather absurdly termed *artificial fecundation*, we owe to the ingenuity of M. Jacobi, (*Mem. de l'Acad de Berlin*, 1764, p. 55.) It is founded on a knowledge of the mode of reproduction in oviparous fishes, and in its turn serves to illustrate the function of generation in fishes. In those places where the fish are easily taken, a female is secured, with roe nearly ready for exclusion, and a proper box with water being prepared, the fish is held by the head, with its tail downwards, and gently squeezed on the belly. The eggs, which are perfect, readily run out into the vessel. A male fish is next procured, and being held in a similar situation, the milt is poured upon the eggs. The eggs thus impregnated, are conveyed to a proper situation as in the second method, and protected from enemies.

The advantages which result from the translation and feeding of fishes, though felt and appreciated in other countries, have been, in a great measure, overlooked in our kingdom. In Scotland and Ireland, and we may likewise include England, there are multitudes of ponds and lakes, which are at present mere useless wastes, but which, if properly stocked with fish, would greatly contribute to the prosperity of the country, by furnishing an additional supply of food. To our forefathers we owe the naturalization of two useful species of fish, namely, the *carp*, which was translated (probably from France or Spain) into England, about the year 1496, and the *pike*, which was introduced about the beginning of the fifteenth century. The gold and silver fishes of China have likewise been naturalized in England, as objects of beauty. We wish this catalogue had been more extensive ; we fondly hope that it will soon increase.

The formation of ponds for salt-water fish, has often been the subject of speculation, but in few instances has it ever been reduced to practice. Indeed the motives for constructing such a pond must originate chiefly in curiosity, as those who are situate on a sea-coast, where such ponds can only be constructed, have access to that great storehouse of life, and may at all seasons derive from it an inexhaustible supply. Besides, there are few situations favourable for the construction of such a pond, and even where most favourable, an expensive barrier must be constructed to separate it from the sea. Some ponds of this kind have been constructed in Scotland. These are well described by Mr NEILL in the Scots Magazine for June 1816, p. 412. *

---

* " A good many years ago, a small fish-pond, into which sea-water could be easily introduced, was constructed by an enterprising individual at Peterhead, in Aberdeenshire, (Mr Arbuthnot). A few sea-fish were occasionally kept in it ; but it soon fell into disuse, and it has of late been neglected. This, however, was, as far as we know, the first attempt of the kind in this country.

" Since that time two sea-fish ponds, of greater dimensions, have been formed by private gentlemen in Scotland, for the conveniency of supplying their families. One of these is at Valleyfied, the seat of Sir ROBERT PRESTON, Bart., on the shore of the Frith of Forth ; the other is situate in Wigtonshire, in an inlet called Portnessock, on the peninsular ridge of country called the Rins of Galloway, nine or ten miles south from Portpatrick, and is the property of Mr MACDOWALL of Logan.

" At the spot where it is formed, there had originally been a small natural basin, communicating with the sea by means of a narrow sinuous fissure, or perhaps an empty vein in the rock. This basin has been enlarged and deepened, by working away the solid rock, which is grey-wacke slate. At flood-tide, the water covers, to the depth of two or three feet, a ledge or walk which passes round an interior or deeper pond, and, at this time, allows tolerably ample space for the rapid motions of the fish,

FISHES, in a domesticated state, are subject to various maladies, the cause and cure of which have not been suc-

---

" The pond is replenished with fishes by the keeper, whose house is hard by. In easy weather, this man rows out in his fishing coble, to the mouth of Logan Bay, in which the inlet of Portnessock is situate. For catching the fish, he uses the common hand line, and the usual baits. He is provided with a wide tub, into which he puts a convenient quantity of sea-water: to this tube he immediately commits such part of his capture as happen to be little hurt by the hook. He finds it necessary, during summer, to cover the tub with a cloth; and in sultry weather he experiences difficulty in keeping the fishes alive in the tub till he reach the shore. This, it seems evident, cannot be ascribed either to mere heat, or to the exhausting of the air contained in the water, by the respiration of the fishes. In all probability, it depends on the influence of the electric fluid of the atmosphere. DE LA CEPEDE, in his essay on the culture of fresh-water fishes, particularly mentions the powerful effect of this fluid on them, when confined in small portions of water, in the course of their transference from one place to another.

" As might naturally be supposed, the fishermen prefer for the pond young fish, or at most those of middle size, to those of large growth. In selecting cod-fish, for example, he rejects all that exceed 6lb., giving the preference to what he styles *lumps*, or young cod-fish, weighing 4lb. or 5lb. In the pond, the fish are not only preserved alive till wanted for use, but, being regularly fed, are found to be fattened. They are taken for use, however, merely by the line and hook; and it is probable, that the fish in best condition will not always be the first to catch at the bait.

" The fishes we observed in the pond were the following:

" 1. Cod, (*Gadus morhua*). They were lively, and caught greedily at shell-fish, which we threw into the pond. They kept chiefly, however, in the deep-water, and, after approaching with a circular sweep, and making a snatch at the prey, descended out of sight to devour it. It has often been doubted, whether the red-ware codling of Scotland was the young merely of the common cod, or a distinct species, *Gadus callarias*. Here one would think the question might easily be decided. Upon describing this red-ware codling, we were assured that it occurs on the coast of Galloway, and that it had sometimes been caught and placed in the pond; but that, after a year, it became as large and as pale in colour as a common grey cod. This accords with our own observations made in less favourable circumstances.

" 2. Haddock (*G. æglefinus*). These, contrary to expectations, we found to be the tamest fishes in the pond. At ebb tide they come to the inner mar-

cessfully ascertained. Trouts, carps, and perches, are subject to various cutaneous diseases. During severe winters, when

---

gin, and eat limpets from the hand of a little boy, the son of the keeper. They appeared white, and rather sickly. One was diseased about the eyes.

"3. Coalfish, (*G. carbonarius.*) Some of these were of a large size, exceeding in dimensions the largest cod in the pond. No fish has received so many different names as the coalfish. When young, it is called at Edinburgh, podley; in the northern islands, sillock; in Galloway, blochan. When a year old, it is styled cooth or piltock, in the north; and glasson in the south-west of Scotland. When full grown, it is named sethe in the north, and stenlock in the south-west. Accordingly, we were now told, that these stenlocks were mere blochans when they were put in. They were become of a fine dark purple colour. They were bold and familiar, floating about slowly and majestically, till some food was thrown to them: this they seized voraciously, whether it consisted of shell-fish or ship biscuit. We were informed, that they, too, occasionally approach the margin, and take their food from the keeper's hand.

"4. Whiting, (*G. merlangus.*) These were scarce in the pond, and very shy.

"5. Pollack, (*G. pollachius.*) This was pretty common, and has been found to answer very well as a pond fish. It is generally called Layde or Lythe.

"Besides these five species of gadus, we are told that the ling (*G. molva,*) had occasionally been kept in the pond.

"6. Salmon. (*Salmo salar*). This was the wildest and the quickest in its motions of all the inhabitants. When a mussel or limpet, freed from the shell, was thrown on the surface of the water, the salmon very often darted forward and took the prey from all competitors, disappearing with a sudden jerk and turn of the body. I suspected this to be the salmon-trout (*S. truta*); but was assured that it was the real salmon, which is occasionally taken in the bay.

"7. Flat-fish or flounders, of two sorts, were also in the pond; but they naturally kept at the bottom, and we did not see them. From the description given by the people, we concluded that they were dab and young plaise.

"The food given to the fishes consists chiefly of sand-eels and of shell-fish, particularly limpets and mussels. In the herring-fishery season, they cut herrings in pieces for this purpose.

"It is remarkable, that all the kinds of sea-fish above enumerated, seem to agree very well together. No fighting had ever been observed by the keeper, and seldom any chasing of one species by another.

the surface of the ponds in which they are kept are frozen over, the various kinds of fish seem to contract diseases, and, in such cases, great mortality often prevails. This seems to arise from want of air in the water, and can only be prevented by removing the fish to a deeper pond, through which there is a constant current. In some rigorous seasons, the extent of this mortality is most alarming, an example of which occurred between 1788 and 1789, when the inhabitants of some districts of France lost nearly all their stock of carp, pike, and tench. *Journal de Physique*, November 1789. *

" None of the fish have ever bred ; indeed, no opportunity of breeding is afforded to them. A warm and shallow retreat, laid with sand and gravel, would have to be prepared for some species ; and large stones, with sea-weed growing on them, would have to be transferred to the pond, and placed so as to be constantly immersed in the water for the use of others. The dimensions of the present pond, however, are two circumscribed to admit of its being used as a breeding place. An addition for this purpose might, without much difficulty, be formed, and here some curious observations might be made. The spawn of various sea-fishes is frequently accidentally dredged up by fishermen, and, could, therefore, no doubt, be procured by using a dredge : its degree of transparency indicates whether it will prove prolific. This might be placed in a protected corner of the breeding pond, and its progress watched. On this branch of the natural history of sea-fishes, little is known."

* In the beginning of the year 1789, Mr Baker states, that carp were taken out of a pond where the ice was broken, frozen crooked and stiff, without the least motion, and ice hanging about them ; but being laid on dry straw in a cellar they all recovered."—Phil. Trans, 1791. p. 92. In the same year, an epidemic distemper affected even those fish which live in the sea, as the following fact, communicated by the late Mr Creech of Edinburgh, in the Appendix to the sixth volume of the *Statistical Account of Scotland*, satisfactorily proves : " On Friday, 4th December 1789, the ship Brothers, Captain Stewart, arrived at Leith from Archangel. The captain reported, that on the coast of Lapland and Norway, he sailed many leagues through immense quantities of dead haddocks floating in the sea. He spoke several English ships, who reported the same fact." Other evidence is stated by the Rev. Cooper Abbs. Phil. Trans. 1792, p. 367.

Fish, considered as an article of food, is regarded as light, and easily digested, and, therefore, well suited for the young, the weak, and the sedentary. But for the same reason, it is unsuitable food for those engaged in laborious occupations. Among the Romans, he who fed on fish was regarded as effeminate. It has often been considered, though, perhaps, without cause, as tending to promote the fertility of the human species; and the immense population of China has been ascribed to the abundant use of this kind of nourishment. Its tendency, however, to encourage diseases of the skin appears to be universally acknowledged, and is, indeed, very evident in the remote islands of this country, as also of Faroe, of Iceland, and of Norway, where fish forms so great a proportion of the food of the inhabitants.

For dietetical uses, fishes have frequently to undergo some sort of preparation, varying according to the situation, the necessities, or the taste of the consumers. Where circumstances permit, they are in general used in a *fresh* state; and even in large cities, where the supply must be brought from a distance, various expedients are resorted to, to prevent the progress of putrefaction. By far the best contrivance for this purpose is the well-boat, in which fish may be brought to the place of sale even in a living state. Placing the fish in boxes, and packing with ice, is another method, and has been extensively employed, particularly in the supply of the capital with salmon.

In many maritime districts, where fish can be got in abundance, a species of refinement in taste, or at least a departure from the simplicity of nature, prevails, to gratify which, the fish are kept for some days, until they begin to putrefy. When used in this state, they are far from being disagreeable, unless to the organs of smell. Such fish are termed by the Zetlanders *blawn-fish*.

Where fish are to be procured only at certain seasons of the year, various methods have been devised, to preserve them during the periods of scarcity. The simplest of these processes is to *dry* them in the sun. They are then used either raw or boiled, and not unfrequently, in some of the poorer districts of the north of Europe, they are ground into powder, to be afterwards formed into bread.

But by far the most successful method of preserving fish, and the one in daily use, is by means of salt. For this purpose they are packed with salt in barrels, as soon after being taken as possible. In this manner are herrings, pilchards, cod, and salmon preserved, as well as many other kinds of esculent fish.

The fish, in many instances, after having been salted in vessels constructed for the purpose, are exposed to the air on a gravelly beach, or in a house, and dried. Cod, ling, and tusk, so prepared, are termed in Scotland *salt-fish*. Salmon in this state is called *kipper*; and haddocks are usually denominated by the name of the place where they have been cured.

After being steeped in salt, herrings are in many places hung up in houses made for the purpose, and dried with the smoke of wood. In this state they are sent to the market, under the name of *red-herrings*.

Although salt is generally employed in the preservation of fish, whether intended to be kept moist or to be dried, vinegar in certain cases is added. It is used, in this country at least, chiefly for the salmon sent from the remote districts to the London market. It can only, however, be employed in the preservation of those fish to which this acid is served as a sauce.

The flesh of any fish is always in the highest perfection, or *in season* as it is called, during the period of the ripening of the milt and the roe. After the fish has deposited the spawn, the flesh becomes soft, and loses a great deal of its peculiar flavour. This is owing to the disappearance of the oil or fat from the flesh, it having been expended in the function of reproduction *. When in season, the thick muscular part of the back, as it contains the smallest quantity of oil, is inferior in flavour, or richness, to the thinner parts about the belly, which are esteemed by epicures as the most savoury morsels.

There are some kinds of fishes, especially those which inhabit the shores of warmer countries, which are reputed *poisonous*. These are, the Tetraodon ocellatus, sceleratus and lineatus, the Sparus pagurus, and a few more. It is generally supposed, and with some probability, that the poisonous quality of these fish proceeds from the food on which they have subsisted. This conjecture is supported by the history of the mussel and the oyster, which owe their occasional noxious qualities to the zoophytes on which they feed. In some cases, the poisonous portion is the liver; in others the whole body, when used after spawning. Perhaps the poisonous quality of these fishes might be considerably diminished, if not entirely removed, were the intestines carefully taken away, and the fish placed for a short time in salt-brine *.

---

* The superiority of *deep-sea* herrings over those caught near the shore and in bays, arises from this circumstance. The former are fat, while the latter have either recently spawned, or are nearly ready for spawning, and, consequently, lean.

† Phil. Trans. 1776, p. 544.

The most convenient distribution of fishes which has yet been proposed, is that which depends on the conditions of the osseous system. In one division, denominated *Cartilaginous*, the bones are soft, and destitute of fibres; while in the other, termed *Osseous*, the bones are more solid, and of a fibrous structure.

ORDER. I.

Cartilaginous Fishes.

In this group, the cartilaginous basis seems to prevail, and the calcareous matter is never distributed in perceptible filaments. The cranium forms only one connected plate destitute of sutures, and all the articulations of the different parts of the skeleton are indistinct. The maxillary and intermaxillary bones are wanting or imperfect, and their place is supplied by a development of the palatine bones and vomer.

1st Subdivision.

Chondropterygious Fishes.

Branchiæ fixed.

1st Tribe.

Lips fitted to act as a sucker.

The skeleton of the fishes of this tribe, is very imperfectly developed. The lips are strengthened by a cartilaginous ring. The branchiæ are placed on the walls of lenticular cavities. The labyrinth of the ear is in the interior of the cranium. The nostrils open by one aperture.

A. Seven branchial apertures on each side, and the same number of cavities.

(1.) *Maxillary ring armed with teeth.*

1. Petromyzon. Lamprey. Margin of the mouth destitute of a beard. *P. marinus* and *fluviatilis.*

2. Homea. Margin of the mouth bearded.

I have ventured to name this genus in honour of Sir EVERARD HOME, who has so successfully investigated the aërating and reproductive organs of the tribe to which it belongs, and who has pointed out its distinguishing internal characters *. The trivial name is due to the late illustrious BANKS, by whom the species was brought to this country from the South Seas. *H. Banksii.*

(2.) *Maxillary ring without teeth.*

3. AMMOCETES. Lip semicircular. *A. branchialis.*

B. Two branchial apertures on each side, and internally only six cavities.

4. MYXINE. This is the Gasterobranchus of BLOCH. *M. glutinosa.* The *Gasterobranchus Dombey* of LACEPEDE, is probably a Homea.

2d Tribe.

Lips unfit to act as a sucker.

This tribe comprehends the old genera Chimæra, Squalus, and Raia. The palatine and post-mandibular bones form the jaw, for the support of the teeth. This jaw is either

---

* " In an animal brought from the South Seas by Sir JOSEPH BANKS, intermediate between the lamprey and myxine, but differing so much from both, as to form a distinct genus, the respiratory organs resemble those of the lamprey in the number of the external openings, and the number of bags; but these organs, and many other parts, differ in the following particulars, in which they agree with those of the myxine. There is no appearance, whatever, of thorax, nor is the pericardium cartilaginous; the bags are flattened spheres placed perpendicularly, their cavities are small, their coats elastic, and the internal orifices communicate directly with the œsophagus, which is small. The œsophagus does not terminate in a valvular slit, but in a loose membranous fold; there are two rows of teeth on each side of the tongue, bent downwards, long, and pointed. There is a posterior nostril, and an appearance resembling an uvula. There is a gall bladder, a row of large mucous glands on each side of the belly, and there is a mesentery to the intestine." Phil. Trans. 1815, p. 258, tab. xii. f. 1.

joined directly to the cranium, or by the intervention of a single bone or stalk, to which the os hyoides is also attached.

A. Pectoral fins free, not coalescing with the snout. The branhcial openings are at the sides.

(1.) *Mouth terminal. Eyes with a dorsal aspect.*

5. SQUATINA. Body depressed. No anal fin. Temporal orifices. *S. vulgaris,* or angel shark.

(2.) *Mouth under the snout. Eyes with a lateral aspect.*

a. Snout produced, and armed with lateral teeth.

6. PRISTES. Sawfish. Teeth of the snout regular, and imbedded in sockets. *P. antiquorum,* a species which the late Dr WALKER in his MS. says, " is found sometimes in Loch Long."

b. Snout abbreviated and unarmed.

This group comprehends the great family of SHARKS, the external and internal characters of which have been very imperfectly investigated.

(A.) Furnished with temporal orifices.

(I.) With an anal fin.

a. With two dorsal fins.

(AA.) Teeth conical or pointed. In the two first genera, the anterior dorsal fin is situated nearly above the pectoral fins, while in the third genus, it is placed nearly over the ventral fin.

6. SQUALUS. Branchial opening embracing nearly the circle of the neck. Teeth without notches. Tail forked. *S. maximus,* or basking shark, and *S. Lelanonius,* or Lochfine shark.

7. GALEUS. Tope. Branchial opening short. Teeth notched. Tail-fin irregular. *G. vulgaris,* the *Squalus Galeus* of LIN.

8. SEYLLIUM. Nostrils near the mouth, with a groove to the lips. Teeth notched. *S. catulus,* and var. *canicula.*

The observations of Sir E. Home, indicate this species to be ovoviviparous *.

9. Cestracion. (Cuvier.) A spine in front of each of the dorsal fins. *C. Phillippi.*

(bb.) Teeth blunt and closely set.

10. Mustelus. Dorsal fins unarmed. *M. vulgaris.*

b. With one anal fin.

11. Notidanus. Resembling the genus Galeus. *N. griseus.*

(II.) Destitute of an anal fin. Two of the genera have the dorsal fins armed with spines, the third is unarmed.

12. Spinax. Teeth small, with a cutting edge. The dorsal spines in front of the fins. *S. acanthias.* This species is ovoviviparous †.

13. Centrina. Under teeth edged, upper ones pointed. The dorsal spines contained in the fins. *C. vulgaris.* Block, Ichthyologie, t. 115.

14. Scymnus. This includes the *Squalus Americanus* of Gmelin.

(B.) Destitute of temporal orifices.

15. Carcharias. Snout conical. This genus includes the following British species. *C. vulgaris*, White Shark, *C. vulpis*, Thresher. *C. glaucus*, Blue Shark, and *C. cornubicus*, Probeagle Shark.

16. Zygana. Balance Fish. Snout truncated, and spread out on each side like the head of a hammer. The eyes are placed terminally on these productions. *Z. Blochii.* Bloch, tab. 117.

A. Pectoral fins coalescing with the snout. The body is depressed. The eyes and temporal orifices are on the dorsal surface of the snout; the mouth, nostrils, and gill-openings on its ventral surface.

---

* Phil. Trans. 1810, p. 211.   † Ibid.

(1.) *Gill openings five on each side.* This includes all the species formerly included in the genus Raia of LINNÆUS.

a. Tail fleshy and of ordinary proportions. The Torpedo and Rhinobatus electricus are furnished with electrical organs. The last species ought probably to constitute a separate genus.

17. RHINOBATUS. Snout with the sides angular. *R. vulgaris.*

18. TORPEDO. Snout with the sides rounded. *T. vulgaris*, or Cramp fish.

b. Tail hard and slender.

(A.) Tail armed with a long serrated spine.

19. TRYGON. Head uniting to form with the pectoral fins an obtuse angle. *T. pastinaca,* Stinging Ray.

20. MYLIOBATIS. Head protruded beyond the pectoral fins, which are much produced laterally. *M. aquila.*

21. CEPHALOPTERA. Head truncated, and the pectoral fins extending in a process on each side. *C. vulgaris* *.

(BB.) Tail without the long serrated spine, but frequently covered with numerous prickles. The important observations of MONTAGU, point out the necessity of exercising great caution in the establishment of species, and even of genera, in this group, either from the armature of the body, or the form of the teeth †.

---

* In Europe, the different species of rays exclusively reside in salt-water. In the Ganges, however, it would appear from the valuable work just published, " An account of the Fishes found in the Ganges and its Branches," by FRANCIS HAMILTON (formerly BUCHANNAN), M. D. 1. vol. 4to, with a vol. of plates, Edinburgh, 1822, that some species nearly allied to the T. pastinaca are common, not only in the estuary, but very far removed from the sea. " For I have seen them (Raia fluviatilis of H.) at Kanpur, more than a thousand miles above the extent of the tide." P. 1.

† After speaking of the appendages at the base of the tail, he says, " Accompanying this truly masculine distinction, are series of large re-

22. RAIA. Ray or skate. Disc rhomboidal. There are several species natives of Britain.

clined hooked spines, never to be found on the other sex, and which begin to shew themselves early in all the species hitherto examined; these are placed in four distinct series, one on each shoulder or fore-part of the wing or pectoral fin, and one on each angle of the wing. These spines are complete hooks resembling those used for fishing, and lie with their points reclined inwards in two or three, and sometimes four parallel lines, but the number of rows, and the number in each row, depends on age; for in very young specimens, I have noticed only four or five spines in a single row. For what purpose this formidable armoury is given exclusively to the males, is not known; but as the hooks are extremely sharp, and lie partly concealed, with their points a trifle reflected, the fishermen's hands are frequently lacerated by incautiously handling the fish. These formidable spines, peculiar to the masculine gender, have occasionally been fixed on as a specific character; and as it does not appear to be generally known that it is only a sexual distinction, it has been thought proper to notice it for the advantage of others who may be pursuing the same track. There is another circumstance, which perhaps, in the discrimination of species, requires more attention than usual; that is, the teeth of both sexes of each species. The necessity of this is particularly evinced by the great difference observable in the teeth of the two sexes of the thornback, *Raia clavata*.

" In search of both sexes of this species, I was naturally led by the usually described essential character of the teeth being blunt, and I was not a little surprised when, amongst several hundreds examined, not one male could be found; but I noticed a ray, not unfrequently taken with the thornback, that was in every other respect similar, except that the wings were generally not so rough, and sometimes quite smooth about the middle. A variety also of this fish had an oblong dusky spot, surrounded with white, in the middle of each wing. The teeth of these fishes were not above half the size of those of the female thornback, and, except a few of the outer series on the lips, were sharp pointed. For a long time I was puzzled to discover to what species of raia these belonged, till, after an examination of a great number, I began to be as much surprised at not finding a female amongst such a quantity of these, as I was at not finding a male amongst those with blunt teeth. These circumstances naturally induced me to conclude, that the sexes of *clavata* had not been accurately defined, and that the leading character of blunt teeth might have been drawn from the female only The fishermen had not noticed the distinction of the teeth in these fishes, and had considered all of them to be thornbacks. After much attention to the

(2). *Gill openings, one on each side.*

23. CHIMÆRA. Snout simply conical. *C. monstrosa.*

24. CALLORHYNCUS. Snout ending in a fleshy hoe-shaped process. *C. antarcticus.*

At the conclusion of this account of the cartilaginous fishes with fixed branchiæ, it appears expedient to direct the attention of the reader to the Sea-snake which was cast ashore on Stronsa, one of the islands of Orkney, in September 1808. The anatomical characters, furnished by the mutilated fragments which were sent to the Wernerian Society, seem to point out a connection with the genus Squalus. But the articulated fins on the sides, the form of the dorsal fin, and the lengthened neck, if the accounts of those who saw the animal can be confided in, prove the propriety of the new genus constituted for its reception, and termed Halsydrus, or Sea-snake. *Scots. Mag.* 1809, p. 7. —The structure of the vertebræ of this animal has been explained with great precision by that celebrated anatomist Dr BARCLAY, and figures of some of the parts have been published from the accurate drawings of Mr Syme. These, and the various descriptions of the animal sent from Orkney, have been given to the world in the first volume of the *Memoirs of the Wernerian Natural History Society of Edinburgh.*

2d Subdivision.

Branchiæ free. This group includes the sturgeons, which have gills like the remaining genera.

---

subject, and after having offered a premium for a male thornback with blunt teeth, an intelligent fisherman assured me, he had examined a vast number since I pointed out the distinction of the teeth, and that he could not find one instance of a male with blunt teeth, nor a female with sharp teeth. It may therefore be fairly inferred, that the sexes of the thornback actually differ in this particular, and that the male has probably been described as a different species, but under what title, it is difficult to ascertain, unless it be Raja fullonica of some authors."— Mem. Wern. Soc. vol. ii. p. 414.

24. ACCIPENSER.   Snout obtusely conical.   *A. sturio.*
25. SPATULARIA   Snout laterally enlarged.   *S. vulgaris.*

ORDER II.

OSSEOUS FISHES.

THE bones of the fishes of this order possess a greater degree of hardness than the cartilaginous kinds, and the earthy matter is obviously arranged in a fibrous form. The articulations are more distinct. The bones of the head are divided by sutures. The gills are always free.

1st Subdivision.

Jaws imperfect.

This includes the *Plectognathes* of CUVIER. The bones are comparatively soft. The intermaxillary bones form the jaw, to which the maxillary bone is firmly fixed. The palatine bones are fixed to the cranium by a suture. These conditions restrain all relative motion among the bones. The ribs are imperfect. There are no ventral fins. Intestinal canal large, without cœca.

1st Tribe.

Jaws exposed and covered with ivory.

The jaws are here formed to act like the beak of a bird.

There are no true teeth, but the jaws being produced and hooked, supply their place. Gill-flap, with six rays.

A. Body capable of being inflated at pleasure. The inflation is produced by air sent from the gills, into a sac formed of a duplicature of the peritoneum, and from thence into the abdomen. The inflation aids the animal in rising in the water, and as the abdomen is covered with spines, it brings these organs of defence into a more favourable position for resistance. The air-bag has two lobes. The gills are peculiar in being only three in number on each side. Nostrils with double feelers. Body tapering behind.

26. DIODON. Beak above and below single. *D. atinga.*

27. TETRAODON. Beak above and below double, in consequence of a suture. *T. stellatus.*

B. Body incapable of being inflated. No tail, the body terminating abruptly, as if part had been cut off. Body smooth.

28. ORTHAGORISCUS. Beak single. No air-bag. The *ductus communis choledocus* opens into the stomach. *O. Mola* and *truncata.*

2d Tribe.

Jaws covered, and supporting ordinary teeth. The snout is produced from the eyes, and terminates in a small mouth, armed with a few distinct teeth on each side. Skin rough, or defended by large scales or plates.

29. BALISTES. Body scaly. The first dorsal fin consisting of one or more moveable serrated spines, the mouth has a single row of eight teeth in each jaw. M. CUVIER distributes the species into four subgenera. 1. Balistes (*B. capriscus*). 2. Monacanthis (*B. chinensis*). 3. Aleuteres (*B. monoceres*). 4. Triacanthes (*B. biaculeatus*).

30. OSTRACION. Body covered with regular fixed osseous plates. The vertebræ are ossified. The jaws have ten or twelve teeth. *O. Triqueter*, the French-fish. Bloch. tab. 130.

2d Subdivision.

Jaws perfect.

In this group, the maxillary and palatine bones, from their developement and articulation, enjoy separate motion, giving to the mouth a greater or less degree of facility in arranging its form.

1st Tribe.

Gills in the form of tufts, disposed in pairs along the gill arches.

This tribe constitutes the *Lophobranches* of Cuvier. The gill-flap is large, but the opening is very small. The body is angular, with a coat of mail. Air-bag large. Intestines without cœca.

A. Mouth terminal.

The snout is much produced, and the aperture of the mouth is subvertical.

(1). *No ventral fins.*

31. Syngnathus. Pipe-fish. Body little tapering and the angles plain. The species of this genus have never been determined in a satisfactory manner. Even the British species are in confusion. There are eight species described, and some varieties; but it is probable that the characters of the sexes have been hastily considered as marks of distinct species. It appears from many observations, that the species of this genus belong to the ovoviviparous division of fishes.

32. Hippocampus. Tail more slender than the body. The angles elevated into spinous ridges. *H. vulgaris* (*Syngnathus hippocampus*, Lin.).

(2). With ventral fins.

33. Solenostomus. Ventral fins united along with the body into an apron, probably connected with the reproductive system. *S. paradoxa.*

B. Mouth under the snout.

34. Pegasus. Snout depressed. Ventrals large. *P. draconis.* Bloch. p. 109., f. 1.

2d Tribe.

Gills disposed on the arches in continuous pectinated ridges.

A. MALACOPTERYGIOUS FISHES.

The fins are supported by cartilaginous articulated rays. M. CUVIER distributes the genera of this division into abdominal, thoracic (*Subrachiens*), and apodal; the two first possessing ventral fins, which are wanting in the last.

(1). *Ventral fins abdominal.*

a. Upper jaw formed by the intermaxillary and maxillary bones. The intermaxillary bones are sessile. The maxillary form a considerable portion of the sides of the mouth, and support teeth when these are present.

(AA). Two dorsal fins, the posterior one fleshy or destitute of rays.

35. SALMO. This genus contains many species, which, by various naturalists, have been distributed into different genera, from the number of the rays of the gill flap, the condition of the teeth, and the size of the mouth. More recently, M. CUVIER has taken into consideration the conditions of the bones of the face. These, however, furnish characters which are destitute of precision. The following subgenera are enumerated by CUVIER. 1. Salmo (*S. salar*). 2. Osmerus (*S. eperlanus*). 3 Coregonus (*S. thymallus*). 4. Argentina (*A. sphyræna*, LIN.). 5. Characinus (*S. argentinus*, GM.) 6. Curimates (*S. edentulus*, BLOCH). 7. Anostomus (*S. anostomus*). 8. Serrosalmus (*S. rhombeus*). 9. Tetragonopterus (*T. argenteus*, ARTEDI). 10. Myletes (*Cyprinus dentex* of LIN.). 11. Hydrocynus (*S. falculus*, BL.). 12. Citharinus (*S. niloticus*). 13. Scopelus (*S. crocodilus*, RISSO). 14. Aulopus (*S. filamentosus*). 15. Gasteropelicus (*G. sternicola*, Hatchet-belly, BLOCH. Tab. 97.). 16. Sternoptrix (*S. draphana*).

(BB). One dorsal fin. In this group, M. CUVIER enumerates the eight following principal genera.

36. CLUPEA. Herring. Gill-flap with about eight rays. Intermaxillaries narrow and short. Maxillaries tri-

trusile. Belly compressed and denticulated. The following subgenera have been indicated: 1. Clupea (*C. Harengus* *). 2. Megalops (*C. cyprinoides*). 3. Engraulis (*C. encrasicolus*). 4. Mystus (*C. atherinoides*). 5. Gnathobolus (*G. mucronatus*). 6. Pristigaster, CUVIER. 7. Notopterus (*Gymnotus notopterus* of PALLAS).

37. ELOPS. About thirty rays in the gill-flap. Belly not denticulated. A strong spine on the dorsal and ventral edges of the tail. *E. saurus.*

38. CHIROCENTRUS. The two middle teeth above and all below much produced. Above each pectoral fin a long pointed scale. Belly sharp. *C. vulgaris.*

39. ERYTHRINUS. Head without scales, hard. Dorsal and ventral fins opposite. Teeth irregularly large. Gill-flap of five rays. *E. malabaricus.*

40. AMIA. Like the preceding, but with twelve flat rays in the gill-flap. *A calva.*

41. SUDIS. Dorsal and anal fins opposite. Two species are known.

42. LEPISOSTEUS. Body covered with osseous scales. Gills united at the throat, and the gill-flap with three rays. *L. osseus.*

43. POLYPTERUS. Gill-flap with one ray. Body with osseous scales. *P. bichir.*

b. Upper jaw formed by the intermaxillaries. The maxillaries are without teeth, and concealed in the lips. M. CUVIER divides this group into three families. The first, represented by the pike and carp, is indicated by the presence of scales, while in the two last the skin is either naked, or furnished with osseous plates.

ESOCIDÆ. Furnished with strong teeth.

---

* A figure of *Clupea alosa* or Shad, is given in Plate III. f. 1.

44. ESOX. Pectoral fins of the ordinary size. M. CUVIER has given indications of the following sub-genera.
1. Esox (*E. lucius*). 2. Galaxias (*E. truttaceus*). 3. Microstoma (*M. vulgaris*, the Serpe microstome of RISSO).
4. Stomias (*E. boa*). 5. Chauliodus (*E. stomius*). 6. Salanx (one new species, Cuv.). 7. Belone (*E. belone*).
8. Scomberesox (*E. saurus*). 9. Hemiramphus (*E. brasiliensis*).

45. EXOCETUS. Pectoral fins greatly developed, and fitted for temporary flight. *E. volitans*.

46. MORMYRUS. Snout produced, mouth small. Gill-flap hid. *M. anguilloides*.

CYPRINIDÆ. Jaws and teeth feeble. The plates of the pharynx thickly set with teeth. Stomach destitute of a pouch, and the intestines without cœca.

47. CYPRINUS. Mouth small, jaws, tongue and palate without teeth. The gill-flap with three rays. The following subgenera have been indicated by CUVIER: 1. Cyprinus (*C. carpio*). 2. Barbus (*C. barbus*). 3. Gobio (*C. gobio*). 4. Tinca (*C. tinca*). 5. Cirrhines (*C. cirrhosus*). 6. Abramis (*C. brama*). 7. Labeo (*C. niloticus*). 8. Leuciscus (*C. leuciscus*). 9. Gonorhynchus (*C. gonorhynchus*).

48. GOBITIS. Lips fleshy, and fit to act as a sucker. Air-bag contained in an osseous case. *G. barbatula*.

49. ANABLEPS. Iris double, ovoviviparous. *A. tetrophthalmus*.

50. POCEILIA. Three rays in the gill-flap. *P. vivipara*.

51. LEBIAS. Five rays in the gill-flap. *L. pacifica*.

52. CYPRINODON. Four rays in the gill-flap. *C. variegatus*.

SILURIDÆ. The maxillaries are greatly reduced or lengthened into filaments.

53. SILURUS. Skin naked. Dorsal fin radiated. The following subgenera are indicated by CUVIER: 1. Silurus (*S. glanis*). 2. Mystus (*S. mystus*). 3. Synodontis (*S. clarias*). 4. Pimelodes (*S. nodosus*). 5. Bagrus (*S. bagre*). 6. Ageneirosus (*S. militaris*). 7. Doras (*S. costatus*) 8. Heterobranchus (*S. anguillaris*). 9. Plotosus (*Platystachus anguillaris* of BLOCH.). 10. Callichtys (*S. callichtys*).

54. MALAPTERURUS. One fleshy dorsal fin placed near the tail. With elecrical organs. *S. electricus*, LIN.

55. PLATYSTACUS. Bones of the operculum ossified and fixed. *Silurus asprido*, L.

56. LORICARIA. Mouth with a ventral aspect. Body mailed. 1. Hypostomus (*L. plecostomus*). 2. Loricaria (*L. cataphracta*).

(2). *Ventral fins thoracic or jugular.*

a. Sides of the body similar. An eye on each side.

(AA.) Furnished with a sucker.

(I.) Sucker thoracic.

57. LEPADOGASTER. Sucker double. *L. ocellatus* and *bimaculatus* of DONOVAN [*], are natives of Britain.

58. CYCLOPTERUS. Sucker single. This genus has been subdivided into, 1. Cyclopterus, having the body furnished with ridges of tubercles, as *C. lumpus*; and, 2. Leparis, having the body smooth, as *L. vulgaris* and *Montagui*.

(II.) Sucker coronal.

59. ECHENEIS. Sucker in the form of an oval disc on the head, with transverse folds. *E. remora*.

---

[*] "The Natural History of British Fishes," 5 vols. 8vo. London, 1808. The figures are in general accurate.

(BB.) Destitute of a sucker.

(I.) Head covered with naked skin or minute scales.

60. GADUS. The ventral fins are jugular and pointed. Gill-flap of seven rays. This extensive genus has been subdivided into, 1. Morrhua (*G. morrhua*), 2. Merglangus (*G. merlangus*), 3. Merluccius (*G. merluccius*), 4. Lotus (*G. lota*), 5. Gadus (*G. mustela*), 6. Brosma (*G. brosme*), 7. Phycis (*Blennius phycis*), 8. Batrachoides (*G. raninus*, MULLER).

(II.) Head covered with hard scales. The scales of the head in the two first genera are irregular, while on the third they are in the form of polygonal plates, similar to those on the head of serpents.

61. LEPIDOLEPRUS. The ventral fins jugular. The suborbitar and nasal bones united to form a snout, under which the upper jaw still executes its motions. *L. trachyrhinchus* of RISSO *.

62. MACROURUS. Ventral fins thoracic. Under lip bearded. *M. rupestres* of BLOCH., t. 177.

63. OPHICEPHALUS. Head depressed. Gill-flap with five rays. *O. punctatus*.

b. Sides dissimilar. One side, in some the right, and in others the left, represents the back, and another the belly, on the latter of which they rest. The eyes are on the dorsal side.

64. PLEURONECTES. With pectoral fins. This genus includes, 1. Pleuronectes (*P. platessa*), 2. Hippoglossus) (*R. hippoglossus*), 3. Rhombus (*P. maximus*†), 4. Solea (*P. solea*).

---

* " Ichthyologie de Nice," Paris, 1810, p. 197. tab. vii. f. 21.

† A figure of Rhombus punctatus, from a Scottish specimen, is given in Plate III. f. 2.

65. ACHIRUS. Destitute of pectoral fins. *Pleuronectes achirus.*

(3). *Ventral fins wanting. Apodal.*

a. Gill opening small, and the gill cover concealed by the skin.

(AA.) With dorsal fins.

66. MURÆNA. In this great natural group the following subgenera have been indicated : 1. Anguilla (*M. anguilla*). 2. Conger (*M. conger*). 3. Ophisurus (*M. serpens*). 4. Muræna (*M. helena*). 5. Sphagebranchus (*S. imberbis*). 6. Apterichtes (*M. cœca*). 7. Synbranchus (*S. marmoratus*). 8. Alabes, CUV. In these subgenera many important characters, for their methodical distribution, present themselves to those who have access to specimens or figures of the different species. In the Anguilla the pectoral fins are large, in the Muræna they are wanting. In some the anal and dorsal fins are obliterated. In the Anguilla, the gill openings are lateral; in Sphagebranchus they are jugular, and in Synbranchus and Alabes, they are jugular and united.

67. LEPTOCEPHALUS. Body compressed like a ribband. *L. Morrisii**.

(BB). No dorsal fin.

68. GYMNOTUS. Anus placed anteriorly, and the anal fin extending along the greater part of the belly. 1. Gymnotus (*G. electricus*). 2. Carapus (*G. macrourus*). 3. Sternarchus (*G. albifrons*).

b. Gill opening and gill-lid apparent, and of the ordinary form. Articulated rays of the dorsal fin simple.

69. OPHIDIUM. Anal, dorsal, and caudal fins united.

---

* Wern. Mem. ii. p. 436, tab. xxii. f. 1.

Tail pointed. *O. barbatum* and *imberbe* are British species.

70. AMMODYTES. Anal, dorsal, and caudal fins separated. *A. tobianus.*

B. ACONTHROPTERYGIOUS FISHES.

The fishes of this division have the first rays of the dorsal, ventral, and anal fins, supported by simple spinous rays. In some cases, the rays of the first dorsal fin are naked. The genera are numerous, and exhibit so many points of resemblance, that it has hitherto been found impracticable to assign distinct limits to each. M. CUVIER has endeavoured to distribute them into different familes and tribes; but, in this case, he has not reached his ordinary degree of success. It would have been more advantageous to the progress of ichthyology, had definite artificial divisions been formed, rather than what are termed natural groups, which consist of genera brought together, before the links which connect them have been determined. But, as no methodical distribution having higher claims has been offered, we shall here give a very brief outline of its divisions.

I. TŒNIOIDÆ. Body lengthened and much compressed, with a dorsal fin along the whole back.

a. Snout short. Maxillaries distinct.

71. CEPOLA. Caudal fin distinct. *C. rubescens.*

72. LOPHOTES. Body pointed behind. *L. lepedianus.*

73. REGALECUS. Body pointed behind. The rays of the ventral fins separate and produced. *R. Hawkenii* has been found in Cornwall.

74. GYMNETRUS. Pectoral fins very small. Upper jaw very extensile. (*G. cepedianus*).

75. TRACHYPTERUS. Distinct caudal fins. Rays of the dorsal fin round, and the first ones denticulated. *Cepola trachyptera.* GM.

76. GYMNOGASTER. No ventral or anal fins. *G. arcticus.*

  b. Snout pointed. Gape wide.

77. TRICHIURUS. Skin with indistinct scales. No caudal fin nor filiform termination. *T. lepturus.* Bloch. tab. 158.

78. LEPIDOPÚS. Caudal fin distinct. *L. tetradens.* Montagu, Mem. Wern. I., p. 82., tab. ii. iii.

79. STYLEPHORUS. Tail ending in a long filament. *S. chordatus.*

II. Gobioidæ. Dorsal fins slender and flexible. Intestine equal, large, and without cœca. No air-bag.

80. BLENNIUS. Ventral fins jugular, and consisting of two rays. Body lengthened compressed: The following subgenera have been instituted : 1. Blennius (*B. Galerita*). 2. Pholis (*B. pholis*). 3. Salarias (*S. quadripennis*). 4. Clinus (*B. mustelaris*). 5. Murænoides (*B. gunnellus*). 6. Ophistognathes (*O. Sonnerati*).

81. ANARRHICAS. No ventral fins. *A. lupus.*

82. GOBIUS. Ventral fin, thoracic, and united like a funnel, by some, erroneously considered as a sucker. 1. Gobius (*G. niger*). 2. Gobioides (*G. lanceolatus*). 3. Tenioides (*G. Schlosseri*). 4. Periophthalmus (*G. Koehlreuleri*). 5. Eleotris (*G. Pisonis*).

83. SILLAGO. Two dorsal fins, the first high. Snout produced. Head scaly. Gill-lid spinous. *S. acuta.*

84. CALLIONYMUS. Gill-openings reduced to a small hole on each side of the neck. Ventrals, jugular, and larger than the pectorals. 1. Callionoymus (*C. dracunculus*). 2. Trichonotus (*T. setigerus*). 3. Comephorus (*C. Baicalensis*).

III. Labroide. Body oblong, scaly. One dorsal fin anteriorly, with strong spines, and terminal filaments. Lips fleshy. Three toothed pharyngian plates; two above and one below. Stomach simple, with two or without cœca. A large air-bag.

85. LABRUS. Lips double, one supported by the jaws, the other by the suborbitars. Gill-flap with five rays. 1. Labrus (*L. vetula*). 2. Julis (*L. Julis*). 3. Crenilabrus (*Lutjanus rupestris* of BLOCH.). 4. Coricus (*L. virescens*). 5. Cheilinus (*Sparus faciatus* of BLOCH.). 6. Epibolus (*Sparus insidiatos*). 9. Gomphosus (*G. coeruleus*).

86. NOVACULA. Body descending suddenly towards the mouth. *Coryphæna novacula*.

87. CHROMIS. Teeth minute. Vertical fins with filaments. *Sparis chromis*.

88. SCARUS. Intermaxillary and premandibular bones convex, rounded, with teeth-like scales on the margin, with young ones for future use. *S. Abildgaardi*.

89. LABRAX of Pallas. Several rows of pores like so many lateral lines.

IV. Spinous portion of the dorsal fin capable of being depressed into a groove, between the scales, on each side at the base. Intestines large, with cœca. A large air-bag.

a. Sparoidæ. Dorsal fin single.

90. SMARIS. Jaws protrusile. Teeth fine. *Sparus mœnu.*

91. BOOPS. Jaws with a single row of cutting teeth. *Sparus salpa.*

92. SPARUS. Teeth on the sides round, with flat summits. Jaws nearly fixed. 1. Sargus (*S. Sargus*). 2. Sparus (*S. aurata*). 3. Pagrus (*S. pagrus*).

The remaining genera form two groups.

(A.) Gape wide. Front-teeth hooked, with thick set fine ones behind. The genera are divided chiefly from the characters furnished by the gill-lid.

93. DENTEX. Gill-lid without spines or notches. *Sparus dentex.*

94. LUTJANUS. L. Lutjanus of BLOCH. From this genus having the opercle plain and the preopercle denticulated, there have been separated the following genera: 1. Diacope (*Holocentrus bengalensis*). 2. Cirrhites (*C. tacheté* of LACEPEDE).

95. BODIANUS. Preopercle plain, and the opercle spinous. *B. guttatus.*

96. SERRANUS. Preopercle denticulated, and the opercle spinous. *Holocentrus gigas.*

97. PLECTROPOMA. Base of the preopercle, with spines directed anteriorly. *Holocentrus calcarifer.*

(B.) Teeth thick set and fine.

98. CANTHARUS. Mouth narrow, teeth in bands. *Sparus cantharus.* The following genera have likewise been instituted in this group: 99. CICHLA (*C. ocellaris*). 100. PRISTOPOMA (*Lutjanus hasta*). 101. SCOLOPIS. (New sp. CUV.). 102. DIAGRAMMA (*Anthias diagramma* of BLOCH). 103. CHEILODACTYLA (*C. fasciata*). 104. MICROPTERA (*M. Dolomieu*). 105. GRAMISTES (*G. orientalis*). 106. PRIACANTHES (*P. macrophthalmus*). 107. POLYPRION (*P. americanus*). 108. HELOCENTRUS (*H. sago*). 109. ACERINA (*Perca cernua*). 110. STELLIFER (*Bodianus stellifer*). 111. SCORPÆNA. This genus has been subdivided into, 1. Scorpæna (*S. porcus*), 2. Synanceia (*S. horrida*), 3. Pterea (*S. volitans*), 4. Tænianotus (*T. triacanthus*).

b. Percaidæ. Dorsal fins separate.

(A.) Head without armature. Dorsal fins widely separated. The following genera belong to this group: 112. ATHERINA (*A. hepsetus*). 113. SPHYRÆNA (*Esox sphyræna*). 114. PARALEPIS (*Coregonus paralepis* of RISSO, MS.). 115. MULLUS (*M. barbatus*). 116. POMATOMUS (*P. teliscopus*, RISSO, Ich. p. 301, Tab. ix. f. 31.). 117. MUGIL (*M. cephalus*).

(B.) Head armed with spines.

118. PERCA. Gape wide. Ventrals thoracic. 1. Perca (*fluviatilis*). 2. Centropomus (*P. nilotica*). 3. Enoplosus (*Chœtodon armatus*). 4. Prochilus (*Sciæna macrolepidota*). 5. Lucioperca (*L. vulgaris*). 6. Terapon (*Hocentrus servus*). 7. Apogon (*Mullus imberbis*).

119. SCIÆNA. Nasal and suborbitar bones swollen and cavernous. 1. Cingla (*Perca zingel*). 2. Umbrina (*S. cirrhosa*). 3. Lonchurus *L. barbatus*). 4. Sciæna (*S. umbra*).

The following genera likewise belong to this group: 120. Pogonias (*P. fasciatus*). 121. Otolithes (*Johnius ruber*). 122. Ancylodon (*Lonchurus ancylodon*). 123. Percis (*Sciæna cylindrica*). 124. Trachinus (*T. draco*).

c. Head armed with a coat of mail, by the extension and hardness of the suborbitar bones.

125. URANOSCOPUS. Head nearly cubical, and the eyes with a dorsal aspect. *U. scaber.*

126. TRIGLA. This genus has been subdivided. 1. Trigla (*T. hirundo*). 2. Peristedion (*T. cataphracta*). 3. Dactylopterus (*T. volitans*). 4. Cephalacanthus (*Gasterosteus spinarilla*).

The following genera belong to this group: 127. MONOCENTRIS (*M. carinata*). 128. COTTUS. 1. Cottus (*C. gobio*). 2. Aspidophorus (*Cottus cataphractus*). 3. Platycephalus (*P. spalula*).

d. Skin smooth, bones soft, and the pectoral fins supported on stalks. Gill openings small behind the pectorals.

129. LOPHIUS. 1. Lophius (*L. Piscatorius*). 2. Antennarius (*L. histrio*). 3. Malthe (*L. vespertilio*).

(V.) Scomberoidæ. Scales small, often scarcely perceptible, unless at the extremity of the lateral line, where they sometimes form a ridge. In other cases, this ridge is formed by a protuberance of the skin, supported by the transverse processes.

a. Two dorsal fins, the first undivided.

130. SCOMBER. Ridge on each side the tail. Anal and second dorsal fins subdivided. 1. Scomber (*S. scombrus*). 2. Thynnus (*S. thynnus*). 3. Orcynus (*S. germon*). 4. Caranx (*S. trachurus*). 5. Citula (New sp. CUVIER). 6. Seriola (*Caranx Dumerili*). 7. Nomeus (*Gobius gronovii*).

131. VOMER. Body compressed. Front truncated. Teeth small. 1. Selene (*S. argentea*). 2. Gallus (*Zeus gallus*). 3. Argyreiosus (*Zeus vomer*). 4. Vomer (*V. Brownii*). M. CUVIER adds to this tribe, with doubt, the Tetrago nurus of RISSO, Ich. p. 347. t. x. f. 37.

b. Separate spines in place of the first dorsal fin.

132. RHYNCHOBDELLA. Body lengthened. No ventral fins. *R. orientalis.*

133. GASTEROSTEUS. With ventral fins. 1. Gasterosteus (*G. aculeatus*). 2. Spinachia (*G. spinachia*). 3. Centronotus (*G. ductor*). 4. Lichia (*Scomber arnia*, BLOCH.). 5. Blepharis (*Zeus ciliaris*).

c. One dorsal fin. Teeth thick set.

134. ZEUS. Body oval, compressed, jaws protrusile. Teeth fine. 1. Zeus (*Z. faber*). 2. Equula (*Zeus insidiator*). 3. Mene (*Z. maculatus*).

The following genera likewise belong to this group: 135. ATROPUS (*Brama atropus*). 136. TRACHICHTHYS (*T. australis*). 137. LAMPRIS (*Zeus Luna*). 138. 1. Ziphias (*Z. gladius*). 2. Istiophorus (*Scomber gladius*). 139. CORYPHÆNA, 1. Centrolopus (*C. niger*). 2. Oligopodis (*O. ater*). 3. Coryphæna (*C. hippurus*). 4. Pterachs (*Coryphæna velifera*).

d. One dorsal fin. Teeth cutting, in a single row. This

includes the following genera: 140. AMPHACANTHUS (*Scarus siganus*). 141. THEUTIS (*T. hepatus*). 142. NASEUS (*chætodon unicornis*).

VI. Scaly finned. The soft anal and dorsal fins are frequently much concealed by the scales. The intestines are long, and the cœca numerous.

a. Teeth like bristles, or shorter and thick set. The following genera may here be enumerated: 143. CHŒTODON (*C. striatus*). 144. PSETTUS (*P. falciformis*). 145. OSPHRONEMUS (*O. olfax*). 146. TOXOTES (*Labrus jaculator*). 147. KURTUS (*K. indicus*). 148. ANABAS (*Perca scandens*). 149. CÆSIO (*C. coerulescens*). 150. BRAMA (*Sparus Raii*).

b. Teeth like hair in a single row. 151. STROMATEUS (*S. niger*). 152. FIATOLA (*S. fiatola*). 153. SESERINUS (*Seserinus* of ROND.) 154. PIMELEPTERUS (*P. Boscii*). 155. PLECTRORHYNCHUS (*P. chetonoide* of LACEPEDE). 156. GLYPHISODON (*Chætodon maculatus*). 157. POMACENTRUS (*Chætodon aruanus*). 158. AMPHIPRION (*A. ephippium*). 159. PREMNAS (*Chætodon biaculeatus*).

c. The two dorsal fins in this group have less of the thick covering of the scales, than in the former ones. 160. TEMNODON (One species, CUV.) 161. EQUES (*E. Americanus*). 162. POLYNEMUS (*P. paradiscus*).

VII. Mouth flute-shaped.

The tubular mouth is produced by the elongation of the ethmoid, vomer, preopercles, interopercles, pterygoid and tympanal bones, at the extremity of which the intermaxillaries, palatines, maxillaries and mandibularies are placed.

163. FISTULARIA. Body cylindrical. 1. Fistularia (*F. tabacaria*). 2. Aulostomus (*F. chinensis*).

164. CENTRISCUS. Body oval and compressed. 1. Centriscus (*C. scolopax*). 2. Amphisile (*C. scutatus*).

Various methods have been practised in the preservation of fishes for a museum. The simplest method consists in dividing the fish vertically and longitudinally, taking care to preserve, attached to one side, the anal, dorsal, and caudal fins. From this side the flesh is then to be scraped off, the bones of the head reduced in size, the base of the fins made thinner, and the specimen stretched out on pasteboard and dried. By this process a lateral view of the fish is preserved; and if the fins and gill-flap are cautiously spread out, the specimen will furnish sufficient marks for recognising the species. A collection of such fishes may be kept in a portfolio, similar to an herbarium.

Many species may be well preserved, by extracting the contents of the body at the mouth, or skinning the fish, with the skin entire from the mouth towards the tail, in the same way as eels are prepared for cooking. Let the skin be restored to its former position, fill the whole with fine sand, and having spread out the fins, let it be dried with care. Almost all wide-mouthed, cylindrical, or tapering fishes may be preserved in this manner. Some recommend filling the skin with plaster of Paris, while others employ cotton. Preserved fishes are usually covered with a coat of varnish, to restore in part the original lustre. But by no means of this sort can we retain many of the brilliant colours which the animals of this class possess when alive; and even the form of some of the soft parts cannot be preserved. Hence fishes are in general preserved in bottles of spirits of wine. In this way, it is true, they take up much room, but they can be subjected to examination at pleasure, and all their characters satisfactorily exhibited, except those depending on colour.

II.

INVERTEBRATA.

Invertebral Animals.

CHARACTERS.—ANIMALS DESTITUTE OF A SKULL AND VERTEBRAL COLUMN, FOR THE PROTECTION OF THE BRAIN AND SPINAL MARROW.

The invertebral animals have few characters of a positive kind, which they possess in common. The skin consists only of a corium and cuticle, both of which, according to circumstances, are furnished with appendices, in the form of shells, crusts, scales, or hairs. These generally supply the place of the osseous system, serving as a protection to the viscera, and as supports to the muscles. The blood, in those cases where a circulating fluid can be detected, is usually of a white or grey colour, seldom inclining to red. When there are both systemic and pulmonic ventricles, they are not united, as in the vertebral animals. With the exception of the genuine viviparous mode of reproduction, the invertebral animals exhibit all the other modifications of that function.

In attempting the division of invertebral animals into subordinate groups, the condition of the nervous system furnishes characters of importance. In one extensive class, which, from the starry form of the species, has been termed RADIATA, the nervous matter appears to be disseminated among the different organs, and never appears in the form of a brain, with its connected filaments. In another class, equally extensive, and which, with propriety, may be denominated GANGLIATA, the brain appears in the form of a collar, surrounding the gullet, near its entrance into the stomach, and sending out filaments, which, in their course, expand into ganglia.

I.
GANGLIATA.

In this group are included the classes *Annulosa* and *Mollusca*. The essential characters, by which these two classes may be distinguished, depend on the condition of the nervous filaments proceeding from the brain, the principal of which, in the former, constitute a knotted cord proceeding to the extremity, while, in the latter, they separate irregularly.

Independent of this internal character, molluscous animals are distinguished from those of the annulose division, by the absence of articulated feet, or the cuticular processes which supply their place, and by the body not being divided into joints or rings.

MOLLUSCA.

Brain surrounding the gullet, and sending out nervous filaments which separate irregularly.

Molluscous animals exhibit very remarkable differences, both in their form and in the number and position of their external members. Neither head nor foot can be observed in some, the principal organs being enclosed in a bag pierced with apertures for the entrance of the food, and egress of the excrementitious matter. In others, with an exterior still remarkably simple, cuticular elongations, termed Tentacula, surround the mouth, and a foot, or instrument of motion, may likewise be perceived. This last organ is in some free at one extremity, in others attached to the body throughout its whole length. In many species there is a head, not, however, analogous to that member in the vertebral animals, and containing the brain and organs of the senses, but distinguished merely as the anterior extremity of the body, separated from the back by a slight groove, and containing the mouth and tentacula.

In many of the animals of this division, the different members of the body are in pairs, and are arranged, in reference to a mesial plane, into right and left. In some, part only of the organs respect a mesial plane, other parts being single, or in unequal numbers. In other species, the organs, which are not in pairs, are arranged round a central axis, and give to the external form a radiated appearance. But these characters are exceedingly variable and uncertain, as indicating the limits of particular tribes; since, in different parts of the same animal, modifications of all these forms may be readily distinguished.

The *skin* of molluscous animals is more simple in its structure, than the same organ in the vertebral animals. The *cuticle* is here very distinct; and, as in other classes, it is thick and coarse where much exposed, but thin and delicate in its texture, where it lines the internal cavities. A *mucous* web may be detected in the cuttle fish and slug, but of great tenuity. The *corium* is destitute of a villous surface; and on its central aspect it is so intimately united to cellular substance, that its fibrous structure can scarcely be distinguished. The *muscular* web, may, in general, be readily perceived. Its fibres proceed in various directions, according to the kind of motion to be executed, and extend or corrugate the skin at pleasure.

The appendices of the skin in this class of animals ought to be carefully studied, as they furnish the most obvious marks for distinguishing species, and for constructing divisions in their systematical arrangement. The appendices of the cuticle are few in number, and perhaps ought to be considered as limited to *hairs*. These, in some species, invest the surface regularly and closely, and may be observed on those which live on land, as well as those which reside in water. In some cases the hairs may be considered as united, and forming continuous crusts or ridges. These hairs, as well as the cuticle, are liable to be worn off, and in some places can seldom be perceived, unless in early age.

The most important appendix to the skin appears to be the *shell*. This part is easily preserved, exhibits fine forms and beautiful colours, and has long occupied the attention of the conchologist. The matter of the shell is secreted by the corium, and the form which it assumes is regulated by the body of the animal. It is coeval with the existence of the animal, and appears previous to the exclusion from the egg; nor can it be dispensed with during the continuance of existence. The solid matter of the shell consists of carbonate of lime, united with a small portion of animal matter, resembling coagulated albumen.

The mouth of the shell is extended by the application of fresh layers of the shelly matter to the margin, and its thickness is increased by a coating on the inner surface. These assertions are abundantly confirmed by the observations of REAUMUR*, whose accurate experiments have greatly contributed to the elucidation of conchology. If a hole be made in the shell of a snail, and a piece of skin so glued to the inner margin as to cover the opening, the shelly matter will not ooze out from the broken margin of the fracture, and cover the outside of the skin, but will form a coating on its inner surface, thus proving it to have exuded from the body of the animal. When a portion of the mouth of the shell of a snail is broken off, and a piece of skin glued to the inner margin, reflected outwardly, and fixed on the body of the shell, the defective part is again supplied, and the matter added to the inner surface of the skin, thus leaving the interposed substance between the new formed portion and

---

* Mémoires de l' Académie des Sciences, 1709.

the fractured edge. Similar experiments, repeated on a variety of shells, both univalve and bivalve, by different naturalists, leave no room to doubt that shells increase in size by the juxtaposition of shelly matter from the common integuments.

Each calcareous layer is more or less enveloped in the animal matter which we have already stated as being present; so that the different layers of successive growth may, by various processes, be distinctly exhibited. If the shell has been exposed for a short time in the fire, the animal matter will appear charred, and its black colour, contrasted with the white earthy matter, will indicate the arrangement of the different strata; in the same manner as the ivory and enamel of a tooth can be distinguished, when subjected to similar treatment. The same satisfactory results may be obtained by a different process. If the shell be steeped in weak muriatic acid, the earthy matter will be dissolved, and the flakes of albumen will remain as the frame-work of the edifice.

The layers of growth may often be distinguished on the surface of the shell, in the form of striæ or ridges, more or less elevated, but parallel to the margin of the aperture. Other inequalities may likewise be observed on the surface, at right angles to the layers of growth, such as ridges, knobs, and spines. These last derive their origin from the inequalities of the skin on which they have been moulded.

In some univalve shells, the layers of growth, parallel to the opening cannot be discerned; when exposed in the fire, there is little darkening of colour; and when dissolved in acids, but a feeble trace of animal matter remains. In the fire, these shells crack in various directions, but exhibit no trace of a scaly structure. By careful management with the file, the shell may be separated into a central layer contiguous to the skin, and a peripheral layer, both similar in structure, though frequently differing in colour. The shells exhibiting such characters have been termed

Porcellaneous, from their dense structure, and the fine polish which their surface presents. The formation of shells of this kind must take place in a different manner from those of the first kind which we have noticed.

If we attend to the form of a young shell belonging to the genus Cypræa of LINNÆUS, we may perceive that an addition of shelly matter to the margin of the aperture, in the manner in which it is applied in other shells, would not enlarge the cavity, but completely close the aperture. The increase of the shell (accompanied with a corresponding increase of its inhabitant), must take place, therefore, either by absorption of the accumulated shelly matter of the mouth, and an elongation in the direction of the greatest curvature of the shell; or the old shell must be thrown off, and a new one produced, suited to the size of the animal. The former supposition has not been entertained, the latter is now generally received by naturalists. The inner coat of such shells appears to be a transudation from the body of the animal, and the outer one to be laid on the surface by the loose reflected lobes of the cloak. In many other shells, portions of matter, more compact than the other parts, may be observed spread on the pillar, and applied to the margin of the mouth by a similar process. Mr PLATT, in support of REAUMUR's opinion, that shells are formed by juxtaposition, against the objections of Mr POUPART *, erroneously considers the different sizes of the Cypreæ as depending on the thickness of the shell increasing according to age, without admitting a corresponding increase of the dimensions of the contained animal, or cavity for its reception.

The shells of the first kind which we have noticed, from being formed of cones or layers applied to the inner edge of the margin, and extending beyond it, have an *imbricated*

---

* Phil. Trans. vol. liv. p. 43.

structure. Those of the second kind, consisting of layers regularly superimposed, have consequently a *laminated* structure; but between the two kinds there are numerous intermediate links, formed by a combination of the two processes.

In some cases, the hard parts of the skin are not entitled to the appellation of Shell, but may rather be considered as Horn. Such are the coverings of the mandibles of the Cuttle-fish, the branchial lid of the Aplysia, and the operculum of the Welk. The two last appendices, however, though horny in some species are shelly in others.

The position of the shell, with respect to the constituent layers of the integuments, exhibits very remarkable differences. In some it appears instead of a cuticle, or at least without an external membrane investing it. In general, however, it occurs between the cuticle and the skin; a position which induces CUVIER * to consider it as analogous to the mucous web of the vertebral animals. Its intimate connection with the muscular system of the animal, and the protection which it affords, seem adverse to such a conclusion. In many species the testaceous substance occurs in folds of the corium, or inserted in its substance. In this position it never acquires the solid texture which shells exposed, or covered only by the cuticle exhibit. Those which are thus concealed are in general white; those which are more exposed are frequently coloured. The colouring, however, does not depend on the direct exposure to the light, as some have imagined, for many shells which are destitute of a cuticle are white, while many of those covered with a dense cuticle are finely variegated beneath.

Between the skin and the shell neither vessels nor nerves have been traced; and the manner in which the latter is

* Leç. d'An. Comp. xiv. 11.

formed, forbids us to expect their existence. Yet the shell cannot be considered as dead matter, so long as it remains in connection with the living animal. In those animals in which the shell is external, there are muscles which connect the animal with its internal surface, and the bond of union being a substance soluble in water, the muscle can be detached by maceration. The analogy between shell and bone is here obvious, although in the one case the connection between the muscle and the bone is permanent, in the other, between the muscle and shell, temporary, or frequently changed during the life of the animal. But the vitality of the shell, if I may use the expression, is demonstrated, from the changes which it undergoes when detached from the animal: The plates of animal matter harden: the epidermis dries, cracks, and falls off; and in many cases the colours fade or disappear. We confess ourselves unable to point out the means employed by the animal to prevent these changes from taking place, by any process similar to circulation.

When the shelly covering consists of two or more pieces, they are joined together, as the articulated bones in the higher classes of animals, by ligaments. These, in some cases, are of great thickness and strength, and, in consequence of their elasticity, assist in the motion of the different parts.

In the mulluscous animals the skin secretes a viscous, adhesive substance, differing according to the medium in which the animal resides, but in all cases calculated to resist its influence. It is probably owing to the lubricating agency of this secretion, that both the cuticle and shell are preserved from decomposition. The skin likewise secretes the colouring matter by which the shells are variegated. The glands from which it proceeds vary much in different individuals, and even in the same individual in different periods of growth.

The characters furnished by the skin and its appendices are extensively employed in the systematical arrangement of molluscous animals. Nearly all those characters which distinguish the species, and many of those on which genera are established, are derived from the form of the shell, the tentacula, or the colour. This last character, however, is one on which little dependence should be placed.

There is nothing peculiar in the *muscular system* of this class of animals. Where the muscles are inserted in the skin, as is usually the case, that organ is in some cases strengthened by condensed cellular substance, and even acquires a coriaceous density.

Molluscous animals preserve themselves in a state of *rest*, chiefly by suction and cementation. The organ which acts as a sucker, is in some cases simple, soft, and muscular, as the foot of the snail, while in others it is compound, and strengthened internally by hard parts, as in the arms of the cuttle fish. The force with which some animals adhere is very considerable, and is strikingly displayed, for example, when we attempt to detach a limpet from the rock.

The rest, which is maintained by cementation, in some cases depends on a glairy secretion, which glues the body of the animal to the substance to which it is disposed to be attached. By such an expedient, the shells of snails adhere to rocks, stones, and plants. It is probable that the bivalve shells of the genus Cyclas, which readily adhere to the side of a glass, secure their temporary attachment by means of their glutinous cuticle. In other animals threads are produced (termed a *Byssus*) from particular glands, and while one extremity is glued to the rock, the other remains in connection with the animal. But there is an attachment more durable than any of these, which takes place in some shells, they being cemented to rocks or stones by calcareous matter, and retained in the same position during the whole term of their existence.

The locomotive powers of the mollusca are confined to creeping and swimming. The former action is performed by alternate contraction and relaxation of the foot, or muscular expansion, which serves as a sucker, and is analogous to the motion of serpents. The motion of swimming is executed either by the serpentine undulations of the foot and the body, or by the action of tentacula, or expanded portions of the integuments. Many species are aided in swimming, by being able to vary the specific gravity of their body at pleasure, and either rise or sink in the water as circumstances may require. In some, as the Janthina, there is a cellular organ peculiarly destined for this purpose, which may be regarded as in some measure analogous to the air-bladder of fishes. In all these exertions their progress is proverbially slow. Some bivalve shells have the power of *leaping* or shifting their position by a sudden jerk, produced by shutting the valves rapidly. This is strikingly displayed in the common Scallop, and is less perfectly exhibited in the river mussels. In a few instances, especially among the slugs, a thread is formed of the viscous secretion of the skin, by which the animal is enabled to suspend itself in the air from the branches of trees.

Although the progressive motions of molluscous animals are comparatively slow, the other muscular actions are executed with ordinary rapidity. The irritability of some parts, as the tentacula and branchiæ, is so great, that the protecting movements are executed almost instantaneously, and the organs are contracted or withdrawn into the body. But these rapid exertions are only called forth in the moments of danger.

The characters furnished by the muscular system, are of great value in the discrimination of species, and in the construction of genera and higher divisions. They are intimately connected with the habits of the animal, and merit the attentive examination of the philosophical naturalist.

In the molluscous animals the *nervous system* is less complicated in its structure than in the higher classes, and the brain is not restricted in its position to the head. The whole nervous system appears in the form of ganglia and filaments. The principal ganglion, or the one to which the term Brain is usually applied, is seated above the gullet or entrance to the stomach. It sends out nerves to the parts about the mouth, the tentacula, and the eyes. It may be considered as analogous to the cerebrum of the vertebral animals. From this ganglion proceed two filaments, one on each side, which in their descent inclose the gullet, and unite underneath to form a second ganglion. From this last, which has been compared to the cerebellum, numerous filaments are likewise distributed to the parts around the mouth, and to the other regions of the body. These filaments in some cases again unite, and form subordinate ganglia. In many cases the brain and ganglia are of a reddish colour, and granulated structure; while the nerves which issue from them are white and uniform, as in the genus Aplysia. The covering of the first ganglion, which is analogous to the *dura mater*, does not adhere to it closely, but leaves a space filled with loose cellular matter. The tunics of the nerves are equally detached; and as they can be inflated or injected readily, some have been led to suppose that the nerves were hollow, and others, that the tunics were the vessels of the lymphatic system.

The organs of perception common to the higher classes of animals, do not all exist in an obvious manner among the mollusca. The *touch*, that universal sense, is here displayed in many cases with great delicacy; and the tentacula, and the other cuticular elongations which we have already referred to, contribute to augment its resources. The sense of *sight* is by no means universally enjoyed by the inhabitants of this class. In a few species, the eye is constructed on the plan of the same organ in the vertebral animals. In general, however, it appears only as a black point, the peculiar functions of which can only be inferred from analogy. In many animals there is no trace of an eye, consequently they cannot possess that varied information which the higher animals derive from that organ. Where eyes exist in this class, they are uniformly two in number. In one tribe only, namely the cuttle-fish, the rudiments of the organs of *hearing* have been detected. The organs adapted to *smelling* cannot be exhibited, but the existence of the sense is demonstrated by the facility with which they discover suitable food, when placed within their reach. The sense of *taste* exists, but it is difficult to point out the particular parts of the mouth fitted for its residence. As they, however, select particular articles of food in preference to others, it may be concluded that taste regulates the choice.

In the classification of the mollusca, the characters furnished by the nervous system, from the difficulty of their detection and exhibition, have never come into use. But those furnished by the organs of perception are highly prized. Of these, the eye is the most obvious and constant. It varies in position in different species; but, among individuals of the same species its characters are constant.

In the cutaneous, muscular, and nervous systems, traces of a general plan may be observed, according to which they have been constructed in the different tribes. In the organs which remain to be considered, there is less uniformity of structure, each family, almost, being constructed according to a model of its own.

The time when molluscous animals feed has not been carefully attended to. Those which live in the water are beyond the reach of accurate observation. Those that reside on land usually shun the light, and creep forth in the evenings to commit their depredations. During warm, dry weather, they stir not from their holes.

The animals under consideration, feed equally on the products of the vegetable and animal kingdom. Those which are *phytivorous* appear to prefer living vegetables, and refuse to eat those which are dried. We are not aware that putrid vegetable matter is consumed by them, although many of the snails and slugs are found under putrid leaves and decayed wood. In these places, there is shelter from the sun, together with dampness, so that it is difficult to determine, whether they sojourn in an agreeable dwelling, or a well-stored larder. Those mollusca which are *carnivorous*, prey on minute animals in a living state, and many of them greedily attack putrid matter.

The means employed to bring the food within the reach of the organs of deglutition, are exceedingly interesting, both on account of their variety and success. Some are provided with tentacula for securing their prey, and conveying it to their mouth, as the cuttle-fish; others protrude a lengthened proboscis, or an extended lip or tongue, and thus bring their food into the mouth. By many, however, which are fixed to the same spot during the continuance of existence, or only capable of very limited locomotive power, successful efforts are made to excite currents in the water, whereby fresh portions of it are brought in contact with the mouth, and its animal or vegetable contents separated. Where part only of any kind of food is taken into the mouth at once, the lips are possessed of sufficient firmness to cut off the requisite portions, or there are corneous mandibles to perform the office.

In the mouth, there is scarcely any process performed

analogous to that of mastication, in the higher orders of animals. When the food is in the mouth, or entering into the gullet, it is mixed with saliva, as in the more perfect animals. The *salivary glands* in which it is secreted, are in general of considerable size, divided into lobes, and, in some cases, separated into distinct masses. In many species the existence of a gullet is doubtful, as the food seems to enter the stomach immediately; while, in others, there is a portion of the intestinal canal which has some claim to the denomination.

The stomach, in many instances, is membranaceous, and can scarcely be distinguished from the remaining portion of the intestinal canal. In some cases, however, it is strong and muscular like the gizzard of a bird, and even fortified with corneous knobs for the reduction of hard substances. In some species, the stomach opens laterally into the pylorus, and, in a few instances, possesses a spiral cœcum attached to it.

The *liver* is usually of large dimensions, and seated close to the stomach, which it, in many cases, envelopes. It is divided into numerous lobes, and receives numerous blood-vessels. There is, however, nothing analogous to the *vena portarum* of quadrupeds. The *bile* is poured, in some, into the stomach, and, in others, into the pyloric extremity of the intestine by different openings. There is no gall-bladder.

There is no division of the canal into small and large intestines, as in the higher classes; or rather, among the mollusca, the relative size of the different parts is reversed. Here the pyloric extremity is usually the largest, while the anal is more slender. The intestine, as in fishes, is short in proportion to the length of the body, and, in its course, is subject to few turns. The anus is, in some, placed on one side of the body; in others, it is terminal, while, in a few, it opens on the back.

The digestive system is thus more simple in its structure than in the higher classes It possesses neither pancreas, spleen, nor mesentery. The calls of hunger are often at distant intervals, and the power of abstinence is great.

The characters furnished by the digestive system are extensively used in the inferior divisions of molluscous animals. The form of the lips, the position of the mouth and anus, and the structure of the stomach, deserve to be attentively considered, as indicating the habits of the species.

Circulating System.—The process by which the food is converted into chyle, has not been satisfactorily traced, nor has the existence of lacteals for the absorption of the chyle been demonstrated. In this class of animals, the veins seem to perform the offices both of lacteals and lymphatics. The blood is white, or rather of a bluish colour. Its mechanical and chemical constitution yet remains to be investigated.

The circulating system of molluscous animals, exhibits very remarkable differences in the different classes. In all of them, however, there is a systemic ventricle; but the other parts of the heart are not of constant occurrence.

The circulating system furnishes few characters which can be employed in systematical arrangements. The structure of the systemic and pulmonary vessels does not appear to be co-ordinate with any particular plan of external configuration and manner, as we see in the case of the pteropoda and gasteropoda. In these, the organs of circulation are very much alike, while the external forms exhibit very obvious differences.

The molluscous animals which respire by means of *lungs*

are few in number, and form a very natural tribe, which Cuvier has termed *gasteropodes pulmones*. In them, the respiratory organ is simple, consisting of a single cavity, on the walls of which the extremities of the pulmonary artery are spread. This cavity communicates externally by an aperture which the animal can open or shut at pleasure.

The mollusca which breathe by means of gills, exhibit very remarkable differences, in their number, structure, and position. In some cases, there is a single cavity communicating by an aperture, through which the water enters. The walls of this cavity exhibit an uneven surface, disposed in ridges, which are the gills, and on which the pulmonic artery is expanded. This structure exhibits itself in the Gasteropoda pectini-branchia. In many cases the gills, though seated in a cavity, like the former, and equally exposed to the contact of the surrounding element, are two in number, one on each side, as in the Cephalopoda. In the Bivalvia, they are four in number, two on each side like leaves, and extend the whole length of the body. In these, the water is admitted at the pleasure of the animal.

The gills of other mollusca are seated externally, and consist either of arborescent productions, or simple cuticular elongations, within which the pulmonary artery terminates. In some of these, as the Pteropoda, the branchial surface is constantly exposed to the action of the surrounding water; while in others, the cuticular expansions, which are analogous to gills, are retractile at the will of the animal.

By means of the characters furnished by the circulating and respiratory systems, the molluscous animals may be divided into several distinct classes. But as we shall employ these characters in the construction of the different divisions to be employed, it is unnecessary, in this place, to enter into their details.

Peculiar secretions.—The molluscous animals are considered as destitute of organs for the production of urine, but they possess various organs for the secretion of peculiar fluids or solids, some of which are useful in the arts.

The coloured fluid, which is secreted by the Cephalopoda and some of the aquatic gasteropoda, appears to consist chiefly of a peculiar mucus, united to a pigment, the properties of which have not been sufficiently investigated. The animals which furnish this secretion, eject it when in danger or irritated, and thus envelope themselves in a dark cloud, and elude the pursuit of their foes. A milky secretion is poured forth over the surface of the skin of some slugs when irritated. Other coloured secretions may likewise be detected in the mollusca, to which we shall afterwards advert. The threadlike secretions, termed a *byssus*, with which some molluscous animals, especially among the Conchifera, fix themselves to other bodies, appear to be of an albuminous nature. A few species in this division have the power of secreting a *luminous* fluid. Its nature, and the organs in which it is elaborated, have not been investigated. It is probable that some animals, as those which have the faculty of raising or lowering themselves in the water, have likewise the power of secreting *air* into those organs which contribute to their buoyancy.

Morbid secretions likewise occur among the animals of this division, chiefly, however, among the Conchifera. The most important of these are *pearls*, so much prized as ornaments of dress.

The organs of generation, which will be noticed afterwards in detail, furnish many important characters for classification. The external openings are those which are detected with the greatest facility, but the structure of the internal organs exhibits more varied and discriminating marks.

Condition of the Mollusca.—Molluscous animals, in reference to their condition, are divided, according to the situation in which they reside, into three groups, which may be termed terrestrial, fluviatile, and marine. Those that inhabit the land belong exclusively to the gasteropoda. Among these, some prefer open pastures, others the rubbish of old walls, while not a few reside in woods or among dead leaves and putrid plants. All the animals of this group respire by means of a pulmonary cavity.

The fluviatile mollusca, or such as reside in fresh waters, include not only many gasteropodous genera, but likewise a few belonging to the Conchifera. Among these, some breathe air by means of a pulmonary cavity, and come to the surface to respire. Such species frequent the more shallow ponds and lakes. Others, respiring by means of gills, are less dependent on the shallowness of the water, and consequently reside in different depths.

The marine mollusca include genera of all the classes. Some burrow in the sand, or adhere to the rocks which are left dry by the receding tide. These are termed *littoral* species. Others, however, which have been denominated *pelagic*, reside in the deep, and are seldom obtained but by dredging, or when thrown ashore during storms.

The effect of temperature in regulating the distribution of molluscous animals, has not been investigated with any degree of care or success. Over the terrestrial and fluviatile species, it probably exercises a very powerful controul, greatly limiting their geographical range. In proof of this, it may be stated, that the south of France possesses several species not to be found in England, while in England, there are a few which have not been detected in Scotland. But, among the marine mollusca, the influence of climate is not felt in the same degree. Living in an element, the bulk and motions of which guard it equally from the extremes of heat or cold, these animals, like the sea-weeds, have a very extensive latitudinal and longitudinal range. Thus, some are common to Greenland and the Mediterranean, others to Britain and the West Indies. The mollusca of the tropical seas, however, differ widely as a whole from those of the temperate regions. Some of the forms appear to be peculiar to warm regions, and, in general, the intensity of colour decreases as we approach the poles. But as there have been few cultivators of this branch of science, the geographical distribution of the species has been but imperfectly explored. Few parts of either England or Scotland have been surveyed by the eye of the helminthologist, so that many species, the range of which is considered as limited, may soon be found to be extensive.

If the observations are few and imperfect, which have been made on the influence of temperature, in regulating the physical distribution of mollusca, we are still in greater ignorance with regard to the power of habit. In the flœtz rocks, the relics of marine and fluviatile mollusca are found mixed in the same bed. This circumstance gave rise to the inquiry, how far the mollusca of fresh water can be habituated to sea-water, and *vice versa*. In the account of the proceedings of the National Institute of France, for the year 1816, it is stated, that M. BEUCHANT, professor at Marseilles, has directed his attention to this subject. He found, that all these animals die immediately, if we suddenly change their place of abode; but that, if we gradually increase the proportion of salt in the water for the one set, and diminish it for the other set, we can, in general, accustom them to live in a kind of water which is not natural to them. He found, however, some species which resisted these attempts, and which could not bear any alteration in the quality of the water in which they reside. Before much confidence can be placed in the accuracy of these results, it would be desirable that the experiments were repeated and varied by other observers. There are, indeed, many sources of error to be guarded against. When we change animals from fresh to salt water, or from salt water to fresh, we must necessarily derange their motions, by compelling them to reside in a medium of a different degree of density from the one in which they have been accustomed to dwell, and to which the arrangement of the different parts of the body is adapted. By such a change of place, it would be difficult for those which breathe air to come to the surface, and descend again in their new situation. In those with gills, the application of a new kind of fluid to the surface of such delicate organs, would considerably influence the function of respiration. The change of situation would likewise be accompanied by a corresponding change of food, and consequently, not merely the organs of locomotion and respiration, but likewise those of digestion, would suffer a derangement in their operations. We know that the power of *suffering* in the animals of this class is very great, and that they survive, though sadly mutilated. Some of the snails will live in a quiescent state for years, without food, and almost without air. Unless, therefore, the animals subjected to these experiments of a change of situation, have been observed to thrive on the food which it spontaneously yields, to execute their accustomed motions, and above all, to propagate their kind, we shall be disposed to conclude, that patient suffering has been mistaken for health, and vivaciousness for the power of accommodation.

Molluscous animals divide themselves into two great divisions, distinguished from each other by well defined characters. In the first, the presence of a head may be recognised, together with eyes and even ears, in some of the groups. In the other, containing animals much less perfect in their organization, there is no head, nor vestiges of eyes or ears in any of the groups. The former have been termed *Mollusca cephala,* the latter *Mollusca acephala.* This arrangement was first employed by the celebrated CUVIER, and afterwards by LAMARK, and other modern systematical writers. In the last work of the former naturalist, this method is departed from, and six orders are now constituted of equal rank, instead of being placed subordinate to the two primary divisions, under which, notwithstanding, they can be suitably distributed.

DIVISION I.

MOLLUSCA CEPHALA.

Head distinct from the body, bearing the lips or jaws.

The head, or the anterior part of the body on which zoologists have bestowed that denomination, possesses more or less freedom of motion, and, on the dorsal aspect, supports either tentacula or eyes, frequently both. The animals of this division exhibit so many modifications of form and structure, in all the series of organs, that the positive characters which they possess in common are few in number. They easily admit, therefore, of subdivision into inferior groups, which exhibit well marked characters of distinction. Two of these groups occupy a primary rank, the others being included under them as subordinate sections. The animals of the first of these groups are all inhabitants of the water, and execute their progressive motion through that element by organs fitted for swimming.

They are destitute of any ventral disc on which to crawl. Those of the second group, including as well animals which inhabit the land, as those which live in fresh water or in the sea, execute progressive motion by means of crawling along the surface of objects, the body resting on a ventral disc, termed a foot.

Section 1.—NATANTIA.

Organs of progressive motion fitted for swimming.

The organs of motion are situate near the anterior extremity of the body, and consist either of flexible tentacula or membranaceous expansions. All the species reside in the sea. They are nearly of the same specific gravity with the surrounding fluid in which they float about, having their motions in a great measure regulated by its changes. It is however probable, that, by means of some contractile movements, they are capable of varying their density, and of rising or sinking in the water. They swim slowly, even with their utmost efforts. M. CUVIER has distributed the animals of this subdivision into two classes, which he has termed *Cephalopoda* and *Pteropoda*.

Class 1.—CEPHALOPODA.

Fins in the form of tentacula, surrounding the mouth.

The cephalopoda, in reference to their external appearance, may be regarded as consisting of two parts; the tunic or sac, which contains the viscera, and the head, surrounded by the tentacula. The sac is, in some species, in the form of a purse, destitute of any appendages, while in others, it exhibits fin-like expansions. It varies considerably in its consistence: in some, it is strengthened on the back internally, by corneous ribs or testaceous plates, and in others, it is protected externally by spiral shells. In some species, it is connected with the head by an intervening space, which may be regarded as a neck, but in others, the tunic and head are continuous behind. In all, it exhibits, after death, great changes of colour.

On the summit of the head there is a flattened disc, in the centre of which is seated the mouth. Round the margin of this oral disc, which is strengthened by a band of muscular fibres, are placed the *arms* or tentacula. Beyond this circle of arms, in some species, there are situated two organs, larger in their dimensions than the arms, which may be denominated *feet*. Both the arms and feet are covered on their central aspect with numerous suckers, by which they are enabled to attach themselves to different bodies, and to seize their prey; and in their axis, both a nerve and artery may be observed. These arms and feet are capable of being moved, at the will of the animal, in every direction, and are the organs by which progressive motion is performed. In the space between the head and tunic in front, there is an opening or *funnel* with a projecting aperture. This funnel opens into the cavity of the sac, and serves both to convey water to the gills, and to carry off the different excreted matters.

The brain in the cephalopoda is contained in an irregular hollow ring, in the cartilaginous border of the oral disk. This cartilage is thickest on the dorsal aspect, and contains the parts which have been denominated *cerebrum* and *cerebellum*, the remaining part of the canal being occupied with the collar, which surrounds the esophagus. The nerves, which proceed directly from the brain to the parts which they are destined to influence, are few in number. From the cerebrum a few small nerves issue, which go to the mouth, and the base of the feet—while some proceed to form ganglia at the mouth, and others supply the feet. The cerebellum, besides furnishing the collar which encircles the gullet, contributes to the formation of the large ganglia which supply the arms—the optic and auditory nerves—those for the funnel, the tunic, and the viscera. From the size of the animals, the ganglia of the nerves are very distinctly displayed. The anastomosing branches of the nerves of the arms are likewise conspicuous. Each nerve, at the base of each arm, sends out two filaments, one to the nerve of the arm on each side. In this manner a chain of nerves is formed round the base of the arm, probably calculated to enable them to act more readily in concert. From the abundant distribution of nerves to the different parts, it appears probable that the sense of touch exists in a tolerably perfect manner. There is no proof of the development of organs for the display of the senses of smell and taste.

The cephalopoda are furnished with two eyes, one on each side of the head. The external membrane on the inner side, which may be compared to the *sclerotica*, differs in many particulars from the covering of the same name in the eyes of the vertebral animals. While it surrounds the contents of the eye from the entrance of the optic nerve to the pupil, it is greatly separated from the choroides. Immediately within its cavity, there is a bag, with a peculiar membranaceous covering, which contains numerous glandular bodies, similar to the milt of fishes, by which the eye is supported, and which probably act as secreting organs (although M. CUVIER could not detect any excretory canals), and likewise an expansion or ganglion of the optic nerve. The concave or anterior surface embraces the *choroides*. This membrane, after enclosing the vitreous humour, forms a zone or diaphragm, which may be compared to the ciliary processes, with an aperture in the centre for the reception of the crystalline lens. The circular margin of this aperture is lodged in a circular groove, and intimately united with it, so that the lens is divided into two unequal hemispheres. Its central surface is coated, as in the higher classes of animals, with the coloured mucous pigment which has been denominated *pigmentum nigrum*. In the cephalopoda, however, it is of a purplish-red colour.

The optic nerve, after entering the sclerotica, expands into a large ganglion, from the peripheral surface of which, issue numerous nervous filaments. These pierce the choroides by as many holes, and go to form, by their reunion, the retina. This important membrane extends to the ciliary zone, and, like it, appears to unite itself with the groove of the lens.

The vitreous humour is contained in a peculiar vesicle, having the lens seated in a concavity on its external surface. The lens divides easily into two parts, the line of separation being the groove which receives the ciliary ligament. Each portion consists of a number of concentric layers of variable thickness, composed of radiated fibres, becoming less and less distinct towards the centre, near which the laminated and radiated appearances cease to be perceptible. An imperfect representation of this structure is given by Sir E. Home, probably from preparations by Mr John Hunter, in the Phil. Trans. vol. lxxxiv. tab. 5. p. 26.

The conjunctiva supplies the place of a cornea, and covers directly the crystalline lens, as there is no aqueous humour. This membrane, in some, is continuous with the skin, but in others, there are imperfect eye-lids formed by its duplicature, previous to passing over the lens. The skin, at the opening of the pupil, formed by the sclerotica, in the absence of an *uvea* and *iris*, is strengthened by a membrane which appears to be muscular, and probably assists in the contraction or enlargement of the aperture.

The animals of the cephalopodous class, besides containing complicated eyes, are likewise furnished with ears. These are situate in the annular cartilage which supports the arms. In this cartilage, there are two cavities, in each of which there is a bag filled with a gelatinous, transparent fluid, and containing a calcareous substance, differing in its consistence according to the species, from the brittleness of starch to the hardness of bone. The auditory nerve penetrates the walls of this labyrinth, and ramifies on the membranous bag which it contains. There is no external opening, nor any apparent alteration in the thickness of the investing integuments.

The digestive system of the cephalopoda exhibits several appearances by which it may be distinguished. The arms which surround the mouth, seize the animals which are to serve as food, and bring them to the mouth. The mouth is situated in the centre of the disc, round which the tentacula are arranged. It is surrounded with a slight fold of the skin, which may be compared to lips, and which is rough on the central aspect. Within these are the two mandibles, of a deep brown colour, hard, horny consistence, and in form resembling the beaks of a parrot. Where free, they are conico-tubular, but where covered, they are open at the central side. The under beak, unlike the same organ in birds, is the largest, the most crooked, and embraces the upper, or the one on the dorsal margin of the mouth. These jaws are merely able to open and shut, as they possess no lateral motion. They are supported by the muscular bed of the mouth, which serves as a mould to fill the cavity towards the point. The tongue is situate between the beaks, and is armed with reflected teeth. These teeth, in consequence of the undulatory motion of the substance of the tongue, expedite the progress of the food into the gullet.

The salivary glands are four in number, and are placed in pairs. The first pair, seated on each side of the muscular bed of the mouth, are divided into numerous lobes, the excretory ducts of which pour their fluid into the beginning of the gullet. The second pair, seated lower down and below the eyes, are not so much divided, and send out separate canals, which unite and pour their contents into the mouth.

The gullet is furnished with a lateral expansion, not unlike the crop of gallinaceous birds. The stomach is muscular, like the gizzard of fowls, and the cuticle is thick, and separates easily from the other membranes. At the pyloric opening of the stomach, there is another aperture equally large, which leads into the *spiral stomach*, or cæcum, as it has been improperly termed by some anatomists. It may with greater propriety be denominated the duodenum, as it performs some of the offices of that part of the gut in the higher orders of animals. This stomach is conical, closed at the distal extremity, and performs about a turn and a half, like a spiral shell. Its inner surface is covered with a ridge, which traverses it in a closely spiral direction. The bile flows into it near the apex, and towards its base glandular orifices, pouring out a thick, yellow fluid, may be observed. The intestine, after leaving the pylorus, in some species, makes one or two turns, in others, it proceeds directly to the anus. This opening is seated at the base of the funnel, on its posterior or dorsal side.

The *liver* is of considerable size, of an orange-yellow colour, and of a soft and spongy texture. It gives rise to two hepatic ducts, which proceed to the extremity of the spiral stomach, where, by a common orifice, they empty the orange-coloured bile which they contain.

The organs of circulation consist merely of veins and arteries. The veins which have their origin in the feet, mouth, and annular cartilage, coalesce, and form two branches, which afterwards unite into a common trunk. This vessel, after descending through part of the viscera into the abdomen, divides into two branches, each of which may be considered as a *vena cava*, conveying the blood to the lateral hearts. Each vena cava, at its origin, is joined by an equally large vessel, which empties its contents in a direction nearly at right angles with the former. These veins arise in the stomach, intestines, liver, and organs of generation. The vena cava receives a second large vessel, nearly in the same direction as the first, which has its origin in the tunic and the supports of the branchiæ. From the size of the vena cava, in consequence of the union of these two branches, and the appearance of muscular ridges on its inner surface, it has been compared by some to an auricle.

On each side, in the common cavity of the tunic, and near the gills, an aperture may be observed, the entrance to a bag or cavity. Each cavity is traversed by the vena cava of that side, and in its passage exhibits a curious conformation. The surface of the vein is covered with spongy, glandular bodies of different shapes. These, upon being pressed, pour out an opake, yellow, mucous fluid. Within, these glands communicate by very wide ducts with the cavity of the vein. Indeed, when air is blown into the vein, it readily passes through the glands into the bag, and thence into the cavity of the tunic; and when air is blown into the bag, it likewise penetrates the gland, and passes into the veins. The arteries with which these glands are furnished are comparatively minute.

It appears probable that these glands separate some principle from the blood, and that this is conveyed away by the ejection of the water from these venous bags into the common cavity. Were it practicable to analyse the yellow mucus which these glands contain, some light might be thrown on the subject. Indeed, it appears not improbable, that this arrangement is analogous in its functions to the urinary system in the most perfect classes.

Each vena cava enters its corresponding lateral heart or ventricle, through an intervening valve. Each lateral heart is situate at the base of each gill, is pear-shaped, black, and moderately thick, with numerous pits on its inner surface. Its narrow end terminates without any valvular structure in the pulmonary artery. In the genus Octopus, the lateral hearts are naked; but in the genera Loligo and Sepia, there is suspended from each, by a slender footstalk, a spongy round body, which is concave beneath. The footstalk consists of fibres, which are attached to the surface of the heart, but there is no communication by ducts or vessels. The use of this organ is unknown.

The animals of this class continually reside in the water, and respire by means of gills or branchiæ. These are double, one on each side, corresponding with the lateral pulmonic ventricles. Each gill is connected at its opposite sides to the tunic, by means of fleshy ligamentous bands. Between these, the double leaves of the gills are arranged in an alternate series. Each leaf is supported by a footstalk from the band, and is subdivided into smaller leaves, to expose a greater surface to the water.

The pulmonary artery passes along this band, sends a branch into each footstalk, which, penetrating the substance of the gills, conveys the blood to its different divisions.

The systemic veins depart from the gills at the opposite extremity. These unite at the inferior band, and from each gill a vessel proceeds to the single central or systemic heart or ventricle. In some of the animals of this class the systemic veins are somewhat enlarged, and assume the appearance of auricles. The two pulmonary, or rather the systemic veins, enter the heart at the opposite side, each at the termination being furnished with a valvular organization.

The systemic heart is white and fleshy, and differs according to the genera in its form, being in the Octopus semicircular, but in the Loligo and Sepia lobed. Besides giving rise to a large aorta, or principal artery, two smaller ones likewise proceed from its cavity. These arteries are furnished at their entrance with valves.

The sexes in the Cephalopoda are distinct, the male and female organs being found on different individuals. There is not, however, any external mark by which they may be distinguished. M. CUVIER found that the males of the Octopus were scarcely a fifth part so numerous as the females.

The male organs of generation consist of the following parts: The *testicle* is a large white glandular purse, containing numerous fringed filaments, from which the seminal fluid is secreted. This fluid passes out of the testicle, by a valvular opening, into the *vas deferens*. This canal is slender, and greatly twisted in its course, and opens into a cavity which has been compared to the *seminal vesicle*. The walls of this last cavity are strong and muscular, and disposed in ridges. Near the opening at the distal extremity of this sac is an aperture leading into an oblong glandular body, regarded as exercising the functions of a *prostate gland*. Beyond this lies a muscular sac, divided at the top, where it opens by two ducts, but connected at the base. In this sac are numerous white thread-like bodies, terminated by a filament, but unconnected with the sac. In the interior they consist of a spiral body, connected at each extremity with a glandular substance. When these bodies are put into water, they twist themselves in various directions, and throw out at one of their extremities an opake fluid. These motions are not excited by placing them in oil or spirit of wine, but they may be exhibited by immersing in water those which have been kept for years in spirits.

These bodies, first observed by SWAMMERDAM, and afterwards by NEEDHAM, have been regarded by some as demonstrating the truth of the vermicular theory of generation; by others, they have been considered as analogous to the pollen of plants—that their tunic is in part soluble in water, and when they are thrown into that fluid, they speedily burst, and spread their impregnating contents over the eggs of the female. Although this last conjecture is plausible, and countenanced by the circumstance that these vermicular bodies are only found at the season of reproduction, the subject is still involved in obscurity. Are these bodies produced in the testicle, and only brought to this bag when nearly ready for exclusion; or, if the product of the bag itself, by what means are they nourished?

The male organs terminate in a cylindrical fleshy body termed the Penis. This is hollow within, and ribbed with muscular bands. Near its base it receives one of the ducts of the vermicular sac, continuous with the one from the prostate gland, forming its canal, and toward the apex the other duct. It projects but a short way into the cavity of the great bag, into which it empties its contents. These pass out of the body at the funnel-form opening in the throat.

The female organs of generation consist of an ovarium and oviduct. The ovarium is a glandular sac, to which the ova are attached by footstalks. The opening by which they issue from the ovarium is wide, and the oviduct (in the Octopus vulgaris and Loligo sagittata), after continuing a short way simple, divides into two branches, each having its external aperture near the anus. The oviducts are furnished within with muscular bands and a mucous lining, and encircled with a large glandular zone, destined, probably, to secrete the integuments of the eggs. In the Loligo vulgaris, and the Sepia, the oviduct continues single. Besides these organs, the Loligo vulgaris and sagittata, and the Sepia, have two large oval glandular bodies, divided by transverse partitions, with their excretory ducts terminating at the anus, the use of which is unknown. The eggs, of the peculiar form already noticed, pass out of the funnel, after which they are supposed to be impregnated by the male, according to the manner of fishes.

The *inky fluid* now remains to be considered, as the most remarkable of the productions of this tribe of animals. The organ in which this fluid is secreted is spongy and glandular. In some species it is contained in a recess of the liver, which has given rise to the opinion, that the coloured fluid which it secreted was bile. In other species, however, this gland is detached from the liver, and either situate in front or beneath that organ. The excretory canal of this gland opens in the rectum, so that the fluid escapes through the funnel. It mixes readily with water, and imparts to it its own peculiar colour. When dried, it is used as a pigment, and is considered as the basis of China ink.

The Cephalopoda are all inhabitants of the sea. They are widely distributed, occurring in the arctic as well as the equatorial seas. In the latter, however, they grow to the largest size. It is reported, that in the Indian Seas, boats have been sunk by these animals affixing to them their long arms, and that they are dreaded by divers.

The two Linnæan genera, Nautilus and Sepia, comprehend all the animals which are at present considered as belonging to this class.

Order I. Nautilacea.

Furnished with a multilocular shell.

This order is involved in the greatest obscurity. None of the recent species have been subjected to an accurate examination, so that their connection with the order Sepiacea may still be considered doubtful. Enough is known of the animals of two of the genera, to furnish some hints for those who are fond of classifying animals from their analogies. These genera are SPIRULA and NAUTILUS.

In the *Spirula*, the shell, which is concealed under the skin of the back, is spiral, with the whorls separate, the mouth orbicular, the chambers perforated by a pipe, and the last cell produced into a tube. The position and use of this terminal tube are unknown. The *S. vulgaris* is the most common species, and inhabits the seas in the West Indies. In the restricted genus *Nautilus*, the shell is supposed to be external, and the body of the animal to be lodged in the last chamber, and to be fixed by a ligament which descends into the central pipe. In the shell itself, the turns of the spire are contiguous, and the last whorl embraces the others on the sides. The *N. pompilus* of RUMPHIUS is the only species in which the animal has been detected.

The other genera which have been formed in this order depend exclusively on the characters furnished by the shells; and the resemblance which these bear to the preceding genera, constitutes all their claim to be included in the present order *.

Order II. SEPIACEA.

Destitute of a multilocular shell.

The sac is strengthened by horny or testaceous plates, unless where the habits of the animal render such support unnecessary.

---

* Since the days of LINNÆUS, our knowledge of the *Multilocular* testacea has been greatly enlarged. He contented himself with arranging all the species with which he was acquainted under one genus, but, in consequence of modern industry, even the genera exceed the number of Linnæan species. Many recent species have been discovered, by the aid of the microscope, among the sand on the sea-shore, and a still greater number in a fossil state among the calcareous strata. These newly discovered kinds exhibit many different characters, and have compelled Conchologists to institute so many new genera for their reception, that the genus Nautilus of LINNÆUS appears rather as the head of a family or order, than as a separate genus of univalve shells. In this department the names of BRUGUIERE, LAMARK, MONFFORT, PARKINSON, and SOWERBY, deserve respectful notice; and it is from their writings that the following remarks concerning the multilocular testacea have been extracted. The multilocular testacea may be divided into three sections: the first including those which are obviously spiral; the second, those which are produced; and the third, those which are of a globular or lenticular form. These sections are merely provisional, and are only intended to render more obvious and intelligible our notices of the genera.

1. *The spiral multilocular testacea.* At the head of this first division stands the modern genus NAUTILUS, in which the turns of the spire are contiguous, and the last whorl incloses the others. The partitions are perforated by a tube. We possess on our shores several species of this genus, of which the N. *crispus* is the most common.

In form, the genus LENTICULINA is nearly related to the former. The margin of the mouth reaches to the centre of the shell on both sides, and the partitions are destitute of a syphon. LAMARK is in possession of a recent shell of this species from the sea near Teneriffe.

The shells which Mr SOWERBY, in his *Mineral Conchology*, has figured under the genus ELLIPSOLITHES, have the whorls conspicuous, although the mouth clasps the body whorl. But it is easily distinguished from the other genera with which it is related by its elliptical form.

1. *Head surrounded with eight arms and two feet.*

The two feet are nearly similar in their structure to the arms, or tentacula, but considerably larger in their dimensions. They have their origin on the ventral side of the

---

The genus Discorbis of Lamark (formerly called by him Planulites), bears a considerable resemblance to the Nautilus in form, but the whorls are all apparent, and the partitions entire. The following species of the genus Nautilus of Montagu, may be inserted in this genus, viz. *crassulus, inflatus, carinatulus, Beccarii* and *Beccarii perversus*. Were we acquainted with the position of this last shell in the animal, we might, on account of its sinistral whorls, consider it as belonging to a new genus.

In the genus Rotalia, the spires approach to a conical shape, and the marginated trigonal aperture is reflected towards the base of the shell. It consists of shells which are now found in a fossil state.

The Nautilus spirula of Linnæus has afforded characters for the construction of a new genus termed Spirula. The whorls are separate, the mouth orbicular, the partitions perforated by a tube, and the last turn of the spire prolonged to a straight line. This last character was unknown to Linnæus, who had only seen the spiral body of the shell.

The genus Spirolina has the last turn of the spire produced like the preceding, but the whorls are contiguous. The partitions are perforated by a tube. The Nautilus *semilituus* and *subarcuatulus* of Montagu are of this genus.

The genus Lituola is allied to the spirula and spirolina in the production of the last whorl. The spires of the body are contiguous, and the partitions are pierced by a number of holes.

In the preceding genera the inner walls of the cavity are simple; but in the two following, the walls are formed into joints by sinuous sutures. The first of these is the Ammonites, including those shells which have been termed *cornua ammonis*. The origin of this name is sought for in their resemblance to the horns of a ram, and to their having been found near the temple of Jupiter Ammon, in Upper Egypt. By the Indians, the *Ammonites sacer* is considered as a metamorphosis of the god Vishnu, and termed by them *salgram* or *salgraman*. It is found among the pebbles of the Gandica, where it joins the Ganges. In this genus the whorls are contiguous, spiral, depressed, and obvious.

The Orbulites of Lamark differs from the Ammonites in the circumstance of the last whorl embracing and concealing the others. In both the

mouth, between that organ and the funnel. The suckers are pedunculated, with their margin strengthened by a

---

syphon is marginal. The Ammonites *discus* of Sowerby appears to be of this genus.

Nearly allied to the preceding is the Turrilites of Montfort. It is similar in internal structure, but while the shells of the former are spirally discoid, those of the present genus are spirally turreted, resembling a Turbo or Turricula. Four species are figured by Sowerby in his *Mineral Conchology*, Vol. I.

The genus Scaphites, formed by Parkinson, possesses very peculiar characters. It commences with a depressed volution, the last turn of which, after being enlarged and elongated, is diminished and reflected inwards.

2. *Multilocular testacea with the shell produced.* It must be confessed that the genera of this section are but imperfectly understood. The recent kinds are too small to admit of any investigation of the animal, so that we are left entirely to conjecture.

The genus Hippurites is of a conical form, and either straight or crooked. Within it is transversely chambered, and furnished with two lateral, longitudinal, obtuse, converging ridges. The last chamber is closed by an operculum.

In the Orthocera the shell is straight or slightly bent, and conical. The chambers are distinct, and pierced with a tube. We possess on our shores many minute species of this genus.

The genus Baculites of Faujus St Fond, possesses a structure similar to the Ammonites, the inner walls being articulated with sinuous sutures, and the partitions perforated. The shell is fusiform, or bent into two parallel limbs. Mr Parkinson has contributed greatly to our knowledge of this genus, and has termed it *Hamites*. We prefer the name of the original discoverer to that of our English naturalist, which is very faulty: For, according to Pliny, "Hammites ovis piscium similis est."

In the fossil genus of *Belemnites*, the shell is straight and conical, the apex is solid, with a groove or fold on one side, and at the thick end there is a conical cavity filled with a shell divided into chambers, all of which are penetrated by a pipe. If we regard this body as the remains of a cephalopodous animal, we may consider the exterior solid extremity to have been a corneous covering, and the chambered alveolus the seat of the body of the animal, which likewise enveloped the base. That the solid base was hard,

corneous ring, furnished with teeth. The sac is furnished with fin-like expansions, and strengthened internally by

and not muscular, like the sac of the Sepiæ, is obvious, from the serpulæ which have been found adhering to its surface, and which probably took up their residence after the death of the animal, and the destruction of the soft covering. That the solid part was different in its nature from shell, appears probable, from the circumstance that the latter, when mineralized, is usually converted into calcareous spar, while the former appears of a fibrous structure.

The alveoli of the belemnite bear so near a resemblance to the species of Orthocera, that some have concluded that the latter were originally parts of a belemnite. Several circumstances, however, militate against this opinion. Orthoceratites are frequently found fossil, where there are no vestiges of belemnites, and even appear to occur in older rocks. Many recent species of the genus Orthocera have been found on our own shores, without the vestige of an external covering. Had they possessed any such solid apex, like the belemnite, it is probable that it would have been detected in the recent kinds, since it is sufficiently durable to retain its form in the solid strata. The shell of the Belemnite was probably, in some respects, internal,—that of the Orthocera was probably external, or covered only by the common integuments. The views here given do not greatly differ from those of Mr PLATT, in the *Philosophical Transactions*, vol. liv. p. 38.

The AMPLEXUS of SOWERBY belongs to this division. It is nearly cylindrical, divided into chambers by numerous transverse septa, which embrace each other with their reflected margins. It contains one species from the limestone rocks of Ireland.

3. *Multilocular testacea of a globular form.* The first genus of this section is the MILIOLA. The shell is composed of three or four oval cells, turning round an axis parallel to their longest diameter. Many recent species of this genus are common on our shores: they were included by MONTAGU in his genus Vermiculum.

In the RENULINA the cells are narrow, linear, unilateral, curved into a part of a circle, and all situate on the same plane. The smallest cell forms a little arch round a marginal axis, and the others are placed contiguous to this on the same side. The species are all fossil.

The GYROGONA is a shell of a spheroidal form, composed of linear, curved, grooved pieces, terminating in two poles. The external surface is obliquely spiral, the spires terminating at each pole. Found only in a fossil state.

The

corneous or testaceous ribs or plates. The head is divided from the sac on all sides by a neck. The margin of the anus is surrounded with tentacula.

1. SEPIA. The sac is furnished on each side throughout its whole length with a narrow fin.

The suckers are irregularly scattered on the arms and feet. The back is strengthened by a complicated calcareous plate, lodged in a peculiar cavity. This plate has been long known in the shop of the apothecary under the name *Cuttle-fish bone*, which was formerly much prized in medicine as an absorbent, but is now chiefly sought after for the purpose of polishing the softer metals. It is somewhat ovate, flatly convex on both sides, and thickest where broadest. The superior half, or the one next the head, is the longest, rounded at the extremity, and thin. The inferior portion becomes suddenly narrow, and ends in a point. It may be considered as consisting of a dermal plate, concave on the central aspect, having its concavity filled up with layers which are convex on their central aspect.

The shells of the genus NUMMULITES are remarkable for their lenticular form. The external surface is smooth, and the cells are concealed; but internally the transverse cells are disposed in a spiral discoid form. The cells are imperforate; they are the Camerinæ of BRUGUIERE,—the Helecites of GUETTARD,—and the Discolithes of FORTIS. This last author supposes, that they are formed in the interior of an animal analogous to the Sepia. The same opinion may be entertained of many other genera of multilocular testacea. FAUJAS ST FOND found a recent specimen of a nummulite among the fragments of the Corallina officinalis, brought from the island of Corsica.

It is probable that the genus LAGENA, formed from the serpulæ *lagenæ* of WALKER (*Testacea minuta Rariora*), belongs to the multilocular testacea; as in some of the species we have observed the appearances of internal divisions.

According to our observations, the dermal plate appears to consist of three different laminæ, arranged parallel to one another. The external or dorsal layer is rough on the surface, and marked by obscure, concentric arches towards the summit, formed by minute knobs, which become larger towards the base, where they appear in the form of interrupted transverse ridges. It is uniform in its structure, and the tubercles possess a polish and hardness equal to porcellaneous shells, although they blacken speedily when put in the fire, and contain a good deal of animal matter. On the central side of this layer there is one flexible and transparent, similar to horn, and smooth on the surface. The third layer is destitute of lustre; and, in hardness and structure, resembles mother-of-pearl shells.

The layers which fill the concavity of this dermal plate are slightly convex on the central aspect, and are in part imbricated. Each layer is attached to the concave surface of the dermal plate, by the upper extremity and the two sides, while the inferior or caudal extremity is free. The inferior and first formed layers are short, occupy the base and middle, and rise from the plate under a more obtuse angle than the new formed layers, which are both the longest and the broadest.

Each layer, which is about one-fiftieth of an inch in diameter, consists of a very thin plate, the dermal surface of which, when viewed with a magnifier, exhibits numerous brain-like gyrations. From the ventral surface of this plate arise numerous perpendicular laminæ, which, when viewed laterally, appear like fine parallel threads, but when examined vertically, are found to be waved, and fold upon themselves. Next the plate they are thin, and not much folded; but towards their other extremity they become thicker, striated across, and more folded, with irregular margins. On the thick, tortuous, even ends of these lamina, the succeeding plate rests, and derives from them the peculiar markings of its surface. These laminæ are closely set, irregularly interrupted, and occasionally anastomose. M. CUVIER states, erroneously *, that these laminæ are hollow pillars disposed in a quincunx order.

The term *bone* has been improperly applied to this complicated plate; " for," according to Mr HATCHETT †, " this substance, in composition, is exactly similar to shell, and consists of various membranes, hardened by carbonate of lime, without the smallest mixture of phosphate."

The most remarkable species of this genus is the *Sepia officinalis,* which is distinguished from the others by its smooth skin. It inhabits the British Seas, and although seldom taken, its bone is cast ashore on different parts of the coast, from the south of England to the Zetland isles.

2. LOLIGO. Calamary. Sides of the sac only furnished partially with fins.

The suckers are disposed on the arms and feet in a double row. The dorsal plate is flexible and corneous, imbedded in the substance of the sac, and is multiplied with years. The following species occur in our seas 1. *L. vulgaris.* 2. *Sagittata.* 3. *Media.* 4. *Sepiola* ‡.

---

* Mem. sur la Seiche, p. 47. † Phil. Trans. vol. lxxxix. p. 321.

‡ Dr LEACH has described three new species of the genus Loligo, which were collected by Mr CRANCH during the voyage to the Congo, in that unfortunate expedition under the direction of Captain TUCKEY. These species belong to a group which have the suckers produced into hooked processes. In two of these species, *L. leptura* and *Smithii,* the suckers on the arms, as well as the feet, are produced into hooks, while, in one species, *L. Banksii,* the feet only are armed with hooks. The

Head surrounded with Eight Arms without Feet.

The suckers have soft margins. The sac is destitute of fin-like expansions, and is either simple or strengthened in the interior by two short corneous processes. The head is united with the sac behind, without the intervention of a neck.

a. *Arms all equal in Size.*

3. OCTOPUS.—Suckers arranged in a double row.

The suckers are sessile. The oviduct is double. The margin of the anus is simple. The *Sepia octopodia* of LIN. is the type of the genus.

4. ELEDONA.—Suckers on the arms disposed in a single row.

M. LAMARK has figured and described two species of this genus, in the *Mem. de la Soc. d'Hist. Nat.* One of these is a native of the Mediterranean, and is remarkable for giving out an odour like musk.

b. *Arms unequal.*

5. OCYTHOE.—Two of the arms at their inner extremities furnished with membranaceous expansions.

In this genus, which was instituted by M. RAFINESQUE, the suckers are in a double row, and supported on foot-

---

The same distinguished naturalist has instituted a new genus nearly allied to Loligo, from two species collected during the same voyage. The following characters are assigned to it.

" Genus *Cranchia*.—Body oval, sac-shaped ; fins approximating, their extremities free ; neck with a frenum behind, connecting it with the sac, and with two other frena, connecting it with the sac before.

" Sp. 1. *Cranchia scabra.*—Sac rough, with hard, rough tubercles.

" Sp. 2. *Cranchia maculata.*—Sac smooth, beautifully mottled with distant ovate spots."—" Narrative of an expedition to explore the river Zaire, usually called the Congo in South Africa, in 1816, under the direction of Captain J. K. TUCKEY, R. N." London, 1818, p. 410.

stalks. In the specimens of the *Ocythoe Cranchii*, procured during the expedition to the Congo, Dr LEACH observed " four oblong spots on the inside of the tube, resembling the surfaces for the secretion of mucus, two inferior and lateral, and two superior, larger, and meeting anteriorly. On the rim of the sac, immediately above the branchiæ, on each side, is a small, short, fleshy tubercle, which fits into an excavation on the opposite side of the sac."—*Phil. Trans.* 1817.

This animal was long considered as the fabricator of the shell termed Argonauta or Paper Nautilus. The observations, however, of Mr CRANCH, the Zoologist to the Congo Expedition, have demonstrated that the shell is merely the temporary residence of this animal, which it quits at pleasure. The body of the animal does not conform in shape to the cavity of the shell, nor to all its irregularities of surface; neither is there any muscular attachment between them. " On the 13th of June" (says Dr LEACH, when publishing the notes of Mr CRANCH), " he placed two living specimens in a vessel of sea-water; the animals very soon protruded their arms, and swam on and below the surface, having all the actions of the common *Polypus* (octopus) of our seas; by means of their suckers, they adhered firmly to any substance with which they came in contact, and when sticking to the sides of the basin, the shell might easily be withdrawn from the animal. They had the power of completely withdrawing within the shell, and of leaving it entirely: One individual quitted its shell and lived several hours swimming about, and showed no inclination to return into it; and others left the shells as he was taking them up in the net. They changed colour, like other animals of the class Cephalopoda; when at rest, the colour was pale flesh-coloured, more or less speckled with purplish; the under parts of the arms were bluish-grey; the suckers whitish." The specimens which furnished an opportunity for making the preceding observations, were met with in the Gulf of Guinea, and afterwards on the voyage, swimming in a small argonauta, on the surface of the sea. The reader, who is desirous of farther information on this subject, may consult Dr LEACH's *Observations on the Genus Ocythoe of Rafinesque*, and Sir E. HOME on the *Distinguishing Characters between the Ova of the Sepia, and those of the Vermes Testacea that live in water*, in the *Philosophical Transactions* for 1817, art. xxii. and xxiii., (both of which are added to the appendix of Captain TUCKEY's *Narrative*), and a paper by Mr SAY, on the Genus Ocythoe, in the *Phil. Trans.* 1819, art. vii.

Class II.—PTEROPEDA.

Fins formed of membranaceous expansions.

This class was instituted by CUVIER, for the reception of a few genera, the peculiar characters of which indicated the impropriety of suffering them to remain in any of those categories which had been previously established. All the species are small in size; and the attempts hitherto made to investigate their internal structure, have, in a great measure, failed in explaining the functions of the organs which are exhibited. The valuable papers of CUVIER, on the Clio, Pneumodermon and Hyalea, include nearly all the accurate information on the subject, of which naturalists are in possession.

The general form of these animals is somewhat ovate. The tunic appears in some genera, as the Clio and Pneumodermon, to be double, the external one soft and thin, the internal exhibiting a fibrous structure, corresponding to the muscular web of the skin of the higher classes. In these animals, however, these two layers are unconnected throughout the greater part of their expansion. In some, as the Cymbullia, the tunic is cartilaginous, while in others it is strengthened by a shell. In these last, the shell in the Limacina is a spiral univalve, covering the abdominal viscera, and in the Hyalea, where it serves the same purpose, it approaches in character a bivalve shell. It is, however, destitute of a hinge, the two valves being united together at their caudal margins, and there is no appearance of a transverse adductor muscle.

The organs of motion in all the genera consist of two fins, or membranaceous expansions, one being seated on each side of the head. They have no foot wherewith to crawl, nor any suckers by which they can adhere to objects. They are, therefore, free animals moving about in the water by means of their fins, and probably possessing, at the same time, a power of varying their specific gravity, as they are capable of varying, to a certain extent, the form of their bodies, and of enlarging or reducing their dimensions. There is nothing peculiar in their nervous system.

The organs of digestion differ greatly from those of the cephalopoda, which we have already considered. They are generally regarded as destitute of eyes and ears. Their tentacula are either seated on the head, forming two complicated branches of filaments, or spread along the margin of the tunic. There are no arms for seizing the food. The mouth, however, is furnished with lips; and, in some, an appearance of a tongue at the entrance of the gullet. The salivary glands are two in number, lengthened, descending a considerable way into the abdomen, and pouring their contents, by means of their excretory canals, into the cavity of the mouth. The gullet, after being encircled by the nervous collar, suffers an enlargement, which has been termed a Crop, contiguous to which is the stomach. Both these cavities exhibit muscular ridges on the inner surface. The liver surrounds the stomach, is intimately united with its contents, and pours in its bile by numerous pores.

The intestine is short, and, after making one or two turns, ascends and terminates in the neck near the mouth.

The circulating system in this class has been but very imperfectly investigated. The pulmonic vessels are unknown, but systemic veins, a single auricle, ventricle, and aorta, have been detected. The heart, in some, is situate on the left, in others, on the right, side of the body.

The aërating organs exhibit very remarkable differences. In the Clio they are in the form of a fine network on the surface of the fins; in the Pneumodermon they are conjectured to form leaf-like ridges on the caudal extremity of the body; or if these ridges are to be considered as particular kinds of fins, the gills may be sought for on the membranaceous expansions of the neck. In the Hyalea the branchiæ form a complex band on each side of the body, at the lateral opening of the shell.

The animals of this class are all hermaphrodites. There is a common cavity, a vesicle, penis, vas deferens and testicle, together with an oviduct and ovarium. These open near the mouth on its ventral margin. There is nothing known with respect to the appearance of the eggs, the period of propagating, or the form of their young.

All the animals of this class inhabit the sea. Some of those, as the Clio and Limacina, frequent the arctic regions, and afford the whale a great part of its sustenance. None of the species of the class have hitherto been detected in the British seas.

M. CUVIER divides the animals of this class into cephalous and acephalous. In the latter division he places the genus hyalea. The head of the animal of this genus, with its inferior neck, may, however, be sufficiently recognised to remove all doubt of its existence.

The characters which may be employed in the classification of this group are numerous; but the influence which their different forms exercise on the habits of the species is still unknown. The following disposition of the genera, though it has no claims to a natural division, may be useful to the student in his investigations.

(1.) *Tunic Strengthened by a Shell.*

6. LIMACINA.—Posterior extremity of the body covered by a spiral shell.

The shell, which is very tender, makes one turn and a-half, is flat on one side, with a large pillar cavity on the other. The fins are two in number, one on each side of the neck. When the animal swims, the head with the fins are protruded.

This genus was instituted by CUVIER, for the reception of the *Clio helicina* of Captain PHIPPS, or Arganauta arctica of FABRICIUS. According to Mr SCORESBY, it is found in great quantities near the coast of Spitzbergen.

7. HYALEA.—Posterior extremity of the body protected by two connected shelly valves.

In the animals of this genus, the body is lodged between two plates or valves, united at the base, where they inclose the caudal extremity. The ventral valve is nearly flat, with an uneven margin, narrow anteriorly, but expanding behind, and terminating in three projecting points. From the middle point four ribs diverge forward, and a muscle arises, which, fixed in the superior viscera, enables the animal to withdraw into the shell. The dorsal valve is shorter than the preceding, the margin flat and circular, and the middle convex outwardly. In the space between the lateral margin of the two valves, on each side, the branchiæ are situate, in a duplicature of the tunic, the sides of which are furnished with filaments The fleshy neck supports the two membranaceous expansions; between which and the base the mouth is situate, surrounded by two lips, and strengthened within by two fleshy cheeks. The opening of the anus and oviduct are at the base of the right fin.

The *Hyalea tridentata*, the best known species of the genus, was first noticed by Forskal, in his *Descriptiones Animalium*, p. 124, as an anomia, and inhabiting the Mediterranean. The same species was likewise taken in abundance in the Gulf of Guinea, by the expedition under Captain Tuckey.

(2.) *Tunic destitute of a shell.*
A. Fins double.
Posterior extremity with leaf-like ridges.

8. Pneumodermon.—Head with two bundles of tentacula.

The body is oval, with a narrow neck, and a fin on each side. The mouth is nearly terminal, furnished on each side with a fleshy lip, and beneath, with a fleshy chin. Each tentaculum consists of a filament, with a tubercle at the end, pierced by a small hole, and considered as exercising the office of a sucker. Cuvier, in his *Memoir sur L'Hayle et Le Pneumoderme*, considered the leaf-like ridges which occur on the caudal extremity of the body, as the branchiæ, and even describes the pulmonary vein which conveys the blood from these to the heart. But, in his *Regne Animal*, he states it as the opinion of his assistant, M. Blainville, that the fin-like expansions of the neck contain the branchiæ on their surface, as in the case of Clio. The rectum and oviduct terminate under the right wing. Cuvier has figured and described the only known species, which he terms *Pneumodermon Peronii*, the trivial name being in honour of the discoverer, M. Peron.

Posterior extremity simple.

9. Clio.—Body ovate, with the tunic elongated and membranaceous.

The head is divided into two lobes, the summits of which are furnished with tentacula. The existence of eyes has not been ascertained. The mouth is transverse, with two lateral longitudinal lips. On each side of the neck arise two blunt, conical, fin-like expansions, with a fine reticulated surface, considered as serving the double purpose of fins and branchiæ. The anus and orifice of generation terminate under the base of the right branchia. The viscera do not fill entirely the cavity of the inner bag. The gut makes only one fold.

The genus Clio, was originally instituted by BROWN in his *Natural History of Jamaica*. It was afterwards embraced and modified by LINNÆUS and PALLAS, in such a manner, as ultimately to exclude the species for the reception of which BROWN originally formed it. It contains two species, the most remarkable of which is the *Clio borealis*. Mr SCORESBY, in his valuable work on the Arctic Regions, states (vol. i. p. 544.) that it occurs in vast numbers in some situations near Spitzbergen, but is not found generally throughout the artic seas. In swimming, it brings the tips of the fins almost into contact, first on one side and then on the other.

10. CLEODORA.—Body covered with a triangular pyramidal tunic.

The fins are membranaceous. The mouth is situate between these, and is furnished with a semicircular lip. This genus was instituted by PERON, for the reception of the Clio of BROWN. The *C. pyramidata* is the best ascertained species. BROWN's *Jamaica*, p. 386, Tab. 43. f. 1. Two other species were taken by the Congo Expedition, in S. Lat. 2° 14′, and E. Long. 9° 55′, and S. Lat. 2° 41′ E. Long. 9° 16′, " both having a spinous process on each side of their shell, near its opening. One species is beautifully sulcated transversely, and the other but slightly so."— *Tuckey's Narrative*, p. 412.

B. Fin single.

11. CYMBULIA.—Tunic cartilaginous and trough-shaped.

The fin is single, divided into three lobes, one of which is small, with two tubercles, and a minute fleshy beard. This genus was instituted by PERON, in *Annales du Museum*, t. xv. t. 3. f. 10. 11.

Section 2.—GASTEROPODA.

Organs of Progressive Motion fitted for Creeping.

This is one of the most extensive groups of Molluscous animals. The marks by which it is distinguished are well defined, and the external and internal characters of the species have been successfully illustrated.

The gasteropoda may be considered as having the body protected dorsally by the cloak, and ventrally by the foot. The *cloak* is either *continuous*, and usually more or less arched, for the reception of the viscera underneath, or it is *interrupted* by a projecting bag, in which are contained the principal digestive and reproductive organs. This projecting bag is tapering and spiral, and always protected externally by a shell. When the cloak is continuous, the surface is variously marked, and frequently exhibits a particular portion, more elevated than the rest, in some cases concealing a testaceous plate, which has been termed the *shield*.

The *foot* situate on the ventral surface, and in opposition to the cloak, exhibits a flat, soft surface, consisting of interlaced muscular fibres. Its central surface serves as a support to the viscera, while externally it constitutes the organ of progressive motion. It is a *sucker* rather than a foot, and enables the animal to adhere to objects when at rest, and to crawl from one place to another by a succession of adhesions, not unlike the leech. It is also used as a fin in swimming.

By the union of the cloak and foot laterally and posteriorly, a sac is formed, which is open in front for the protrusion of the neck and head. The line of junction between the cloak and foot is marked, in general, by peculiarities in the condition of the margins of both.

The *neck* is usually divided from the cloak by a *collar*, or thickened margin belonging to the cloak, or rather to the shield, while in other cases it is continuous. Underneath, the neck is frequently attached to the foot.

The *head* supports the tentacula and eyes, is free dorsally, but frequently intimately connected with the foot on its ventral side. The portion between the tentacula and the mouth is termed the *snout*, (*le mufle* of the French, and its margin *le chaperon*). The mouth exhibits various modifications of fleshy lips and corneous jaws. The inside of the cheeks are covered in some species with reflected teeth to aid deglutition. The *tongue* can scarcely be detected in some of the genera; while, in others, it is a simple tubercle, or a strap-shaped, spiral organ, armed with transverse rows of teeth. This spiral tongue, where it is fixed to the base of the mouth, is broadest, and there also the spinous processes are strongest. The spiral part is narrowest and softest, and folded up behind the pharynx. M. CUVIER conjectures, and apparently with plausibility, that the spiral portion comes forward into the mouth to act as a tongue, in proportion as the anterior part is worn by use and absorbed. (See his *Memoire sur la Vivipare d eau douce*, p. 12).

The organs of respiration exhibit the two modifications of lungs and gills, to enable us to divide the gasteropoda into two classes, which we have termed Pulmonifera and Branchifera. M. Cuvier appears to have been in some measure aware of the importance of the distinction when he instituted his order *Pulmonés*; but he afterwards suffered himself to be more influenced by the presence of an operculum, the shape of the aperture of the shell, and the supposed separation of the sexes, than by the characters of the respiratory organs.

Class 1.—Pulmonifera.

The pulmonary cavity is single and lateral. Its orifice is capable of being closed at the will of the animal. The bloodvessels are spread, chiefly on the walls and roof, like delicate net-work. The opening of the cavity is usually on the right side, with the anus behind it, and the sexual orifice is in the front near the head. In some of the genera, these openings are situate on the left side. The shells of the former are denominated *dextral*; of the latter *sinistral*. This change in the position of the external openings is accompanied by a corresponding alteration in the arrangement of the internal organs. The heart, for example, is always placed on the side opposite the pulmonary cavity. In the dextral shells, therefore, it is sinistral. In both kinds, however, all the organs preserve the same relation to the back and belly, the head and tail. It is impossible, therefore, to conceive a dextral animal changed into a sinistral, by any circumstance which could take place at the period of hatching, as M. Bosc was inclined to believe. This arrangement of the organs must have been not merely congenital, but coeval with the formation of the embryo. In some species all the individuals are sinistral, while in others the occurrence is rarely met with in a solitary example.

MOLLUSCA.

The former are in their natural state, the latter ought to be regarded as monsters. Where the character is permanent, it should constitute a generical difference.

The reproductive system of the animals of this class exhibits the sexual organs, in general, united in the same individual. Mutual impregnation, however, is necessary. All the species are oviparous. The eggs are either naked, as in the terrestrial genera, or enveloped in a gelatinous mass, like the aquatic kinds. The embryo acquires nearly all its members while in the egg, and the shell is of a proportional size previous to hatching. Sir EVERARD HOME, when treating of the distinguished characters between the ova of the sepia, and those of the vermes testacea that live in water (*Phil. Trans.* 1817, p. 297), and when referring to the ova of the vermes testacea, says, " If the shell were formed in the ovum, the process of aërating the blood must be very materially interfered with, for this reason, the covering, or shell of the egg, first drops off, and the young is hatched before the shell of the animal is formed; this I have seen take place in the eggs of the garden snail, but in the testacea that live in water, the young requires some defence in the period between the egg being hatched and the young acquiring its shell, which is not necessary in those that live on land; for this purpose, the ova are enclosed in chambers of a particular kind." The assertion here made, and founded on *a priori* considerations, that the shell is not formed until after the egg is hatched, is opposed by every observation which we have been able to make on the subject; and what is more surprising, it is at variance with his own observations on the garden snail, the very example produced in its confirmation. The eggs of a snail, laid on 5th August 1773, were hatched on the 20th of that month, and their condition at this time distinctly stated.

"On the 20th," he says, "the young were hatched, and the shell completely formed." It is much more becoming in a philosopher to observe how nature operates, than to pronounce what she *must* do.

Order I.—Terrestrial.

The animals of this order reside constantly on the land. When by accident they fall into the water, they appear to be incapable of using their foot as a sucker or as a fin, and die after a few writhings. The species in general prefer moist places, and are seldom very active in dry weather. After a shower they speedily leave their hiding places, and at this time they may be readily collected. The eggs are hatched on land.

1st Subdivision.

Cloak and foot parallel, and containing the viscera between them.

In this group are included those animals denominated slugs in this country. They possess four retractile tentacula, of unequal length. The eyes are two in number, in the form of black points, seated at the tips of the posterior tentacula.

1st Tribe.

Cloak with a shield. This shield is, in general, strengthened internally by a deposition of earthy matter, in the form of grains, or a shelly plate.

A. Shield anterior. In this group the shield is placed nearer the head than the tail. It contains four genera, two of which have compound tails, or furnished with peculiar organs, while in the remaining genera the tails are simple. The mouth consists of lips, which are capable of small extension, and above, the entrance is armed with a concave corneous jaw with a notch in the middle. The tongue is merely armed with soft transverse ridges, pointed before, and terminated by a short cartilaginous cone. There is a sensible dilatation of the gullet, which marks the place of the stomach, at the under extremity of which, is the rudiment of a cæcum at the pyloric opening. The intestine makes several folds, chiefly in the liver, before it reaches the anus. The salivary glands reach to the extremity of the gullet. The liver is divided into five lobes, which give rise to two ducts that open into the pylorus.

The circulating system consists of two venæ cavæ, which give out numerous branches to the pulmonary cavity. The aërated blood is conveyed by several ducts to a simple membranaceous systemic auricle. Between the auricle and ventricle there are two valves. The ventricle is more muscular than the auricle. The arteries, which take their rise from a single aorta, are characterised by a peculiar opacity, and whiteness of colour, as if they were filled with milk.

The organ of viscosity nearly encircles the pericardium. It consists of regularly pectinated plates. Its excretory canal terminates at the pulmonary cavity.

The organs of generation consist, in the female parts, of an ovarium, oviduct, and uterus; and in the male, of a testicle, vas deferens, and penis, together with the pedunculated vesicle; and, as common to both the sexual organs, there is a cavity opening externally, in which, by separate orifices, the uterus, penis, and vesicle, terminate.

1. *Extremity of the Tail Compound.*

12. ARION.—A mucous orifice at the termination of the cloak.

This genus has been recently instituted by M. le Baron D'Audebard De Férrussac, in his *Histoire Naturelle Générale et Particuliére du Mollusques Terrestres et Fluviatiles*, folio, Paris, 1819, 3e. liv. p. 53. The species of which it consists were formerly confounded with those which now constitute the restricted genus Limax. It differs, however, in possessing the mucous pore, in the pulmonary orifice being near the anterior margin of the shield, with the sexual orifice underneath, and in the soft state of the calcareous matter, in the shield. The author now quoted, has described four species, and illustrated their characters by beautiful and expressive figures. The *Limax ater* (together with its variety *rufus*) of British writers may be regarded as the type of the genus.

13. PLECTROPHORUS.—A conical protuberant shell at the termination of the cloak.

This genus, likewise instituted by M. FERRUSSAC, nearly resembles the preceding in form. Three species have been described and figured, which, however, differ remarkably from one another.

2. *Extremity of the Tail simple.*

The genera of this group have neither a terminal mucous orifice nor shell.

14. LIMAX.—Pulmonary orifice near the posterior margin of the shield.

This genus, as now restricted by M. FERRUSSAC, differs from the Arion in the absence of the caudal mucous pore, the position of the pulmonary cavity, and the orifice of the sexual organs placed under the superior right tentaculum. The calcareous matter of the shield is more solid, and appears as a shelly plate. Six species have been described and figured. The *Limax cinereus* of LISTER is the type of the genus.

15. LIMACELLUS.—Pulmonary orifice near the anterior margin of the shield.

This genus was instituted by M. DE BLAINVILLE, in the *Journal de Physique*, December 1817, p. 442, pl. 11, f. 5, nov. His observations were made on a specimen in a shrivelled state, preserved in spirits of wine, which was communicated to him by Dr Leach of the British Museum. Its history is, therefore, necessarily imperfect. The animal is rounded before, and pointed behind, and dorsally protuberant. The whole body is smooth, soft, and of a greyish-white colour. The shield, which adheres on all sides, is destitute of any shelly plate. It is notched near its anterior margin, at the opening of the pulmonary cavity. The foot is broad, and separated from the cloak by a slight fold, which, on the right side, forms a groove, leading from the base of the right tentaculum to the posterior extremity of the body. It is inferred from this appearance, that the sexual organs are disjoined, the female parts being seated in the tail, while the male organs occupy the ordinary position. The only species, *Limacellus lactescens*, is supposed to have been brought from the Antilles.

B. Shield posterior.

In this group, the shield is placed nearer to the tail than in the preceding, and is fortified internally with a subspiral plate. It contains one genus.

16. Parmacella.—Posterior extremity of the shield containing the shell.

The pulmonary cavity is placed underneath the shell of the shield. This arrangement occasions a corresponding posterior position to the heart. Along the back, from the shield to the head, are three grooves, the middle one of which is double. The shield itself adheres only at the posterior portion, the anterior part being free. The internal structure is similar to the slugs. The only marked difference, indeed, consists in two conical appendages of the sexual cavity, by which there is an approach to the species of *Helicix*.

The *Parmacella Olivieri* is the only known species, and was first described, and its structure unfolded, by M. Cuvier. It was brought from Mesopotamia by M. Olivier.

2d Tribe. Cloak destitute of a shield.

In this tribe, the pulmonary cavity is situate near the tail.

A. Tail covered with a single spiral open shell.

17. TESTACELLA.—Pulmonary cavity underneath the shell.

The vent and pulmonary cavity are, from the position of the protecting shell, on which they are dependent, nearly terminal. The foot extends on each side beyond the body. From the manner in which the blood is aerated, the auricle and ventricle are placed longitudinally, the latter being anterior.

This genus at present consists of three species. One of these, *T. haliotoideus*, is a native of France and Spain. It lives in the soil, and feeds on the earthworm. Another species, the *T. Maugei*, was first observed at Teneriffe, by M. MAUGE. It has been figured in the splendid work on the Mollusca, by FERRUSSAC, already referred to, Tab. vii. f. 10—12, from specimens found in the botanical garden of Bristol, and communicated by Dr LEACH. It has been conjectured, that the British examples may have been introduced along with plants from Teneriffe. The third species, *T. Ambiguus*, is established on the doubtful authority of a shell, in the collection of M. LAMARK.

B. Destitute of an external shell.

18. VERONICELLUS.—Cloak fortified posteriorly by an internal shelly plate. Tentacula four.

This genus was instituted by M. BLAINVILLE, for the reception of the *V. lavis*, established from a specimen preserved in spirits, belonging to the British Museum, and communicated by Dr LEACH It is figured by FERRUSSAC, Tab. vii. Fig. 6, 7. The body is somewhat pointed before, and rounded behind. The cloak is large; the foot rather narrow, and plaited on the edges. The opening to the pulmonary cavity is situate on the right side posteriorly, under the shell. A little in advance of this, is another opening, in the middle of which is the funnel-shaped aperture of the anus. The sexual organs are united, and placed at the base of the right tentaculum.

It is not improbable, according to FERRUSSAC, that the *Limax nudus cinereus terrestris* of SLOANE, Jam. ii. p. 190, Tab. 233, f. 2, 3, may belong to this genus, or rather, that the individual brought home by SLOANE, may have been the identical specimen submitted to M. BLAINVILLE's examination.

19. ONCHIDIUM.—Cloak tuberculated. Snout enlarged and emarginate. Tentacula two in number, with eyes at the tips.

This genus was instituted by Dr BUCHANAN (now HAMILTON), in *Lin. Trans.* vol. v. p. 132, for the reception of a species which he found in Bengal, on the leaves of *Typha Elephantina*. It is not, according to this naturalist, " like many others of the worm kind, an hermaphrodite animal; for the male and female organs of generation are in distinct individuals. I have not yet perceived any mark to distinguish the sexes, while they are not in copulation; as, in both, the anus and sexual organs are placed in a perforation (cloaca communis), in the under part of the tail, immediately behind the foot; but, during coition, the distinction of sexes is very evident, the penis protruding to a great length, considering the size of the animal."

2d Subdivision.

Cloak and foot not parallel; the viscera contained in a spiral, dorsal protuberance, protected by a shell.

This group includes the animals usually denominated SNAILS. They bear a very close resemblance to the Slugs. The shield, however, has a thickened margin in front, destined to secrete the matter of the shell. In the part corresponding with the centre of the shield in the slugs, there is (as CUVIER has characteristically termed it) a *natural rupture*, through which the viscera are protruded into a conical bag twisted spirally. In this bag are contained the principal viscera, the liver occupying its extremity. The body of the animal is attached to the pillar of the shell by a complicated muscle, which shifts its place with the growth of the animal. The mouth is furnished above with a thin arched corneous mandible, notched on the edges. The whole body, including the foot and head, are, in general, capable of being withdrawn into the cavity of the shell. In one genus, the aperture is closed by a lid.

1st Tribe.

The foot is furnished with a lid or operculum, for closing the mouth of the shell when the animal withdraws itself into the cavity.

20. CYCLOSTOMA.—Aperture of the shell circular.

The tentacula are linear and subretractile. The primary ones have subglobular, highly-polished extremities, considered by MONTAGU as the eyes. The true eyes, however, are placed at the exterior base of the large tentacula, and are elevated on tubercles, which are the rudiments of the second pair.

The aperture of the pulmonary cavity is situate on the neck. The sexes are likewise separate; the penis of the male being large, flat, and muscular. The mouth is formed into a kind of proboscis, and the upper lip is deeply emarginate. The *Turbo elegans* of British conchologists is the type of the genus.

2d Tribe.

Foot destitute of a lid.

A. Aperture of the shell with a thickened margin.

In all this division, the margin of the shell, while the animal is young, is thin; but, upon reaching a certain period, it becomes thick, and bordered with a ring, after which there is no increase of size.

1. *Last formed Whorl of the Shell greatly larger than the penultimate one.*

21. *Helix.*—Snail. Aperture of the shell lunulated; the width and length nearly equal.

The snails differ from the slugs chiefly in the organs of reproduction. The vagina, previous to its termination in the sexual cavity, is joined by the canal of the vesicle, and by two ducts, each proceeding from a bundle of multifid vesicles. Each bundle consists of a stem or duct, and numerous branches, with blunt terminations. These organs secrete a thin milky fluid, the use of which is unknown.

Connected with the sexual cavity is the bag in which the darts are produced. The bag itself is muscular, with longitudinal grooves, and a glandular body at the extremity. This glandular body secretes the dart, which is in the form of a lengthened pyramid, consisting of calcareous filaments nearly resembling asbestus. Previous to the sexual union, the two snails touch each other repeatedly with the mouth and tentacula, and at last the dart of the one is pushed forth by its muscular bag, and directed against the body of the other, into which it enters, never penetrating through the integuments, and even, in many cases, falling short of its mark. Whether the use of the dart is merely to stimulate, or whether it is subservient to any other purpose, can scarcely be said to be determined.

The species belonging to this genus are numerous, and exhibit, in the form, the markings, and the coverings of the shell, numerous characters for their subdivision.

22. BULIMUS.—Aperture of the shell longer than broad.

The structure of the animals of this genus has not been determined; but analogy would lead us to conclude, that it is similar to the snails. While the shells of the helix are globose, those of bulimus are turreted.

2. *Last Whorl nearly of the same size as the penultimate one, or even less.*

The species which are related to the Turbo *bidens perversus* and *muscorum* of LINNÆUS, constitute a very natural family, which may be termed PUPADÆ, distinguished by the mouth being, in general, furnished with teeth, or testaceous laminæ, and the last whorl nearly the same as the preceding. Perhaps the most convenient way of dividing them is into two sections, the first including the dextral, and the second the sinistral shells.

The dextral pupadæ form three genera. The PUPA, as originally constructed by LAMARK, was equally faulty with many of the old Linnæan genera. As we have restricted it to include dextral shells, with the animal possessing four tentacula, with eyes at the tips of the two longest, we can receive into it the *muscorum, sexdentatus,* and *juniperi* of MONTAGU. The genus Chondrus of CUVIER contains the tridens of MONTAGU. In the genus CARYCHIUM, formed by MULLER, the tentacula are only two in number, with the eyes placed at the base. It is represented by the T. *carychium* of MONTAGU.

The sinistral pupadæ form two genera. The first, which is the Clausilia of DRAPARNAUD, contains sinistral shells, the animal furnished with four tentacula, having eyes at the tips of the two longest. To the pillar there is attached internally a twisted plate. This contains the following British species,—*perversa, nigricans, laminata, biplicata,* and *labiata.* The other genus, called VERTIGO, was formed by MULLER. The animal possesses only two tentacula, with the eyes on their tips. The T. *vertigo* is the type of the genus.

B. Aperture of the shell destitute of a thickened margin.
There is in this group no certain indication of maturity or stationary growth.

1. *Mouth of the Shell at the Pillar entire.*
23. Vitrina.—Margin of the shield double.

The upper fold of the shield is divided into several lobes, which are capable of being reflected over the surface of the shell. The shell itself is not capable of containing the whole body of the animal. The *Helix pellucida* of Muller is the type of the genus. It is a common British species, and was hastily regarded by Montagu as the fry of the *Helix lucida* *.

2. *Mouth of the Shell at the Pillar effuse.*
24. Succinea.—Termination of the pillar rounded.

The mouth is large in proportion to the size of the shell, with the outer-lip thin, and the pillar attenuated. We are at a loss to account for the conduct of Lamark in substituting a new name for this genus, without any apparent reason, and thus adding to the synonimes with which the science is already oppressed. The name first employed by Draparnaud, indicates one of the most striking characters of the type of the genus; whereas the term *Amphibulina*, used by Lamark, is founded on a mistake, and is apt to mislead. The *Helix succinea* (the type of the genus), although found in damp places, is not amphibious. It never enters the water voluntarily. Indeed Muller says, " Sponte in aquam descendere nunquam vidi, e contra quoties eum aquæ immisi, confestim egrediebatur." The same remark is made by Montagu, and we have often witnessed its truth.

25. Achatina.—The termination of the pillar truncated.

---

* A figure of this species is given in Plate IV. f. 1.

The *Buccinum acicula* of MULLER, a native of England, belongs to this genus; and likewise the *Helix octona* of LINNÆUS, erroneously considered as a native of Britain.

Order II.—AQUATIC.

The aquatic pulmoniferous gasteropoda have their residence constantly in the water. They possess two tentacula only. These are usually flattened, incapable of being withdrawn, and having the eyes at the internal base. The food consists of aquatic plants. Respiration can only take place at the surface of the water, to which the animals occasionally ascend, to expel from the pulmonary cavity the vitiated air, and replenish it with a fresh supply. The sexes are united. The spawn, which is in the form of a rounded gelatinous mass, containing many ova, is deposited on aquatic plants under water. Previous to hatching, the fœtus must be aërated by means of some branchial arrangement.

1st Subdivision.

Body protected externally by a shell.

The animals belonging to this subdivision bear a very close resemblance to the snails, in the structure of their body, and the form of their shell.

1st Tribe.

The protecting shell spirally twisted.

A. Shell turreted.

1. *Whorls dextral.*

24. LYMNEUS.—Aperture of the shell having the right lip joined to the left at the base, and folding back on the pillar.

The tentacula are lanceolate and depressed. The mouth is furnished with three jaws; the lateral ones simple; the upper one crescent-shaped, and emarginate. The male and female organs, though intimately connected internally, have their external orifices separated to a considerable distance, the former issuing under the right tentaculum, the latter at the pulmonary cavity. In consequence of this arrangement, the individuals of *L. stagnalis* have been observed by GEOFFROY and MULLER to unite together in a chain during coition, the first and last members of the series exercising only one of the sexual functions, the intervening individuals impregnating and receiving impregnation at the same time. Whether this is the constant or only accidental practice of this species, does not appear to be determined. We know that many other species of the genus are mutually impregnated, as usual, in pairs only.

The species of this genus are numerous. They reside in pools, lakes, and rivers, and furnish a favourite repast to the different kinds of trouts and water-fowl. The following are natives of Britain:—*L. stagnalis, fragilis, palustris, fossarius, octanfractus, detritus, auricularius, putris, glutinosus,* and probably *Helix lutea* of MONTAGU.

With regard to the *Lymneus auricularius*, it would appear, from the observations of DRAPARNAUD, " *Histoire des Mollusques,*" p. 49., that it exhibits a very singular structure of the respiratory organs. We shall quote his own words:—" L'animal est pourvu de quatre filamens ou tubes qui partent de la partie supériure du cou, près du manteau; ce sont des trachées. Ces tubes sont longs, blancs et très transparens, et on ne les distingue bien qu' à la loupe. Leur surface est comme rugueuse, et leur extrémité est un peu renflée. Ils sont rétractiles. L'animal les fait sortir à volonté, un, deux, trois ensemble : il les agite et les contourne sans cesse en divers sens : ce qui fait qu'on les prendroit pour de petits vers. Je présume que par ce mouvement ces organes sèparent de l'eau l'air que y'est contenu et l'absorbent. Cet animal est très sujet, ainsi que les autres gasteropodes fluviatiles, à être infesté par le *nais vermicularis,* qui se loge ordinairement entre le cou et le manteau, au dessous des tentacules, et s'agite sans cesse d'un mouvement vermiculaire." But little doubt, we think, can be entertained that this naturalist had been deceived by some of the parasitical leeches which infest the aquatic pulmonifera, and that, instead of breathing by means of tubular gills, the animal of the *L. auricularius* possesses, like those which it resembles in other characters, a pulmonary cavity.

2. *Whorls sinistral.*

25. PHYSA.—Pillar-lip destitute of a fold.

The external appearance of the animal is similar to the Lymneus; but the margin of the cloak is loose, divided into lobes, and capable of being reflected over the surface of the shell near the mouth. This genus was instituted by DRAPARNAUD. The *Bulla fontinalis* of British authors is regarded as the type of the genus.

26. APLEXA.—Pillar-lip, with a fold.

This genus was instituted by us for the reception of the *Bulla hypnorum* and *rivalis* of British writers. The shell is more produced than in the physa. The cloak of the animal is incapable of being reflected on the shell, and its margin is destitute of lobes.

B. Shells depressed.

The spires revolve in nearly the same horizontal line. The tentacula are long and filiform.

27. PLANORBIS.—Cavity of the shell entire.

This is another sinistral genus; the vent, pulmonary cavity, and sexual organs, being on the left, and the heart on the right side. The *P. corneus*, the type of the genus, pours forth, when irritated, a purple fluid from the sides, between the foot and the margin of the cloak.

28. SEGMENTINA.—Cavity of the shell divided.

Externally, the shell appears similar to planorbis; but internally, it is divided by testaceous, transverse partitions, into several chambers, which communicate with each other by triradiated apertures. It is uncertain whether the animal is to be considered as dextral or sinistral. This genus was instituted by us several years ago, for the reception of the *Nautilus lacustris* of LIGHTFOOT, first described and figured in *Phil. Trans.* vol. lxxvi. p. 160. Tab. 1. f. 1. 8.

2d Tribe.
Shell simply conical.
29. ANCYLUS.—Foot short, elliptical.

The tentacula are short, compressed, and a little truncated. This genus was formed by GEOFFROY, and includes the *Patella lacustris* and *oblonga* of British conchologists.

2d Subdivision.
Body destitute of the external protection of a shell.
30. PERONIA.—Head with two long retractile tentacula. The snout is divided into two broad appendages. Between the tentacula, towards the right side, is the opening for the penis. The anus is terminal, immediately above which is the entrance to the pulmonary cavity; and on the right is the opening to the female organs, from which a groove runs towards the right lobe of the snout.

The mouth is destitute of a proboscis or jaws. The tongue is merely a cartilaginous plate grooved transversely. The gullet is long in proportion, with a villous surface. There are three stomachs, each distinguished by its peculiar characters. The first is a true gizzard, covered internally with a cartilaginous cuticle, and its walls formed of two strong muscles, with connecting ligaments. The second stomach is funnel-shaped, with prominent ridges both on its external and internal surface. These ridges, at their origin internally, are highest, and project considerably into the cavity, acting like a valve in retarding the progress of the food. The third stomach is short and cylindrical, covered internally with equal longitudinal fine ridges. The intestine is nearly of equal thickness throughout, and upwards of twice the length of the body. The salivary glands are much branched, and pour their contents into the entrance of the gullet. The liver, in the animals of this genus, is distributed into three separate portions, each of which may be regarded as a distinct liver, an arrangement which is not known to take place in any other animal. The first liver is situate near the middle of the body, on the right side; while the second is placed near the posterior extremity. The ducts enter the cardiac opening of the stomach, each by a separate hole, and seem to occupy the place of the zone of gastric glands observed in birds. The third liver is placed at the posterior end of the gizzard, into which it pours its contents by a short duct.

The most remarkable feature of the circulating system, is the position of the lungs, at the posterior extremity of the body, which occasions a corresponding arrangement in the connecting organs. The entrance to the pulmonary cavity is immediately above the anus. The vessels in which the blood is aërated, are distributed on the roof and sides of the cavity. The pulmonic veins consist of two receptacles, one on each side, extending nearly the length of the body, which may be considered as venæ cavæ. These receive the blood by numerous vessels, and convey it directly to the lungs. The aërated blood is conveyed by a systemic vein into a large auricle, seated in front of the lungs, of considerable size, with the walls fortified on the interior by branched ligaments. The ventricle is placed at its anterior extremity, and separated by two valves. The aorta arises from the opposite side of the ventricle, its main trunk passing on towards the head.

The male and female organs of generation, although occurring in the same individual, appear to occupy different parts of the body. The opening of the male organs is at the tentacula, which leads to a cavity terminating in two unequal recesses. The anterior is the smallest, and receives the termination of a vessel three or four times longer than the body, which takes its rise at the external base of the cavity, apparently from the cellular substance, and, after a variety of convolutions in the neighbourhood of the mouth, opens into the recess. The second recess is the largest, and the vessel connected with it is most complicated. Its origin is in a mass which occupies a considerable portion of the abdominal cavity, and which consists of a vessel forming a great number of complicated convolutions, liberally supplied with bloodvessels. The duct which proceeds from this mass, undergoes for a short space a sudden thickening of its walls, after which it again contracts, and, before it terminates in a perforated glandular knob in the recess, it contains a pedunculated fleshy body, with a sharp-pointed corneous extremity, probably capable of being protruded into the recess and cavity.

The parts which are considered as forming the female organs, or those which are connected with the sexual cavity on the right side of the anus, consist of an ovarium, divided into two lobes, each of which may be perceived to be again minutely subdivided. The oviduct is tortuous, and passes through a glandular body, which, in the other gasteropoda, is regarded as the testicle. The pedunculated vesicle gives out two ducts, one of which goes to the testicle, the other to the uterus. It is difficult to form even a conjecture regarding the uses of all this complicated sexual apparatus. The subject can only be elucidated by an attentive examination of the condition of the organs at different seasons of the year, and by studying, at the same time, the habits of the animals.

The preceding description of the characters of the genus is taken from the anatomical details of a species found creeping upon the rocks under water in the Mauritius, by M. PERON, which CUVIER referred to the genus Onchidium of BUCHANAN, already noticed. We have ventured to institute the genus, and to name in honour of the discoverer of the first ascertained species. CUVIER conjectures that it breathes free air, and has accordingly inserted it among the *Pulmones aquatique.* Some doubts, however, may reasonably be entertained as to the truth of this supposition. It would certainly be an unexpected occurrence, to find a marine gasteropodous mollusca obliged to come to the surface, at intervals, to respire. It will probably be found that it is truly branchiferous.

Class II.—BRANCHIFERA.

The molluscous animals of this class are more numerous than those of the preceding. They chiefly inhabit the waters of the ocean, a few genera only being met with in fresh water lakes and rivers. The branchiæ which constitute their aerating organs, exhibit numerous varieties of form, position, and protection, and furnish valuable characters for their methodical distribution.

Order I.—BRANCHIÆ EXTERNAL.

The branchiæ are pedunculated, and more or less plumose. They are moveable at the will of the animal, and, in general, are capable of great alteration of form.

1st Tribe.

Branchiæ exposed. In nearly all the genera, the branchiæ are numerous and distributed regularly over the cloaks or sides.

A. Branchiæ issuing from the cloak dorsally.

1. *Body exposed, and destitute of a shell.*

In many species the back is covered with perforated papillæ, which pour out a mucous secretion. All the species are hermaphrodite, with reciprocal impregnation.

a. Anus situate near the posterior extremity of the back, and surrounded with a fringe of plumose branchiæ.

31. Doris. Oral tentacula two, vent without scales.

The cloak is covered with retractile papillæ, and separated from the foot by a distinct duplicature. Towards its anterior margin are placed the two superior tentacula. These are retractile, surrounded at the base with a short sheath, and supported on a slender stem, having an enlarged compound plicated summit. The neck is short, and above the mouth there is a small projecting membrane, connected at each side with the oral tentacula, which are in general minute, and of difficult detection.

The mouth is in the form of a short trunk, leading to fleshy lips, within which the tongue is placed. This last organ is covered with minute reflected hairs, and, from its motion, appears to be destined exclusively for deglutition. The gullet is a simple membranaceous tube, terminating in a stomach, which presents on the interior a few longitudinal folds. It is furnished with a small cæcum, the extremity of which receives the bile from the liver. The stomach likewise receives the secretion of another gland, which is not connected with the liver, in the form of a small bag, the inner surface of which is covered with numerous papillæ. The intestine is lodged in a groove on the surface of the liver, and proceeds directly to the anus.

The liver itself is divided into two lobes, and gives rise to numerous biliary ducts, which proceed to the stomach. But it likewise gives rise to a duct which proceeds to a small bag plaited on the inside, and afterwards opens on the surface at a small hole near the anus. It yet remains to be determined, whether the fluid carried off by this conduit be excrementitious matter, separated by the liver, or whether the gland which produces it be distinct from that organ, but so interwoven therewith as to elude the observation of the anatomist.

It is obvious, from the structure of the digestive organs, that the species subsist on soft food, requiring neither cutting nor grinding, and, in this respect, differ remarkably from the species of the genus Tritonia, which were formerly arranged along with them.

The organs of generation differ little from the other hermaphrodite gasteropoda. The vesicle furnishes two canals, one of which goes to the testicle, the other to the penis. There is likewise a minute bag connected with the canal of the latter. The spawn is deposited on sea-weeds and stones. It is gelatinous, of a white colour, and in appearance resembles the Spongia compressa.

The following species are natives of the British seas: 1. *D. Argo;* 2. *Verrucosa;* 3. *Lævis;* 4. *Marginata;* 5. *Nodosa;* 6. *Quadricornis;* 7. *Nigricans.*

32. POLYCERA. Oral tentacula more than two.

The branchiæ, when withdrawn, are protected by two scales. The superior tentacula resemble those of the doris, the oral ones are more numerous, sometimes amounting to six. *P. flava* and *pennigera* are British examples

b. Anus situate on the right side, and unconnected with the branchiæ.

(A.) Mouth furnished with corneous jaws.

These jaws are in the form of narrow plates, which cut the food by crossing each other like the blades of a pair of scissors.

(1.) Branchiæ disposed along the back or sides, and unconnected with membranaceous expansions.

(*a.*) Tentacula limited to two in number.

33. TERGIPES. Branchiæ furnished with a sheath at the base.

The branchiæ form a single row on each side, and are

qualified to act as suckers. The *Limax tergipes* of Forskael, Des. An. p. 99, is the type of the genus. *T. maculata*, described by Montagu, Lin. Trans. vii. p. 80, t. vii. f. 8, 9, is a British example.

34. Tritonia. Branchiæ destitute of basilar sheaths.

The branchiæ are in the form of plumes, or imbricated productions, placed in a row on each side the back. The tentacula, which are partially retractile, have a sheath at the base. In some of the species there are indications of eyes.

The mouth consists of two lips, which are placed longitudinally, and open into a short canal. The jaws consist of two corneous plates, united at the upper dorsal edge, slightly arched, and meeting at their upper margin, for the purpose of cutting. Within these is the tongue, which differs remarkably from the same member in the Doris. In the latter, the spines with which it is beset are reflected, and draw the food to the gullet, while in the former, the spines are deflected, and serve to keep the food within the reach of the jaws. The tongue of the doris, therefore, serves for deglutition, that of the tritonia for mastication. M. Cuvier describes the functions of both as similar. The salivary glands are placed on each side the gullet, and empty their contents behind the jaws. The gullet has a few longitudinal folds; the stomach is simple, scarcely differing from the gullet; and the intestine proceeds almost directly to the anus, situate on the right side. The liver is small, and situate behind, enveloping the stomach, and intimately united with the ovarium. The organs of generation exhibit nothing remarkable. The pedunculated vesicle has a simple canal. The external opening of the organs of generation is situate a little before and beneath the anus. The *T. arborescens, pinnatifida,* and *bifida,* are examples of British species.

(*b.*) Tentacula four in number.

The branchiæ are simple, tapering, or clavate, and disposed in transverse rows on each. These branchiæ in some species readily fall off, and, as if independent, are capable of swimming about for a short time in the water. This is executed by means of minute hairs with which their surface is covered, and which move rapidly, pushing forward the distal extremity.

35. MONTAGUA. Branchiæ in continuous rows across the back.

This genus, which differs from the other, not merely in the arrangement of the branchiæ, but in possessing a cluster of short papillæ on the right side, probably connected with the anus, we have ventured to name in honour of the late GEORGE MONTAGU, the well-known author of Testacea Britannica, and of several valuable papers in the Linnean Transactions, on molluscous animals. The two species which may be referred to this genus, were detected in Devonshire by this observer. The first, *M. longicornis*, (Lin. Trans. vol. ix. p. 107, tab. vii. f. 1,) is the type of the genus. The other species, *M. cærulea*, (Lin. Trans. vol. vii. p. 78, tab. vii. f. 4, 5,) is probably the type of another genus.

36. EOLIDA. Branchiæ interrupted on the back.

This genus, which was instituted by M. CUVIER, includes the following British examples: 1. *E. papillosa;* 2. *Plumosa;* 3. *Pedata;* 4. *Purpurascens* *.

(2.) Branchiæ disposed on lateral membranaceous expansions.

These expansions serve the double purpose of supporting the branchiæ, and acting as fins.

---

* There is a figure of this species given in Plate IV. f. 2.

37. SCYLLEA. Branchiæ seated dorsally on the fins. Tentacula two.

On each side of the back are two membranaceous expansions, and one on the tail, supporting on their dorsal surface scattered plumose branchiæ. Each of the tentacula is furnished with a large funnel-shaped sheath. The foot is very narrow, with a mesial groove, used in climbing up the stalks of sea-weeds. The mouth is placed at the base of the tentacula, and surrounded with a semicircular lip. The tongue is in the form of a tubercle, with reflected points. The gullet is plaited longitudinally. The stomach is short and cylindrical, with a ring of hard, longitudinal scales. The liver consists of six unequal globules, and the bile is poured into the cardiac extremity of the gullet. The *Scyllea pelagica* has been long known to naturalists, and appears to be very common in the equatorial seas, adhering to the stems of the *Fucus natans*.

38. GLAUCUS. Branchiæ seated on the margin of the fins. Four simple tentacula.

On each side of the body there are three or four membranaceous expansions, the margins of which are fringed with the simple branchial filaments. This genus was instituted by R. FORSTER, and the oldest known species, *G. radiatus*, is figured, Phil. Trans. vol. liii. tab. iii.

(B.) Mouth destitute of corneous jaws.

39. THETHYS. Branchiæ forming a row on each side of the back, consisting of fringed processes, alternately larger and smaller.

The body is ovate, with the cloak and foot continuous. The neck is distinct from the foot, and is narrow. Above, the neck is continuous with the cloak, from which arises a large semicircular expansion, used probably as a fin. The margin of this expansion is fringed with numerous filaments, and on the upper surface, within the border, is a row of conical tubercles. The true tentacula are placed towards the base of this fin near the neck. Each of them consists of a small fleshy cone, striated across, with a semicircular sheath behind. The branchiæ consist of a tapering, fleshy stalk, spirally twisted towards the summit with a series of filaments on one side. They are fourteen in number on each side, alternately and oppositely small and large. The anus opens in front of the third branchia on the right side. The orifice of generation is exhibited under the first branchia of the same side. In front of each of the larger branchiæ, is a small cavity with a small filament in the centre.

The mouth is situate underneath the tentacula. It consists of a large funnel, covered within with soft papillæ, destitute of jaws or tongue. The gullet is short, the stomach simple, fleshy, and covered with a thick cuticle. The salivary glands are slender and branched, and open into the gullet. The intestine is likewise short, and proceeds directly to the anus. The liver pours the bile into the canal at the pylorus; and likewise sends out another duct, which opens externally near the anus. The organs of generation are similar to the doris.

The *T. fimbria* is the type of the genus, a figure of which, with its anatomical details, it given by M. CUVIER, in his *Memoire sur le Genre Thethys*.

2. *Body concealed in a Spiral Shell.*

40. This section includes the genus VALVATA of MULLER, represented by two British species *V. cristata* (*Helix cristata* of MONTAGU), and *V. piscinalis* (*Turbo fontinalis*). These resemble in aspect the aquatic pulmoniferous gasteropoda. The branchiæ appear in the form of a feather, with a central stem, and a row of compound branches on each side, decreasing in size from the base to the free extremity. It issues from the neck near the middle, a

short way behind the anterior tentacula. Near this plume, but towards the right side, is a single simple filament, like a tentaculum. The anterior tentacula occupy the usual position, are setaceous, and have the eyes placed at the base behind. The spiral shell is capable of containing the body, and the aperture can be closed by a spirally striated operculum attached to the foot. The internal structure is unknown.

B. Branchiæ issuing laterally from between the cloak and foot.

This division includes the orders *Cyclo-branchia* and *Infero-branchia* of Cuvier, which we have ventured to bring together, as connected by the common character of the position of the gills.

(1.) *Body protected dorsally by a shelly covering.* *Cyclo-branchia.*

a. Shell simple.

41. Patella. Shell entire. Mouth with tentacula.

This genus differs from the others of this order. The back is covered by a conical shell, within the cavity of which the animal is capable of withdrawing itself. The cloak is large, covering both the head and foot. It is united with the shell along its superior margin. The foot is fleshy, and furnished with numerous muscular filaments, which unite, in the superior part of the cloak, to form a strong muscle, by which the body adheres to the shell. The action of this muscle brings the shell close to the surface to which the foot adheres, or removes it to a distance.

The head is furnished with a large, fleshy snout, supporting at the base two pointed tentacula. The eyes are placed on a small elevation at the external base of the tentacula. A little way behind the head, and below the cloak, on the right side, are two apertures, being the anus and orifice of generation. The gills occupy the same position as in the preceding genera. In some, the branchiæ form a complete circle; in others, the circle is interrupted anteriorly at the head.

Within the trunk, the mouth is fortified by two cartilaginous cheeks, which, at their union anteriorly, support the base of the tongue. This last is a most singular organ. It is longer than the whole body, narrow, and covered with three rows of short reflected spines, interrupted longitudinally and transversely. Its fixed end only can be exercised in deglutition, its free end being coiled up the abdomen. On the upper side of the mouth is a semicircular osseous plate, or upper jaw. The gullet is furnished with a dilatable pharynx. The stomach is elliptical, with the cardia and pylorus at opposite extremities. The intestines are variously folded, and are several times longer than the body. The salivary glands are minute. The liver is intimately united with the stomach and intestines.

The heart is situate on the left side, in the anterior part of the body. The auricle receives the aërated blood from one vein when the circle of the gills is complete, and by two when interrupted. This auricle is placed on the anterior side of the heart. An aorta arises from each side, to convey the blood to the body.

The ovarium is placed underneath the liver; and, as it exhibits some differences of organization, M. CUVIER infers that it likewise contains the male organs.

The species belonging to this genus are numerous, and appear to admit of distribution into sections; the first having the branchial circle complete, the second interrupted *.

b. Shell divided. Mouth destitute of tentacula.

---

* See a figure of P. vulgaris, Plate IV. f. 3.

42. Chiton. Shell constituting a series of imbricated dorsal plates.

The body is elliptical. The cloak is firm and cartilaginous, and variously marked on the margin. The dorsal plates are arched, and occupy the middle and sides of the back, where they are implanted in the cloak, in an imbricated manner, the posterior margin of the first valve covering the anterior margin of the second. The foot is narrow. The mouth is surrounded with a semicircular curled membrane, and is destitute of tentacula. The anus consists of a short tube, placed at the posterior extremity of the cloak. The external orifice of generation has not been detected.

The mouth is capable of forming a short proboscis. The tongue is short, and armed with strong, reflected spines. The gullet is short, and the stomach, which is lengthened and folded, is membranaceous. The intestine is several times longer than the body, and much folded. The liver is divided into numerous lobes, and intimately united with the stomach and intestines.

The heart is situate at the posterior part of the body. The auricle is placed posteriorly, and receives the aerated blood from two veins. Each vein descends along the base of the gills, collecting the aerated blood from the particular side of the body to which it belongs; and, what is most remarkable, when opposite the ventricle, it is suddenly enlarged, and sends off a branch which communicates with it, and again contracts and unites with its fellow from the opposite side, to form the auricle. A single aorta arises from the anterior side.

The ovarium is conical, and divided into numerous lobes. Behind, two ducts seem to arise, and to proceed one to each side; but it has not been determined whether they open externally. No male organs have been detected; nor is there any thing accurately known with regard to the peculiar nature of their hermaphroditism.

43. CHITONELLUS. Dorsal plates not imbricated.

In this genus, instituted by LAMARCK, (Hist. Nat. des Animaux sans Vertebres, vol. vi. p. 316.), the shells are slender, narrow, and are disposed longitudinally, and not in contact, along the middle of the back, leaving the sides of the cloak naked. Two species, *C. lævis* and *striatus* were brought from New Holland, by PERON and LE SUEUR.

(2.) *Body naked. Infero-branchia.*

44. PHYLLIDIA. Anus placed dorsally near the extremity of the cloak.

The body, in the animals of this genus, is ovate. The foot is narrow in front. The cloak is broad, coriaceous, and destitute of a shell. Towards its anterior extremity are two cavities, from which issue the retractile superior tentacula, as in the genus Doris. Nearly at the posterior extremity is another cavity, containing the anus. This opening, though similar in situation to that of the Doris, is merely a short simple tube. The head is immediately above the anterior margin of the foot, above which is the mouth, having a small conical feeler on each side. Under the margin of the cloak on the right side, and about half way between the mouth and the middle of the body, are two openings, in a tubercle, for the organs of generation. The branchiæ consist of slender complicated leaves, which surround the body between the foot and the cloak. The circle is interrupted at the head and at the tubercle of generation.

The mouth is destitute of jaws. The gullet is simple, ending in a membranaceous stomach. The pylorus is placed near the cardia, and the intestine goes directly to the anus. The salivary glands are small, and placed near the mouth. The liver is large in proportion.

The heart is placed in the middle of the back. The auricle is simple, placed on the side next the tail, and supplied by the two systemic veins which collect the aërated blood from the branchiæ on each side. There is a simple aorta arising from the opposite side of the heart.

The organs of generation appear to be similar to those of the preceding class; but they have not as yet been minutely examined. The existence of eyes is not satisfactorily determined.

The animals of this genus appear to be inhabitants of the tropical seas. CUVIER has given descriptions and figures of three species, which differ remarkably from one another in the protuberances of the cloak.

45. DIPHYLLIDIA. Anus placed on the right side.

This genus was formed by M. CUVIER in his *Regne Animal*, vol. iii. p. 395, from an imperfectly investigated animal, in the cabinet of M. BRUGMANS at Leyden. The cloak is pointed behind, with a feeler and small tubercle on each side.

2d Tribe.

Branchiæ simple, and concealed when at rest under a lid. *Tectibranchia* of CUVIER.

A. Head furnished with tentacula.

1. *Tentacula four in number.*

a. Branchiæ lateral.

46. APLYSIA. Branchiæ with a corneous lid.

The body of the Aplysia is ovate, acuminated behind, and produced before to form a neck. The foot is narrower than the body. In the middle of the back is a corneous plate inclosed in a bag in the skin, and on each side, and behind, there is a fold by which this part may be concealed. The head is slightly emarginate, with a feeler on each side. The superior feelers are situate on the neck. In front of each of these is a small black point or eye.

The branchiæ are situate underneath the dorsal plate, on the right side, and exhibit a complicated plumose ridge, capable of expansion beyond the edge of the plate. The anus is situate immediately behind the branchiæ, and before these is the orifice of generation, from which proceeds a groove along the neck to the inferior base of the fore feeler, on the right side, where there is an opening for the penis.

Within the longitudinal lips there are two smooth, corneous plates, the substitutes for jaws; the tongue is rough, as in many of the other gasteropoda. The gullet is short, and suddenly expands into a large subspiral crop, with membranaceous walls. To this, a gizzard with muscular walls succeeds, the interior of which is armed with numerous pyramidal teeth, with irregular summits, of a cartilaginous nature. The connection between these teeth and the integuments is so slender, that they are displaced by the application of the smallest force. They, however, project so far into the cavity, as to offer resistance to the progress of the food. There is yet another stomach, armed on the one side with deflected, pointed, cartilaginous teeth. At the pyloric extremity are two membranaceous ridges, between which are biliary orifices, and the opening into a long narrow cœcum, with simple walls, which is contained within the liver. The intestine is simple, and after two turns ends in a rectum.

The salivary glands are very long, and, as usual, empty their contents into the pharynx. The liver is divided into three portions by the folds of the intestine, each of which consists of several lobes. The biliary vessels are very large, and open at the mouth of the cœcum into the last stomach. The food of the Aplysia consists of sea-weeds and minute shells.

The circulating organs are remarkable. On each side the body, in the region of the dorsal plate, there is a large vessel, which receives blood from different parts of the body, and which likewise, by various openings, has a free communication with the cavity of the abdomen. In this respect there is a resemblance to the spongy, glandular bodies of the venæ cavæ of the Cephalopoda. These two vessels, or *venæ cavæ*, unite posteriorly, and transmit their contents to the gills. The aërated blood is now conveyed to an auricle, of large dimensions, and uncommonly thin walls, situate beneath and towards the front of the dorsal plate, and emptying its contents through a valve, into the right side of the ventricle. The aorta, which issues from the left and anterior side, divides into two branches, the smallest of which proceeds to the liver on the left. The larger branch is again divided, the smaller branch proceeding to the stomach. The largest trunk that remains, before it leaves the pericardium, has two singular bodies attached to it, consisting of comparatively large vessels, opening from this aortic branch. The use of these glands is unknown.

The organs of generation likewise exhibit some remarkable peculiarities. The ovarium is situate in the posterior part of the abdomen. The oviduct is tortuous in its course, passes along the surface of the testicle, and, after uniting with a clavate appendage, opens into a common canal. The testicle is firm, apparently homogeneous in its texture, of a yellow colour, with spiral ridges on its surface. The vas deferens arises from a complex, glandular body, and unites with the common canal. This common duct, before it reaches the external orifice, receives the contents of the pedunculated vesicle, and has attached to it a botryoidal, glandular organ, the use of which is unknown, but which some suppose to be employed to secrete an acrid liquor regarded as venomous.

It is obvious from this structure, that the seminal fluid and eggs must come in contact in the common canal, and at the single orifice, provided they are both ejected at the same time. From the orifice to the right fore-feeler there is a sulcus, leading to the pore containing the retractile penis. This organ, like those of the other mollusca, is solid. It terminates in a small filament. The external groove is the only connection between it and the other sexual organs.

There is a peculiar secretion of a purple fluid, which here deserves to be recorded. It issues from a spongy texture, underneath the free side of the dorsal plate. Connected with this cellular reservoir is a glandular body of a considerable size, which is supposed to secrete the coloured fluid. This gland is supplied by a large branch of the glandular aorta, and gives out two very large veins to the left vena cava.

The purple fluid itself has never been carefully investigated. It is not altered by the air after drying, nor is its colour destroyed by acids or alkalies, although the tint is a little changed, and rendered less pure. Both these reagents precipitate white flakes from the fluid.

This liquor is poured out by the animal when in danger or constrained, and colours the water for several yards around. It ejects it readily when put in fresh water; and, when entangled in a net, several yards of it in the neighbourhood are sometimes stained, greatly to the amazement of the unsuspecting fishermen.

The Aplysia has been long known in the records of superstition under the name of the *Sea Hare*. Its flesh, and the inky fluid it pours out, have been regarded as deleterious to the human frame. Even to touch it was supposed to occasion the loss of the hair; while the sight of it would not fail to subdue the obstinacy of concealed pregnancy. The progress of science has exposed the errors, or perhaps tricks, of the earlier observers, and proved the innocence of an animal formerly invested with every repulsive and noxious attribute. The *A. depilans*, the type of the genus, is of frequent occurrence on the British shores. The *A. punctata* of Cuvier may be regarded merely as a variety *.

47. Notarchus. Lid of the branchiæ soft.

There is an oblique groove from the neck leading to the branchiæ. The structure is similar to Aplysia. M. Cuvier instituted this genus in his *Regne Animal*, vol. ii. p. 395, and vol. iv. tab. xi. f. 1.

b. Branchiæ terminal.

48. Dolabella. Dorsal plate a solid shell.

---

* It is probably at this place where the genus Gasteroplax of Blainville, published by Lamarck, under the ill-judged title Umbrella (*Hist. Nat. &c.* vol. vi. p. 339), should be introduced. The following is the extended character which has been communicated:

" Corpus valde crassum, obovatum, testa dorsali onustum; pede amplissimo, subtus plano, undique prominente, anterius sinu emarginato, postice attenuato. Caput non distinctum. Cavitas infundibuliformis in sinu antico pedis os in fundo recondens. Tentaculo quatuor; superiora duo, crassa, brevia, truncata, hinc fissa, intus transversim sublamellosa; altera duo, tenuia, cristata, pedicellata, ad oris latera. Branchiæ foliaceæ, serratim ordinatæ, infra cutis marginem per totam longitudinem lateris dextri. Anus post extremitatem posticam branchiarum.

Testa externa, orbicularis, subirregularis, planulata, superne convexiuscula, albida, versus medium mucrone, apicicali brevissimo præbita; marginibus acutis: interna facie subconcava; disco calloso, colorato, ad centrum impresso, limbo lævi cincto."

Doubts seem to be entertained whether the shell is to be considered as belonging to the cloak or the foot. Two species are known.

This genus differs from Aplysia, in the dorsal plate being calcareous and hard. The fore part of the body is narrow; behind, it is larger, and obliquely truncated. The disc thus formed is circular, surrounded with a fringe of fleshy filaments. From the centre of this disc, a longitudinal slit extends forward, a little way beyond the anterior margin, and contains the branchiæ. The position and structure of the other organs are precisely similar to those of the Aplysia.

This genus was instituted by LAMARCK, from characters derived exclusively from the dorsal plate or shell. CUVIER afterwards examined a species brought from the Mauritius by PERON, which he considers as the one figured by RUMPHIUS in his Amboinshe Rariteitkamer, Tab. x. No. 6., and which he has consecrated to his memory, naming it Dolabella Rumphii.

2. Tentacula two in number.

49. PLEUROBRANCHUS. Cloak and foot expanded, between which, on the middle of the right side, the branchiæ are placed. The cloak is strengthened in the middle above the branchiæ by a thin expanded subspiral shell. The neck is short, and in some contracted, with the front emarginate, exhibiting the commencement of the inferior tentacula. The upper tentacula are tubular and cloven. The gills occur at the edge of the dorsal plate. In front of these are the orifices of the organs of generation, and the anus is situate immediately behind the gills.

The mouth is furnished with a short retractile, proboscis. The tongue occupies both sides of the mouth, and is covered with spines. The gullet is enlarged into a kind of crop before it enters the stomach: this is folded, and is divided by contractions into three parts. The first stomach has muscular walls of moderate thickness, with a single longitudinal band. The second has membranaceous walls, with longitudinal internal ridges, and the third has thin and simple walls.

The gut is short. The salivary glands are situate at the folds of the stomach, and by two canals empty their contents into the mouth. The liver is placed on the stomach, and empties itself into the lower part of the crop.

The heart is nearly in the middle of the back. Its auricle is on the right side, at the base of the branchiæ; and the ventricle sends out at the opposite side three arteries.

M. CUVIER has figured and described the P. Peronii with its anatomical details. Two species likewise appear to be known as natives of the British seas.

B. Head destitute of tentacula.

50. BULLA. Body of the animal protected by a convoluted shell.

The body is oblong, becoming a little narrower in front. Below, the foot is broad, thin, and waved on the margin, expanded on each side behind, and capable of being turned upwards. At the posterior part of the foot, but separated from it by a groove, there is a broad, membranaceous appendage, a part of which is folded upwards, and a part spread over bodies, like the foot. It assists in closing the mouth of the shell, and in its position and use is analogous to the operculum, in the following order. Above the foot, in front, also, but separated from it by a groove, there is a fat, fleshy expansion, which CUVIER terms the Tentacular Disc, considering it as formed by the union of the inferior and superior tentacula. In the centre of the disc, in the *Bulla hydatis* *, MONTAGU observed two eyes. Between this portion of the back and the posterior extremity, is the dorsal plate or shell, forming the genus *Bulla* of conchologists. In some species, this shell is covered by the integuments, while in others it is exposed. But in all, the part containing it is partially con-

---

* Linn. Trans. vol. ix. tab. 6. f. 4.

cealed by the animal, by means of the reflected margins of the foot, and its appendage. Along the right side of the body there is a groove, formed by the foot and its appendage, on one side, and the dorsal plate and tentacular disc on the other. The branchiæ are situate in a cavity under the shell or dorsal plate, and resemble those of the *Aplysia*. Behind the gills, in the lateral groove, is the anus; and in front of these, the orifice of the united organs of generation. The penis is removed as in the *Aplysia*, and connected by a similar slit.

The mouth is, as usual, in front, above the foot and beneath the tentacular disc, both of which serve as lips. The cheeks are strengthened on each side by a corneous plate. The tongue is well developed in some, as the *B. ampulla*, while in the *B. aperta* it is reduced to a small tubercle. The gullet is large, and in the *B. lignaria* makes two folds before entering the gizzard. This last organ is fortified by three testaceous plates, convex and rough on the inner surface, and attached to strong, muscular walls. These plates exhibit in the different species considerable varieties of form and markings. The intestine, before terminating in the anus, makes several convolutions in the substance of the liver. The salivary glands exhibit considerable differences. In the *B. ampulla*, they are long and narrow, and their inferior extremity fixed to the gizzard. In the *B. aperta* and *lignaria*, they are short, with the extremity free. In the *B. hydatis* they are long, unequal, and the extremity of the one, belonging to the left side, is forked. The liver forms a part of the contents inclosed in the spire of the shell. It envelopes the intestine, and empties the bile into its pyloric extremity.

The auricle and ventricle appear to occupy the same relative position as in the *Aplysia*, but the structure of the arteries is unknown. The organs of generation have also so near a resemblance as to forbid a detailed description. Some species are said to eject a coloured fluid, like the *Aplysia*, from the lid of the branchiæ. A gland is observed in the *Bulla lignaria*, similar to the *Aplysia*, in which it is probable the fluid is prepared.

The species of this genus have not been sufficiently investigated in a living state. When preserved in spirits, it is impossible to form a correct idea of their true appearance, as exhibited when alive in sea-water, since they usually exist as a shapeless mass. CUVIER has given delineations of such preserved species, but they bear no resemblance to the figures of MONTAGU, of the same species, taken from living objects.

M. LAMARCK is inclined to divide this genus into two, distinguishing those which have the shell concealed, by the term *Bullæa*, from such as have the shell in part exposed, which he retains in the genus *Bulla*. The shells of the genus *Bullæa* are thin and white, as *B. aperta*; those of *Bulla* stronger, more opake, and covered with an epidermis, which, after the death of the animal, is easily detached, as *B. lignaria*.

51. DORIDIUM (of Mekel). Destitute of a dorsal plate or shell.

There is a cavity in the cloak, with a spiral turn. The branchiæ, and accompanying organs, are placed far behind. There is here no appearance of a spinous tongue; the gullet is simple, and the stomach is membranaceous. *D. carnosum*, a native of the Mediterranean, is the type of the genus.

Order II.—BRANCHIÆ INTERNAL.

The aerating organs are contained in a cavity, and appear in the form of sessile, pectinated ridges.

1st Subdivision.

Heart entire, and detached from the rectum.

This group, forming the *Pectinibranchia* of CUVIER, includes nearly all the marine gasteropoda which have spiral univalve shells. It likewise contains a few species which inhabit the fresh water.

The foot is usually fortified above, on its posterior extremity, with a corneous plate, which acts as a lid to the shell, when the animal is withdrawn into the cavity. The anterior extremity is in some of the species double. The anterior margin of the cloak forms a thick band, or arch, rising from the foot, behind which is the portion of the body that is always contained in the shell, and which is covered with a very thin skin. Between the margin of the cloak and foot is situate the head, supported on a short neck. The tentacula are two in number, bearing eyes at their base, or on short lateral processes, which have some claims to be considered as tentacula. The hood is frequently emarginate, and sometimes fringed. The mouth is more or less in the form of a proboscis, in some cases armed within with spinous lips, or furnished with a long narrow spiral tongue, armed with spines, as in the common periwinkle. The nature of this kind of tongue, the spiral extremity of which is free and lodged in the abdomen, is not well understood.

The entry to the gills is by a large aperture between the margin of the cloak and neck, at the middle, or towards the right side. These are contained in a cavity on the back of the animal, and consist of leaves arranged in one or more rows, which adhere to the walls of the cavity. At the entrance of this cavity is the anus and oviduct.

The male and female organs are considered not only as distinct, but as occurring on different individuals. The evidence in support of this opinion is in many cases complete. The penis is in some external, and incapable of being withdrawn, while in others it is retractile, and situate in a cavity in the right tentaculum.

The body of the animal is attached to the shell by means of two muscles, which adhere to the pillar near the same place, and shift their position, by an arrangement not well understood, in proportion as the individual increases in size. These muscles terminate in the foot and mouth.

The animals of this order have not been examined sufficiently in detail, to admit of their distribution into natural groups, distinguished by characters founded on important differences of organisation. The form of the shell has been resorted to, with the view of assisting arrangement. The characters thus furnished would be useful and valuable, were they the index of any peculiar internal structure. But, unfortunately, animals, widely different in structure, inhabit shells of the same form, and *vice versa*, so that, however useful the mere conchologist may find the form of the shell to be in his arrangements, it can only be regarded by the zoologist as occupying a subordinate place. Without, therefore, entering into any details regarding the structure of the few species which have been examined anatomically, we shall merely point out the tribes and families which have been contemplated, the characters of which in a great measure depend on the shape of the shell.

1st Tribe.
Shell external.

The shelly covering exhibits all the variations of the spiral form. The internal structure has hitherto been in a great measure neglected, so that the characters employed in the methodical distribution of the species and genera, are derived from the shelly appendage of the cloak. The groups, therefore, are merely artificial, temporary combinations.

A. Aperture of the shell entire.

As co-existent with this character of the shell, the anterior margin of the cloak, at the entrance to the branchial cavity, is found likewise to be entire.

1. *Aperture of the shell closed by a pedal lid, or operculum.*

The three following families appear to be the indications of as many natural groups, the genera of which admit of still more minute arrangement.

Family 1. *Turbonidæ.* Aperture of the shell round or ovate.

This family includes the greater number of the species of the Linnæan genus *Turbo*. The genera into which it is now divided may be distributed into two sections, from the residence of the animals.

Section 1. *Marine.*

The marine Turbonidæ are of frequent occurrence, and compose the genera Turbo, Delphinula, Turritella, Scalaria, Odostomia, Monodonta, Phasianella, and Vermicularia. Some of the species are known to be ovoviviparous, and it is probable that the same kind of reproduction prevails in all of them. Remarkable differences may be observed in the form of the hood, the length of the peduncles supporting the eyes, and the number and distribution of the filaments surrounding the body.

Section 2. *Fluviatile.*

The Fluviatile Turbonidæ are limited in their number, both in regard to genera and species. The genera are only three, Ampullaria, Melania, and Paludina. In the last of these, including the *Helix vivipara*, and *tentaculata*, the sexes are obviously distinct in different individuals.

Family 2. *Neritadæ.* Aperture semicircular, with an oblique, straight pillar-lip.

Section 1. *Marine.*

This includes the genera Nerita and Natica.

Section 2. *Fluviatile.*

This contains only the genus Neritina, including *Nerita fluviatilis* of LINNÆUS.

Family 3. *Trochusidæ.* Aperture of the shell subquadrangular.

All the genera, including Trochus, Solarium, and Pyramidella, are marine. The cloak on each side is usually ornamented with three filaments.

2. *Aperture of the shell exposed.*—The foot destitute of a lid. Marine.

52. JANTHINA. Foot with an adhering spongy body.

In this genus, represented by the *Helix janthina* of LINNÆUS, the spongy body is capable of changing its dimensions, and enabling the animal to sink or rise in the water at pleasure. When irritated, it ejects a purple fluid from the cellular margin of the cloak above the gills, not unlike the Aplysia. This species was added to the British Fauna, by the late Miss HUTCHINS.

53. VELUTINA. Foot simple.

This genus was formed by us for the reception of the *Bulla velutina* of MULLER (*Zool. Dan.* tab. ci. f. 1, 2, 3, 4), the *Helix lævigata* of British writers.

The foot is destitute of lid or appendage, and is broad before, and pointed behind. The tentacula are two in number, short and filiform, with eyes at their external base. The head is broad and short. In addition to these characters given by MULLER, we have been enabled to add the following, from a specimen, somewhat altered, which was found in the stomach of a cod-fish. The animal adheres to the shell by two linear muscles, one on each side the cloak. The branchial cavity is towards the left side. The tongue is spinous, narrow, with its free extremity spiral. Eyes rather behind the tentacula. Penis exserted on the right side of the neck, immediately behind the eye. Cloak large in proportion to the size of the foot. We have termed the genus *Velutina*, bestowing on the species the trivial name *vulgaris*.

B. Anterior margin of the aperture of the shell canaliculated.

This groove in the aperture of the shell is produced by the anterior margin of the cloak being extended over the opening into the gills, for the purpose of acting like a tube or syphon, in conveying the water to and from the branchial cavity. The species are considered as oviparous, with distinct sexes in separate individuals.

1. *Shell convoluted.*—The shell has an oval or linear mouth parallel with its length. The whorls, which are small segments of large circles, are wrapped round the pillar, and the one rising a little above the other, embrace or inclose the preceding ones. The four following families appear to belong to this division.

Family 1. *Conusidæ.* Furnished with a long proboscis, and produced tentacula, with the eyes near the summit on the outside. The lid is placed obliquely on the foot, and is too small to fill the mouth of the shell. The genera Conus and Terebellum form this family.

Family 2. *Cypreadæ.* Cloak enlarged, and capable of folding over the shell. There is no lid. The genus Cyprea is the type.

Family 3. *Ovuladæ.* Both extremities of the aperture canaliculated. The inhabitants of all the genera, Ovula, Calpurna, and Volva, are unknown. The last genus includes the *Bulla patula* of PENNANT.

Family 4. *Volutadæ.* Canal of the aperture abbreviated. Pillar-lip plaited. The foot appears to be destitute of a lid. The genera are numerous; Voluta, Oliva, Cymbium, Marginella, Cancellaria, Mitra, Ancilla, Volvaria, and Tornatella. The last genus contains the *Voluta tornatilis* of British writers.

2. *Shell turreted.* The whorls of the shell, the revolving spire of which is subconical, scarcely embrace one another, but are merely united at the margins. Three families may here be established.

Family 1. *Buccinidæ.* Canal short, scarcely produced beyond the anterior margin of the lip, and bent towards the right. The tentacula are remote, and the head is destitute of a hood. The mouth has a retractile proboscis. The following genera belong to this family: Buccinum, Eburna, Dolium, Harpa, Nassa, Purpura, Cassis, Morio, and Terebra.

Family 2. *Muricedæ.* Canal produced, and straight. The tentacula approach the head and mouth as in the preceding family. The genera are Murex, Typhis, Ranella, Fusus, Pleurotoma, Pyrula, Fasciolaria, and Turbinella.

Family 3. *Cerithiadæ.* Canal short and recurved. Head with a hood. This family contains the marine genus Cerithium, and the fluviatile one Potamidum.

Family 4. *Strombusidæ.* Canal short, and bent towards the right. The outer margin of the aperture becomes palmated with age, and exhibits a second canal, generally near the former, for the passage of the head. The following are the genera: Strombus, Pterocera, and Rostellaria.

2d Tribe.
Shell internal.

This tribe consists at present of only one genus, termed Sigaretus, two species of which are natives of Britain. The foot of the animals belonging to this genus, or rather of the species which constitutes the type, is oval, with a duplicature in front. The cloak is broad, with an indentation on the left side, in front, leading to the branchial cavity. A ring of transverse muscles unites the cloak with the foot. On the back is placed the shell, which does not appear on the outside, as it is covered by a thick cuticle. It is lodged in a sac, and united by a muscle, which adheres to the pillar. The hood is produced, at each side, into a flattened tentaculum, with an eye at the external base. The anus is situate at the branchial indentation on the left side. The penis is situate on the right side of the neck; it is external, with a crooked, blunt, lateral process near its extremity.

The mouth is in the form of a short proboscis. The tongue is armed with spines, and is long and spirally folded. The salivary glands are large. The stomach is membranaceous, giving off the intestine near the cardia. The intestine makes two folds. The liver, with the testicle in the male, and the ovarium in the female, occupy the posterior part of the body, under the spire of the shell.

2d Subdivision.

Heart traversed by the rectum.

This group includes the order Scutibranchia of CUVIER. In general form, and in the structure and position of the branchiæ, the resemblance to the genera of the preceding subdivision is very great. They differ, however, in many particulars. The heart is furnished with two auricles, and is perforated by the intestine. The sexes appear to be incorporated in the same individual, or rather the male organs are unknown. The body is protected by a shell, the aperture of which is wide, and never closed by a lid.

1st Tribe.

Shell ear-shaped, flat, with a lateral, and nearly concealed spire.

Family, *Haliotidæ*, including the genera Haliotis, Padola, and Stomatia. These genera exhibit well-marked characters in the shell. The left margin of the shell in

Haliotis, is pierced by a row of holes. In Padola, these holes are nearly obliterated; but there is an internal groove and external ridge in the line of their direction. In Stomatia, there are neither holes nor ridges. In the Halyotis, the foot is oval and large. The sides of the body all round are ornamented with one or more rows of simple or branched filaments. The shell is placed on the back with the spiral part behind, and the row of holes on the left side, through which some of the filaments are protruded. The animal is attached to the shell by a single large muscle. The entry to the branchial cavity, which likewise contains the termination of the rectum and oviduct, is on the back. The gills are in two ridges, consisting of complicated branched filaments. At the entrance of the cavity, the cloak is furnished with a slit, the left margin of which rests upon the pillar of the shell. The edges of this slit are furnished with filaments, which pass through the anterior holes of the shell. The use of this singular arrangement is unknown. The branchial cavity likewise contains the viscous organ, in common with the Pectinibranchiæ.

The hood is emarginate, with a long tentaculum on each side, behind which, towards the side, is a cylindrical protuberance, bearing the eye at the top. The mouth is in the form of a short proboscis, with two corneous plates as cheeks, and a long narrow tongue extending backwards, and covered with spines. The pharynx is dilatable, with internal folds. The salivary glands are very small. The gullet is very short. The stomach is divided into two portions, the first of which is striated longitudinally with a glandular structure, and receives a biliary duct. The second is separated from the former by a valve, is smaller, with transverse striæ, and a double ridge. It likewise receives bile through two apertures. There is another valve at the pylorus; and the intestine, after making some turns, is surrounded by

the heart. There is an auricle on each side, receiving the aërated blood from each of the gills.

2d Tribe.

Shell conical, simple, or slightly revolute at the apex.

A. Cavity of the shell interrupted by a testaceous plate. This division consists of three genera, each of which may be regarded as the type of a family, although, for the present, they are all included in one.

Family, *Crepiduladæ*. The *marine* genera are, Crepidula and Calyptrea, the latter including the *Patella Chinensis* of British writers. In the former, the gills form a transverse ridge on the roof of the cavity, consisting of filaments extending beyond the margin. The eyes are at the base of the tentacula. There is only one fluviatile genus, termed Navicella.

B. Cavity of the shell entire.

Family 1. *Capuluside*. Shell entire.

This includes the genera Capulus (containing *Patella*, *Hungarica*, and *Antiquata*, of British writers) and Carinaria, represented by the *Argonauta vitrea*. In the Capulus, the foot is complicated on its anterior margin. The shell adheres to the animal by a circular muscle, leaving an opening in front, for the issue of the head and entrance to the branchial cavity. The gills form a single ridge across the roof. The mouth is in the form of an extended proboscis, with a deep groove above. The tentacula, which are two in number, have the eyes at the external base. The anus is on the right side of the branchial cavity. In the Carinaria, the foot appears to be compressed, and formed for swimming. The head is covered with a group of tubercles. The mouth is furnished with a proboscis. Near the middle of the body, the shell is attached. The surface of the body above is closely covered with small tubercles.

It is probable that the species here alluded to is the same with the *Pterotrachea coronata* of FORASKÆL.

Family 2. *Fissurelladæ.* Shell with a slit, or perforation.

In the Fissurella, the apex of the shell is perforated, and united to the cloak by a circular muscle open in front. The cloak forms a duplicature in front for the branchial cavity, which extends to the perforated apex of the shell. The gills consist of two ridges; at the dorsal extremity of which is the anus. It is probable that the excrements are ejected at the perforation in the apex of the shell, and likewise the water which enters the branchial cavity in front. The head is furnished with two tentacula, bearing the eyes at the external base. The *Patella græca* and *apertura* may be quoted as British examples of the genus.

The genus Emarginula differs from the former in the apex of the shell not being perforated. Its place, however, is supplied by a slit on the anterior margin, which is the entrance to the branchiæ and anus. The foot is surrounded with a row of filaments, and the eyes are supported on short foot-stalks, characters in which it approaches the genus Halyotis. The *Patella fissura* of conchologists is considered as the type of the genus.

DIVISION II.

MOLLUSCA ACEPHALA.

Destitute of a distinct head, or neck.

The animals of this division are much more simple in their organization than those of the preceding division. In none of the species are there any rudiments of organs of hearing or of sight. They are destitute of jaws or other hard parts about the mouth. They all inhabit the water, and possess branchiæ. The organs of the two sexes are incorporated in the same individual, and reciprocal union is unnecessary. They are either oviparous, or ovoviviparous. The presence or absence of a shelly covering, furnish characters for a twofold distribution of the groups.

Section 1.—ACEPHALA CONCHIFERA.

The shell in all cases is external and bivalve; and exhibits very remarkable differences in the form, relative size, and connection of the valves. The cloak is likewise in the form of two leaves, corresponding with the valves which protect it.

Order I.—BRACHIOPODA.

Mouth with a spiral arm on each side fringed with filaments.

The genera included in this group constitute the Brachiopoda of CUVIER. The lobes of the cloak are free anteriorly. From the body, between the lobes, the arms have their origin at the margin of the mouth. These arms are capable of folding up spirally. All the species are permanently attached to foreign bodies, and inhabit the sea. Their nervous and reproductive systems have received but little elucidation.

1st Subdivision.

Shell supported on a fleshy peduncle.

54. LINGULA. Valves equal, the apex of both attached to the peduncle.

The peduncle is nearly cylindrical, cartilaginous, and covered with a membrane consisting of circular fibres. The valves are oval, flat, and destitute of teeth, or elastic ligaments. The adductor muscles are numerous, obliquely placed, and appear capable of giving to the valves a considerable degree of lateral motion. The cloak is thin, and has interspersed muscular fibres. Its margin is thickened, and fringed with fine hairs of nearly equal length.

The arms are fleshy in their substance, conical, elongated and compressed in their form, and ornamented on the external surface with thickset fringes or tentacula. The mouth is simple, and situate between the arms at their base, There is no enlargement of the alimentary canal, which can be regarded as a stomach, and the anus is a simple aperture situate on the side. There are marked indications of salivary glands and a liver. The blood is conveyed to the gills by two vessels, which are divided at the separation of the lobes into two branches, one of these going to the half of one lobe, and another to the opposite half of the other lobe. Two systemic veins occupy a similar position, and return the aërated blood to the two lateral systemic ventricles. The gills themselves are arranged in a pectinated form, on the inner surface of each lobe of the cloak. There is nothing known of the nervous or reproductive systems of this animal.

The *Lingula unguis* is the only species of the genus, and appears to be confined to the Indian seas. The valves were first figured by SEBA, together with the peduncle by which they are supported. LINNÆUS having seen only one valve, conjectured that it belonged to Patella, and named it *P. unguis*. CHEMNITZ examined both valves, without the peduncle, and pronounced them connected with the genus Pinna. BUGIERE, aware of SEBA's figure, contemplated the formation of the new genus for its reception, which LAMARCK executed. M. CUVIER afterwards dissected one of the individuals, which SEBA had possessed, and unfolded characters in its organisation, sufficient not only to warrant the construction of a new genus, but a new class.

Some petrifactions have recently been referred to this genus; but, in the absence of all vestige of the peduncle, we do not consider the mere form of the shell as furnishing characters sufficiently obvious and precise to warrant such distribution.

55. TEREBRATULA. Valves unequal; the peduncle passing through an aperture in the largest valve.

The arms are shorter than those of the Lingula, and are said to be forked. They are supported within by numerous arcuated plates

M. LAMARCK divides the recent kinds into two sections.

1. *Shell smooth, or destitute of longitudinal ribs.* The *T. cranium*, a native of the Zetland seas, may be quoted as an example The peduncle is simple *.

2. *Shell ribbed longitudinally.* The *T. aurita* †, which inhabits Loch Broom, is another, though recently discovered British example. The larger valve is broadest in the middle, semicircular in front, and narrowing towards the apex, in consequence of the sides being compressed or bent inwards. The ribs from the beak towards the anterior margin are the most distinct, rounded, and about eight in number; those towards the sides are obsolete. The under valve is nearly orbicular, with the margin at the hinge truncated, or rather obtusely angular, and having the sides depressed, and forming small auricles, as in the genus Pecten, but not produced. The ribs are obsoletely wrinkled across, and the margin is waved by the ribs being concave internally. The inner surface of both valves, especially the largest, is finely punctulated. The hinge is formed by a projection on each side, the proximal margin of the perforation in the large valve entering corresponding depressions in the smaller one. The margin of the perforation itself is completed by the application of the smaller valve.

---

* See Plate IV. f. 4. † Plate IV. f. 5.

The peduncle is short, and consists of numerous unequal-sized tubular threads, attached by a complicated tendino-muscular apparatus, chiefly to the larger valve. The spiral arms seemed to have simple summits, and to be destitute of testaceous plates. The smallness and probable youth of the subject, however, rendered a minute examination of the structure impracticable. Trawled up by us in Loch Broom, near the harbour of Stornoway, 16th August 1821. This species approaches nearest to the *T. truncata*.

The fossil species of this genus are numerous, and occur in the older and newer floetz formations. They furnish obvious characters for the construction of many genera, some of which have been already established.

2d Subdivision.

Shell sessile.

56. CRIOPUS. Under valve cemented to stones.

The under valve is membranaceous, flat, and adhering; the upper is flatly conical, and resembles a Patella, in which genus, from neglecting the structure of the animal, it has usually been placed.

The *C. anomalus*, *Patella anomala* of MULLER (*Zool. Dan.* tab. v. f. 1. 8.), has a branched, double ovarium, with round eggs. It has been described and figured by us as a native of the Zetland seas, under the term *Patella distorta* (*Edin. Encyclopædia*, vol. vii. p. 65. tab. cciv. f. 4.; and by MONTAGU, *Linn. Trans.* vol. xi. p. 195. tab. xiii. f. 5.)

Order II.—BIVALVIA.

Mouth destitute of fringed spiral arms.

The animals of this group form the class Conchifera of LAMARCK, the Bivalvia of the older naturalists. The shells exhibit great variety of form and relative size. They are joined together at the hinge, which is either plain or toothed, and corresponds in position with the back of the animal. The connection of the two valves is secured by the intervention of an elastic horny *ligament*, the office of which is to keep the valves open. It is either external or internal. The valves are closed by means of *adductor muscles*, intermixed with tendons, and, passing transversely through the animal, adhere to the corresponding places in the inside of each shell. By the contractions of these muscles the free edges of the valves are brought into contact, at the same time that the ligament is compressed or stretched, according as it is internal or external. The number of muscular impressions is employed by LAMARCK in the division of the Bivalvia into two orders, *Dimyaires* and *Monomyaires*. This distinction, however, he has not attended to with care, as in his family *Mytilacées*, which he includes in his second order, or those having one adductor muscle, there are obviously two adductor muscles, although the one is certainly much larger and more complicated than the other. Besides these impressions of the adductor muscles, there are others connected with the foot and byssus. The cloak lines the inside of the shells. In some cases it is entirely open, when the border corresponding with the free margin of the shell is thickened, and more or less fringed with contractile irritable filaments. In other cases, the cloak in front is more or less united, and even forms tubular elongations, which are termed *syphons*.

Locomotion is denied to many species of this order. Among these some are immoveably cemented to rocks and stones, as oysters; a few are attached by a cartilaginous ligament, as the Anomiæ; while others are fixed by means of a *byssus*. This last organ consists of numerous filaments issuing from a complicated apparatus in the breast, connected with a secreting gland, and with the shell, by the intervention of tendinous bands. The *foot* is seated a little towards the mouth, is usually tongue-shaped, capable of considerable elongation, with a furrow on its posterior surface. This organ, where a byssus is present, is considered as employed in spinning and fixing the threads. When there is no byssus, it either acts as a sucker, enabling the animal to crawl along the surface of bodies, or as a paw, to dig holes in the sand or mud. None of the species can float in the water; they either crawl or leap, the last kind of motion being effected by suddenly opening and shutting the valves. In securing a residence, some of the species *bore* into different substances by means of a rotatory motion of the shell. It was at one time supposed that the dwelling was formed by a secretion affecting the solution of the surrounding substance. But the very different substances penetrated by the same species, as limestone, slate-clay, and wood, forbid us to entertain such a supposition.

The nervous system is here but little developed. The superior and inferior ganglia, surrounding the gullet, give rise to all the nervous filaments which proceed through the body.

The digestive organs are scarcely less simple. The food is soft and swallowed entire, and either brought to the mouth by accident, or by eddies produced in the water, by the opening and shutting of the shells, aided in some cases by the syphons.

It may be proper, in order to understand the relative situation of the parts, here to state, that, upon laying the animal upon its back, and opening the cloak, the abdomen appears to occupy the middle longitudinally, and the branchia to be arranged on each side. The mouth is situate at the anterior extremity, and consists of a simple aperture entering into the gullet, or rather stomach. It is surrounded by four flattened moveable tentacula, two of which, in some are in part united with the cloak, while in others they are free to the base. In their structure they resemble the branchiæ. The stomach is full of cells, the bottom of each pierced with a biliary duct. A singular organ, termed the *crystalline process*, cylindrical, cartilaginous, and transparent, is found in some species, projecting into the cavity of the stomach. The liver is large, surrounds the stomach, and pours out its contents by numerous openings. The intestine terminates posteriorly by a tubular anus.

The branchiæ consists of two ribbands on each side, extending the length of the body, free on the sides and margin, and striated transversely. These plates are frequently of unequal size. The blood is brought to these by means of pulmonic veins, without the intervention of the heart. The aërated blood is transmitted to a systemic heart, consisting of one or two auricles, and a ventricle.

The most important of the peculiar secretions of the animals of this class is the Pearl. This substance, equally prized by the savage and the citizen, is composed, like shells, of carbonate of lime, united with a small portion of animal matter. Pearls appear to be exclusively the production of the bivalve testacea. Among these, all the shells having a mother-of-pearl inside, produce them occasionally. But there are a few species which yield them in the greatest plenty, and of the finest colour. The most remarkable of these is the *Avicula margaritifera*. This shell, which was placed by Linnæus among the mussels, is very widely distributed in the Indian Seas; and it is from it and another species of the same genus, termed *Avicula hirundo*, found in the European seas, that the pearls of commerce are procured. The *Pinna*, so famous for furnishing a byssus or kind of thread, with which garments can be manufactured, likewise produces pearls of considerable size.

They have seldom the silvery whiteness of the pearls from the Avicula, being usually tinged with brown. But the shell, which in Britain produces the finest pearls, is the *Unio margaritifera*, which was placed by LINNÆUS in the genus *Mya*. It is found in all our alpine rivers. The Conway and the Irt in England, the rivers of Tyrone and Donegal in Ireland, and the Tay and the Ythan in Scotland, have long been famous for the production of pearls. These concretions are found between the membranes of the cloak of the animal, as in the Avicula; or adhering to the inside of the shell, as in the Unio. In the former case, they seem to be a morbid secretion of testaceous matter; in the latter, the matter appears to be accumulated against the internal opening of some hole with which the shell has been pierced by some of its foes. LINNÆUS, from the consideration of this circumstance, endeavoured, by piercing the shell, to excite the animal to secrete pearl; but his attempts, though they procured him a place among the Swedish nobility, and a pecuniary reward, were finally abandoned; the process being found too tedious and uncertain to be of any public utility. The largest pearl of which we have any notice, is one which came from Panama, and was presented to Philip II. of Spain, in 1579. It was of the size of a pigeon's egg. Sir Robert Sibbald mentions his having seen pearls from the rivers of Scotland as large as a bean.

The reproductive organs of the Bivalvia, hitherto examined, consist of an ovarium occupying the sides of the body, and penetrating the membranes of the cloak. They appear to have the organs of both sexes incorporated, and to propagate without intercourse. LAMARCK is disposed to consider impregnation as produced by the male fluid dispersed through the water; a supposition unsupported even by analogy in the animal kingdom. Many species are ovoviviparous; in which case the eggs when ripe pass into the gills, where they are hatched.

The methodical distribution of the Bivalvia appears to be attended with peculiar difficulties, in consequence of the uniformity which prevails in the structure and disposition of their organs. The characters furnished by the shell, though useful in the construction of generic as well as specific distinctions, have been abandoned by those who prefer a knowledge of the structure, to an acquaintance with the form of an animal. The characters derived from the presence of a byssus, a foot, or of syphons, appear to be nearly of co-ordinate importance. M. CUVIER gives the preference to those founded on the appearances of the latter, and distributes the genera into five families, an arrangement which we here propose to follow. These, however, may be considered as occupying a much higher rank, and each as including numerous families.

1st Subdivision.

Cloak open. There are no syphons, the anterior margin of the cloak being as open as the mouth of the shell. When the valves open, the water comes immediately in contact with the branchiæ and mouth. The margin of the mantle has a double fringe of filaments.

1st Tribe.

Valves closed by one adductor muscle.

A. *Pectenidæ.* Animals free or fixed only by a byssus. Furnished with a foot.

Into this family, contemplated by LAMARCK, the following ill assorted genera may be placed: Pecten, Lima, Pedum, Plicatula, Vulsella, Placuna, Gryphæa, Perna, and Crenatula.

In the animals of the genus Pecten, represented by the common scallop, there is a small foot, supported on a short stalk arising from the abdomen. The margin of the cloak is surrounded with two rows of tentacula, some of which, in the external row, have greenish tuberculated summits. The mouth is surrounded with numerous branched tentacula, in place of the four ordinary labial appendages.

B. *Ostreadæ.* Shell cemented to foreign bodies. Body destitute of a foot.

To this family the following genera are related: Ostrea, Spondylus, and Anomia. The last genus ought to form a family apart. It is distinguished by the singular character of the adductor muscle, a portion of which is attached to the corneous or testaceous plate, which passes through the cardinal perforation, and adheres to rocks. There is a small foot, which is capable of being likewise protruded through the cardinal perforation. In the genus SPONDYLUS, the margin of the cloak is fringed with a double row of tentacula having tuberculated summits, and the foot is seated on a short stalk, with a large radiated disk.

2d Tribe.

Shell closed by two adductor muscles.

The two genera, Avicula and Meleagrina (of LAMARCK), form one family of this tribe; the genus Pinna another; and the Arcadæ a third, including Arca, Pectunculus, Nucula, Cucullæa Trigonia, and Castalia.

2d Subdivision.

Cloak more or less closed, forming syphons.

The further division of this group depends on the modifications of the syphons, or aperture of the cloak.

1st Tribe.

The union of the cloak forms only one syphon.

This syphon is situate posteriorly opposite the anus, and serves for the issuing of the excrements. The other large opening allows the water to enter to the mouth and gills.

This tribe may be divided into families. The first, *Mytilusida*, will include the genera Mytilus, Modiolus and Lithodomus, which are furnished with a byssus. The second, *Uniodæ*, will embrace Unio, Ilyiria, Anodonta, and Iridina. These are destitute of a byssus.

M. Cuvier is disposed to place in this group the genera Cardita, Venericardia, and Crassatella.

2d Tribe.

Cloak closed posteriorly, and anteriorly forming three apertures The first serves for the passage of the byssus, and is the largest. The second admits water to the branchiæ and mouth: and the third is opposite the anus. The valves are closed by one adductor muscle. There are only two genera belonging to this tribe, Tridacna and Hippopus.

In the two remaining tribes there are three openings in the cloak. Two of these are posterior, and near each other; sometimes, indeed, they are tubular and united. There is no byssus, but always a foot.

3d Tribe.

Anterior opening large, allowing the water free acess to the mouth and gills, and the feet freedom of motion.

The structure of the animals is yet too imperfectly examined, to enable any one to establish families on permanent characters. The attempt which Lamarck has made may be considered as a complete failure, independent of the wanton changes of nomenclature with which it is chargeable. The following are the principal genera belonging to this tribe: Chama, Iscordia, Cardium, Donax, Cyclas, Corbis, Tellina, Loripes, Lucina, Venus, Capsa, Petricola, Corbula, and Mactra.

In the genus *Chama*, the two posterior apertures are in the form of short tubes, the anterior one is small, and indicates the corresponding size of the foot. The foot of the animals of the *Iscordia* is much larger, and the anterior aperture is large in proportion. In the common cockle (Cardium), prized by many as an agreeable article of food, the foot occupies a large share of the cavity of the shell. It is bent in the middle, with the point directed forwards. In the genera Cyclas, Tellena, Donax, and Venus, the foot is long and tongue-shaped, and the posterior tubes, in general, considerably produced, and more or less united at the base. In the genus Loripes, the foot is small and cylindrical, and the tubes are short and united. In the Mactra the tubes are likewise short, but the foot is compressed.

4th Tribe.

Anterior opening small, and not exposing the mouth or gills.

In this tribe the mantle is closed in front; and even when the valves are open, neither mouth nor gills are visible. The anterior opening serves for a passage to the foot, and the posterior openings, in the form of two long tubes, united by a common membrane, serve for the entrance and exit of the water to the mouth and branchiæ, and the ejection of the fœces, the dorsal syphon serving the latter purpose. The cuticle of the shell covers also the exposed portion of the cloak, so that, when an animal is removed from the shell, it remains as a loose membrane on the margin of the valves, as was first observed by REAUMUR. All the genera prefer concealment, burrowing in sand, mud, or wood, with the head downwards, and the syphons rising to the surface. The following genera belong to this tribe: Mya, Lutraria, Anatina, Glycemeris, Panopea, Pandora, Gastrochena, Byssomia, Hiatella, Solen, Sanguinolaria, Pholas, Teredo, and Fistularia.

Section II.—ACEPHALA TUNCATA.

Covering soft or corriaceous.

The formation of this interesting group of animals was first publicly announced by LAMARCK in his *Histoire Naturelle des Animaux sans Vertebres*, Tom. iii. p. 80. (1816). The labours of DESMARET, LESUEUR, and CUVIER, aided by the descriptions of ELLIS and PALLAS, paved the way for the masterly efforts of SAVIGNY, to whom we owe the most extensive, new, and accurate information yet given, concerning the animals of this group. His observations are contained in his *Recherches Anatomiques sur les Ascidies composées, et sur les Ascidies simples*, inserted in his *Memoires sur les Animaux sans Vertebres*. 8vo. Paris, 1816.

The covering of the animals of this group consists of an external and internal sac or tunic, either entirely united, or unconnected, except at the apertures. The surface is smooth in some, and rough in others, and in a few species defended by an artificial covering of agglutinated shells and sand. The sacs are furnished with muscular bands, and are capable of contraction. Some of the species, by means of contractile movements, float about in the water; others, receiving that element into the branchial cavity, and ejecting it forcibly at the opposite one, push themselves forward. Many, however, are fixed during life to sea-weeds and stones.

The apertures of the tunic are two in number, The one, frequently the largest, is destined for receiving the water into the cavity, to supply the mouth and gills. This is termed the *branchial* cavity. The other is destined for the exit of the water, the eggs, and the fœces, and termed the *anal* opening. These apertures are sometimes placed near each other, at other times at opposite extremities of the body, and variously provided with tentacula or valves.

The mouth is simple, destitute of spiral arms, and opening in the interior of the cavity of the body, between the branchiæ, as in the bivalvia. It possesses neither jaws nor tentacula. The alimentary canal is very simple, and can scarcely be distinguished into gullet, stomach, and intestine. The food is soft, and such as the bounty of the waves bestow. The liver adheres to the stomach, and in many species is divided into distinct lobes.

The circulating system appears to be reduced to a single systemic ventricle. The gills cover the walls of the cavity, in the form of ridges, more or less complicated, and seldom symmetrical.

The reproductive organs consist of an ovarium, either simple or complicated, with some additional glands, the uses of which have not been ascertained. The species are considered as hermaphrodite, and independent of reciprocal impregnation. They appear, in some genera, not only to be oviparous, but to be gemmiparous and compound, many individuals being organically connected, and capable of simultaneous movements. They are all inhabitants of the sea.

1st Subdivision.

Interior tunic detached from the external one, and united only at the two orifices.

The branchiæ are large, equal, and spread on the central walls of the inner sac. The branchial orifice has an inner membranaceous denticulated ring, or circle of tentacula.

1st Tribe.

Body permanently fixed to other bodies.

In this tribe the branchial and anal orifices are not opposite each other, and do not communicate through the branchial cavity which, at its opening, is furnished with tentacular filaments. The branchiæ are conjoined anteriorly.

A. Simple.

This division includes the genus *Ascidia* of LINNÆUS. The individuals are independent of each other, and although they frequently adhere together in clusters, they are destitute of a common covering, or organical connection.

I. *Apertures furnished with four rays.*

The animals of this group have the external tunic coriaceous, dry, opaque, rough, folded, and frequently covered with extraneous bodies, or inclosing such. The branchial orifice has four rays, the anal one the same, or divided transversely. The branchiæ are divided longitudinally into persistent regular deep folds.

a. Body pedunculated.

The peduncle, in this division, may be said to have its rise in the summit of the body, which it serves to suspend. The abdomen is lateral. The meshes of the branchiæ are destitute of papilæ.

57. BOLTENIA. The tentacular filaments of the branchial circle are compound. There is no liver, and the ovarium is compound. Only one species is known. *B. fusiforme.* Savigny, *Mem.* tab. i. f. 1. and tab. v. f. 5. It is the *Vorticella Bolteni* of Lin. and the *Ascidia clavata* of Shaw.

b. Body sessile.

M. SAVIGNY describes this group as a genus, which he terms Cynthia, and which he divides into four sub-genera.

(A). Tentacular filaments of the branchial orifice compound. The folds of the branchiæ more than eight in number. The liver distinct, and surrounding the stomach. Ovarium divided with one division at least on each side the body. The intestine destitute of a rib.

58. CYNTHIA. Meshes of the branchiæ unchanged by the folds. *C. Momus.* Sav. Tab. i. f. 2.

59. CÆSIRA. Meshes of the branchiæ interrupted by the folds. *C. Diona* of Sav. Tab. vii. f. 1. The *Ascidia quadridentata* of Forskael.

(B). Tentacular filaments of the branchial orifice simple.

The folds of the branchiæ eight in number, four on each side, and the meshes uninterrupted. Intestine strengthened by a cylindrical rib from the pylorus to the anus. Liver absent or indistinct.

60. STYELA. Ovarium divided, one division at least on each side. *S. Canopus*. Sav. Tab. viii. f. 1.

61. PANDOCIA. Ovarium single, and situate in the fold of the intestine. The *Ascidia conchilega*, a native species is the type.

2. *Apertures with indistinct rays, or more than four.*

The external tunic is here soft, easily cut, and translucent. The rays (when existing) of the branchial orifice, amount to eight or nine; and those of the anal to six at least. The branchiæ are destitute of longitudinal folds. The tentacular filaments of the branchial circle are simple. Liver indistinct. Ovarium single.

a. Body pedunculated.

The stalk is here placed at the base, and serves to support the body, being of an opposite character from that of the Boltenia.

62. CLAVELINA. Branchial and anal orifices without rays. Angles of the branchial meshes simple. Intestine destitute of a rib. The *Ascidia clavata* of Pallas, and the *A. lepadiformis* of Muller belong to this genus; the latter of these is now recorded as a British species.

b. Body sessile.

The branchial orifice with eight or nine rays, and the anal with six. The angles of the branchial meshes with papillæ. No liver. A cylindrical rib extending from the pylorus to the anus.

(A.) Tunic and branchial cavity straight.

63. PIRENA. The branchial sac as extended as the tunic. Stomach not resting on the intestine. *P. Phusca* of FORSKAEL is the type to which SAVIGNY has added three other species. The *Ascidia prunum* of MULLER, a native species, is probably of this genus.

64. CIONA. Branchial sac shorter than the tunic, and exceeded by the viscera. *C. Ascidia intestinalis*, LIN. is a native example of this genus.

(B.) Tunic turned up at the base.

65. PHALLUSIA. Branchial sac extending beyond the viscera into the pouch of the sac. Stomach resting on the mass of viscera. The *Ascidia mentula* of MULLER, a native species, is the type.

There is one genus, supposed to be nearly related to the preceding, which is involved in great obscurity, the *Bipapilaria* of LAMARCK, which appears to be pedunculated, with two apertures, each furnished with three setaceous tentacula.

B. Compound.

The animals belonging to this division were formerly included in the genus Alcyonium of LINNÆUS, and placed among the Zoophytes. They are compound animals, many individuals united by a common integument, and arranged according to a uniform plan.

In some cases, there is only one system of individuals in the mass, in other cases, there are many, similarly arranged and contiguous. The tentacular filaments of the branchiæ appear to be distinct. They are destitute of the intestinal rib which occurs in some of the preceding genera.

1. *Branchial orifice radiated.*

a. Branchial and anal orifices, with six rays.

(A). Body sessile. The angles of the branchial meshes furnished with papillæ. The thorax, or cavity containing the branchiæ, cylindrical. The abdomen is inferior, with a stalk. Ovarium sessile, and single.

65. DIAZONA. Body orbicular, with a single system of animals disposed in concentric circles.

The substance is gelatinous. The ovarium enclosed in the fold of the intestine. *D. violacea* of SAV. Tab. ii. f. 3.

66. POLYZONA. Body polymorphous, with many systems disposed subcircularly.

The body is subcartilaginous. The individuals are disposed irregularly around the common centre. SAVIGNY inadvertently termed this genus Distoma, a name long pre-occupied among the Intestina. The *Alcyonium rubrum* of Plancus, and the *Distomus variolosus* of GAERTNER, belong to this genus. The last is a native species.

(B.) Body pedunculated.

67. SIGILLINA. Body a solid cone, consisting of a single system of many individuals, irregularly disposed, one above the other.

The thorax is short, and hemispherical. The angles of the branchial meshes destitute of papillæ. The abdomen is inferior, sessile and larger than the thorax. The single ovarium is pedunculated. *S. australis*, SAV. Tab. iii. f. 2. brought from New Holland, by M. PERON, is the only known species.

b Branchial orifice only furnished with six rays.

(A). Body pedunculated. System single, circular, and terminal.

68. SYNOICUM. Anal orifice rayed.

The body is cylindrical. The anal orifice has six very unequal rays; the three largest forming the exterior margin of the central star. The stomach is simple. The angles of the branchial meshes destitute of papillæ. Ovarium single, sessile attached to the bottom of the abdomen, and descending perpendicularly. The *S. turgens* of PHIPPS is the type. In the month of August 1817, we observed at the Isle of May another species, adhering to a rock, and differing from the *turgens* chiefly in the smoothness of its skin.

69. SYDNEUM. Anal orifice simple and tubular.

The body is inversely conical. The stomach surrounded with glands. Intestine spirally folded. Ovarium pedunculated. The *S. turbinatum* is the only known species, and was sent to SAVIGNY by LEACH from the British seas.

(B.) Body sessile, polymorphous.

(*a.*) Each system with a central cavity.

70. POLYCLINUM. Systems numerous, convex, stellular. Individuals arranged irregularly round the common centre. Abdomen inferior, pedunculated and less than the thorax. Ovarium single, pedunculated, attached to the side of the abdominal cavity, and drooping. M. SAVIGNY describes one species from the Mauritius, and five from the Gulf of Suez.

(*b.*) Systems destitute of central cavity, and the angles of the branchial meshes without papillæ.

71. ALPIDIUM. Individuals in a single row round the common centre.

The thorax is cylindrical. The abdomen inferior sessile, and of the size of the thorax. Ovarium single, sessile, placed at the bottom of the abdomen, and prolonged perpendicularly. SAVIGNY divides the genus into two tribes. In the first, the individuals are simply oblong, with an ovarium shorter than the body, as *A. ficus* (*Alcyonium ficus,* LINN.). In the second, the individuals are filiform, with an ovarium longer than the body, as *A. effusum* of SAVIGNY, Tab. xvi. f. 3.

71. Didemnum. Individuals in indistinct systems.

The thorax is short and subglobular. The abdomen inferior, pedunculated, and larger than the thorax. The anal opening is obscure. The ovarium is single, sessile, and placed on the side of the abdomen. *D. candidum* and *viscosum*, from the Gulf of Suez, are the only known species.

II. *Branchial Orifice simple.*

The species appear in the form of thin fleshy crusts on stones and sea-weeds. The individuals are stellularly arranged in distinct systems. The branchial orifice is circular, and undivided. The abdomen is sublateral, and fixed at the bottom of the branchial cavity. The intestine is small, and the anus indistinct. The angles of the branchial meshes are without papillæ.

73. Botryllus. Systems furnished with a central cavity.

The systems are prominent, and consist of one or more regular concentric rows. The ovarium is double, being attached to each side of the branchial sac.

This genus is subdivided by Savigny into *Botryli stellati*, and *Botrylli conglomerati*. In the first, where the individuals are distributed in a single row, there are some species in which the individuals are cylindrical, with approaching orifices, and the limb of the central cavity not apparent after death and probably short, as the *B. rosaceus Leachii* and *Borlasii*. In other species, the individuals are ovoid, with remote orifices, and the limb of the central cavity is always apparent and notched, as *B. Schlosseri*, *stellatus*, *gemmeus*, and *minutus*. In the Botrylli conglomerati, in which the individuals are disposed in several rows, there is only one species, *B. conglomeratus*.

74. Eucælium. Systems destitute of a central cavity.

The individuals are distributed in a single row, and the ovarium is single, sessile, and attached to the side of the abdominal cavity. The *E. hospitiolum* of Sav. Tab. iv. f. 4, is the only known species.

2d Tribe.

Body free and moving about in the water.

75. PYROSONA.

The body is gelatinous, in the form of a lengthened bag open at the widest end. The individuals are arranged perpendicularly to the axis of the central cavity, super-imposed on one another. The branchial orifice is external, without rays, and with an appendage over its upper margin. The anal orifice is opposite, and terminates in the central cavity. Branchial sac destitute of folds, with a membranaceous ring at the entry. The branchiæ are disjoined. The abdomen is inferior to the branchiæ, and not separated by any contraction. Liver distinct, globular, and retained in a fold of the intestine. Ovarium double, opposite, and situate at the upper extremity of the branchial cavity.

M. SAVIGNY divides the species into Pyrosomata verticilla, having the individuals arranged in regular prominent rings, as *P. elegans* of LESUEUR; and Pyrosomata paniculata, having the individuals forming irregular circles unequally prominent, as *P. giganteum* and *atlanticum*.

2d Subdivision.

Inner tunic adhering throughout to the external one.

The body is gelatinous, transparent, and simple. The branchial cavity is open at both ends, communicating freely with the anus. The branchial orifice is in the form of a transverse slit, with one edge, in the form of a valve, to accelerate the entrance of the water into the cavity. The inner tunic is strengthened by numerous transverse muscular bands, which, by contracting, diminish the diameter of the cavity, and eject the water from the anal orifice, thereby propelling the body through the water The digestive organs are situate at the inner end of the cavity. The mouth and rectum are simple; the former placed between the two branchiæ, the latter directed towards the anal orifice. The heart is contiguous to the stomach, at the bottom of what may be termed the branchial sac, and is enveloped in a membranaceous pericardium. The branchiæ are double, not incorporated with the walls of the sac, but with two folds of unequal length. The largest is free in the middle, fixed at each extremity, and opposed to the dorsal groove, and traverse the cavity obliquely. The other extends from the base of the first to the extremity of the dorsal groove. The surface of the branchiæ consists of transverse vessels in a single range in some species, and a double range in others.

When young, many individuals often adhere, and form chains and circles. But the fully grown individuals are always detached and single.

This subdivision comprehends the species of the genus Salpa; they are exceedingly numerous, and appear to belong to many different genera. M. CUVIER has given indications of some of these, chiefly derived from the shape. A few are furnished with an elevated crest or fin, as the Thalia of BROWN; a few have both extremities rounded or truncated, as *Salpa octofera* of CUVIER) others have one extremity produced, as *Holothuria zonaria* of GMELIN; and even both extremities produced, as *Salpa maxima* of FORSKAEL. The *Sulpa moniliformis*, so common in the Hebrides, and first recorded as a native by Dr MACCULLOCH, in his valuable *Description of the Western Isles*, Vol. ii. p. 188, and imperfectly figured in its young state, at Tab. xxix. fig. 2., appears to be closely allied to the *S. maxima* of FORSKAEL, and but very remotely with the *S. polycratica* and *confederata* with which it is compared.

This observer states, that " It cannot bear to be confined in a limited portion of water, as it died even in a ship's bucket in less than half an hour. With us, in similar circumstances, those taken in the evening were alive at noon on the following day."

The preparation of molluscous animals for exhibition in a museum is attended with peculiar difficulty. The shells, indeed, need only to be cleaned with a soft brush, and the marine kinds to be steeped in fresh water, to extract all the saline ingredients, and dried, when they are fit for the cabinet. The soft parts, however, can seldom be distended by any substance, and dried. They are usually, therefore, preserved in spirits of wine, where but too frequently they appear a shapeless mass. The animal should be permitted to die slowly, that the different parts may become relaxed, otherwise the examination of the form of the body, at a future period, becomes impracticable. A quantity of spirits should be injected into the stomach, or other cavities of the body, immediately after death, to prevent putrefaction, as it frequently happens when the body is immersed in spirits, without such precaution, that the viscera become unfit for examination while the integuments have been preserved in a sound state.

II.

ANNULOSA.

Brain surrounding the gullet and sending out a knotted filament to the posterior extremity of the body.

The longitudinal filament, which issues from the ventral side of the nervous collar, is frequently double at its origin, and in some cases continues distinct throughout its whole length. At each ring of the body it forms a ganglion, and where the filament is double a union takes place of both. From these ganglia nerves are sent to the neighbouring parts.

Besides these conditions of the nervous system, there are others connected with form and structure, which distinguishes the annulose animals. In general their shape is lengthened, approaching to cylindrical. The body is divided into rings by transverse strictures more or less distinctly marked. Even the different appendices of the skin frequently exhibit their annulose structure.

The annulose animals form two groups, which, while they indicate the closest affinity in the nervous system and divisions of the body, exhibit at the same time remarkable distinctive characters. These characters indeed are so obvious, as to have been perceived by all naturalists, and to have induced them to assign to each a separate place in their systems. The first group, the species belonging to which have articulated limbs, is identical with the fifth class in the LINNEAN system, termed Insecta. The second group, containing animals destitute of articulated limbs, includes genera of the different orders of the class Vermes of the same system. As the remarks which appear necessary to be made on one group, must differ widely from those applicable to the other, it will be proper to consider each apart.

Subdivision I.—ANNULOSE ANIMALS WITH ARTICULATED LIMBS FOR LOCOMOTION.

The SKIN of the animals of this group serves the double purpose of protection and support, and represents the cutaneous and osseous systems of the vertebral animals. Its structure appears much more simple than in the higher classes, as it can neither be said to possess a mucous or cellular web or true skin. It bears the nearest resemblance to the cuticle of the skin of the higher classes, or rather, all the laminæ of perfect skins are here incorporated into one uniform plate. It exhibits very remarkable varieties of texture. In some genera it is soft and pliable; while, in others, as some of the weevils, it approaches the consistence of bone, or appears as a calcareous crust in the crabs. In some species it is elastic, in others brittle.

The appendices of the skin consist of spines, hairs, and scales. The spines are merely projecting portions of its substance, and are usually distributed over certain parts of the feet, to aid the locomotive powers. Hairs are often distributed over the whole body; and, while they pass into spines on the one hand, they become, on the other, so exceedingly fine as to require the aid of a powerful magnifier to trace their character. These spines and hairs, being merely elongations of the skin, are not easily rubbed off. It is otherwise with scales. Some of these are inserted into their skin at their proximal, and are free at their distal extremity, and in some insects are so feebly connected, as to fall off, in many species, by touching them with the finger. These scales, in the butterfly, bear a remote resemblance to feathers in their form, and are very extensively used as pleasing objects for the microscope.

In those insects, which undergo several changes of form during life, the cutaneous system, in the first periods, possesses considerable powers of production. In the caterpillar state, the skin is cast or changed, several times, and along with it, the spines and hairs by which it is covered. That which is cast off bears a resemblance to the cuticle in the perfect animals; but the skin which remains to supply its place, is similar in its structure. When the insects arrive at their last stage, or that of maturity, the reproductive power of the skin does not seem to be exerted. Neither holes in the wings, nor fractures in the joints, appear to be repaired. Among the crabs and spiders, the casting of the skin takes place periodically.

Comparative anatomy has hitherto failed in detecting any glands subservient to the functions of the skin. As the aquatic insects, however, are never wet with water in which they reside, it is probable that the skin is besmeared with some unctuous matter. In some instances, indeed, the skin resists being wet, even after the death of the animal has taken place for some time, but previous to becoming dry.

The MUSCLES of insects appear to possess the same internal structure as the corresponding organs in the higher classes. They are nearly all simple, and more or less transparent and whitish. All those which are concerned in the production of locomotion have their origin and insertion in the skin, apparently without the intervention of tendons. The different members appear to be connected with each other, by the intervention of a more transparent, tough substance, than the skin in other places, to which the name of ligament may be applied. The action of the muscles will be best understood by a description of the different parts of the body, and the motions which these perform.

The HEAD contains the organs of the external senses and the mouth. It is joined to the trunk behind, and has its motions regulated by its mode of connection. Where the head is united to the trunk by a cylinder of ligament, the motion of which it is susceptible is various, limited chiefly in the dorsal direction by the superior margin of the trunk. When the articulation is effected by the immediate contact of the more solid surfaces, the three following modifications of joints present themselves. In the first, there are two or more rounded, smooth tubercles, received into corresponding cavities in the trunk. The motion is consequently either backwards or forwards. In the second mode of articulation, the head is rounded posteriorly, and received into a socket in the thorax. In this manner great liberty of motion is obtained. In many cases, however, it is restrained by projections of the trunk, which limit it in one direction. In the third, the articulation takes place by the contact of two flat surfaces. The head is frequently contracted behind, and the trunk in front, to diminish the uniting surfaces, but such joint admits but of very imperfect motion.

The muscles of the head take their rise near the abdominal edge of the trunk, and, entering the occipital hole, become attached to its margin. Those which move the head upwards take their rise on the upper part of the trunk; while those which depress it arise from the under side. These last are the largest. Those which arise from the lateral parts of the trunk give to the head its lateral motions.

The characters derived from the head, which are used in classification, are chiefly taken from the markings of its surface, the inequalities of its margin, its size, and shape. They are, in general, obvious and permanent.

The TRUNK, as an organ of support to the other members, may be considered as the most important organ of the body. The terms, however, which are employed to designate its different parts are neither appropriate nor well defined. LINNÆUS, in reference to this organ, gives the following enumeration of its parts: " Truncus, inter caput et abdomen, pedatus, thorace supra dorso, postice scutello, subtus pectore sternoque." By modern authors, it is usually divided into the thorax and breast.

The *thorax* is the second ring of the body, and is united on the fore part with the head, and behind with the third ring or breast. In many kinds, this part is minute, while, in others, it occupies a large portion of the body. On its ventral aspect in insects, it bears the fore legs, or first pair, and between these is the *thoracic sternum*, frequently in the form of a keel, and terminating behind by a spinous process, which rests upon the pectoral sternum, as in the genera Elater and Dytiscus. It may be considered as the only fixed part of the body, giving origin to the muscles of the head, the fore legs, and frequently, also, to the breast and abdomen.

The characters furnished by the thorax are extensively employed in the arrangement of the genera and species. These are chiefly taken from its appearances on the back of the animal.

The *Breast* or third ring of the body, in insects, is frequently so much incorporated with the thorax, as to appear as one organ. In other cases it seems to form a part of the abdomen. In the back, it frequently exhibits a horny process, termed *scutellum*, analogous in consistence to the thorax, to which it is united behind. At this place of the breast, the wings have their origin. On its ventral side it supports the middle and hind legs. Between these, at the base, is the *pectoral sternum*, which, in some species, expands into a cover for the first joint of the hind legs, and in others is produced to cover a portion of the abdomen. The breast contains the muscles for moving the middle and hind legs, the wings and abdomen; and it likewise contains some very strong muscles, passing from its ventral to its dorsal surface, and calculated to bring these, when required, nearer together. This motion is probably facilitated by the number of sutures with which its skin is traversed. The characters for classification are chiefly taken from the scutellum and sternum.

The ABDOMEN is the last portion of the body. In some species, it is sessile, and intimately united with the breast; in others it is divided by a stalk. It consists of rings, varying in number in different genera. These rings are joined together in many species by a simple adhesion of the margin, while, in others, the posterior margin of the one includes the anterior margin of the other. The motion of the whole abdomen takes place by means of muscles, which arise in the breast, and, in those where it is sessile, the motion is very limited. The rings themselves are likewise susceptible of a little motion, especially when they are included. The muscular fibres which change their position are longitudinal, and pass from the posterior margin of one joint to the anterior of the other, and are able to draw the ring to one side, or pull it within the one which immediately precedes it.

The classical characters derived from the abdomen are chiefly taken from the number of the rings, their connection, and the condition of their surface. The anus and external orifice of generation are situate at the termination of the abdomen. These have frequently appendices, which we will afterwards notice.

These three portions of the body, in the different tribes, exhibit very remarkable combinations. In some of the crustacea, the head, trunk, and abdomen, are incorporated on the dorsal surface of the body. In some of the Arachnidæ, the head and trunk are united, while, in others, the head appears to be distinct, while the trunk and abdomen are incorporated. These modifications are extensively employed in the methodical distribution of the groups.

The MEMBERS of the articulated Annulosa are of two kinds, wings and feet.

The wings, which exclusively belong to insects, vary in their number, structure, and appendices. In one tribe, even among these, they are wanting, and hence termed Apterous; in another they are two in number, but the greater number have four. These wings are either membranaceous, and supported by corneous ribs, which form a net-work in their substance; or, where the wings are four in number, the upper pair are sometimes crustaceous, obtain the name of *elytra*, and serve as a covering to the inferior ones. The ribs of the wings, improperly denominated nerves, in the manner of their distribution, the hairs by which they are covered, and the form of the vacant spaces, exhibit great regularity in the individuals of the same species. The manner in which the membranaceous wings are folded up, when at rest, is various. In some they are folded longitudinally, in others transversely, and in others obliquely. Each of these arrangements prevails throughout extensive groups. In the Diptera there is, under or behind each wing, a stalk terminating in a small knob. These are termed *halteres*, or *poizers*, and are considered as the rudiments of the second pair of wings. Between each poizer and the base of the wing, one or two spoon-like scales are found, termed winglets. They have likewise been observed in a coleopterous insect, *Dytiscus marginalis*. In some of the diptera they are absent. The use of these appendices of the wings has not been satisfactorily determined. The muscles which move the wings take their rise in the breast, and are capable of executing their functions with great celerity. The elytra perform no other motion than elevation and depression, and serve merely to protect the wings when at rest, not to assist them when flying.

The characters employed in the classification of the primary divisions of insects are, in a great measure, derived from the wings. Their presence or absence—their number and appendices—their texture and consistence, together with their size, position, and manner of folding up, yield marks which are of easy detection, and which experience has found to be permanent.

The *legs* never fall short of six in number. The first pair, when six only are present, take their rise under the thorax, and the second and third pairs under the breast. When more numerous, some of them are attached to the abdomen. They consist of the five following parts, the hip (*coxa*), the thigh (*femur*), the leg (*tibia*), the toe (*tarsus*), and the claw (*unguis*).

The *hip*, or coxa, serves to unite the limb with the body. It is usually short, or nearly as broad as it is long. It is imbedded in the body, and is limited in its motion by the mode of insertion. Sometimes its proximal extremity is globular, and received into a corresponding cavity of the body, giving to it a very extensive degree of motion. In other cases, the coxæ are consolidated with the skin.

The *thigh* or femur is usually united with the coxa, in such a manner as only to admit of motion backwards and forwards. At its coxal extremity, the femur has, in some tribes, one or two eminences, in some cases produced into spines, which are termed *trochanters*, and as they are hollow within, they are considered as furnishing suitable insertions to the muscles for particular motions. When the thigh is slender and cylindrical, the motion of the legs is confined to walking, but where leaping is required, or even swimming, the thigh is thick, and bellied, to give room to the requisite muscles.

The *leg*, or tibia, is articulated to the femur in such a manner as to admit of motion only in the same plane in which it moves. In those species which swim, this joint is long and flattened, while, in those which dig holes in the ground, it is strong and serrated on the margin.

The *toe*, or tarsus, consists of several joints, which are articulated more loosely than the preceding parts of the limb, and admit not only of motion outwards and inwards, but likewise in a lateral direction. Its strongest muscle, as in the other parts of the leg, is the flexor. It is only by means of this part that the feet can apply closely to any object. The joints vary in number, length, and size; and, in many species, are furnished with very singular appendices. Those in insects which deserve particular notice are denominated cushions and suckers, and are situate on the under surface of the joints. The cushions are either soft and smooth on the surface, or consist of an enlargement closely covered with short hairs. These, by their elasticity and resistance, aid the animal in climbing and leaping. In some animals they are single in each joint, in others double, and either extending to all, or confined to a few of the joints. The suckers vary greatly in number, shape, and position. In all, they are capable of being applied to the surface of a smooth body. In some, the whole disc is applied, in others only the margin. These suckers are either sessile or seated on footstalks; sometimes they occupy the tibial joints of the tarsus only; in other cases they are confined to the last joint, and are from one to three in number *.

The *claw* (unguis) is attached to the distal edge of the last joint of the tarsus. In a few species it is single, or capable of being opposed to a projection, serving the purpose of a thumb. In others, the claws are double, and either move in the same plane, or act in opposition. The muscles of the different parts are all included in the limb, except those of the coxæ, which originate in the trunk. The former gives origin to those of the tibia, the latter to the first joint of the tarsus, and this again to the one which succeeds. These muscles are chiefly flexors and extensors, the former placed on the ventral, the latter on the dorsal surface of the limb.

These divisions of the limbs are frequently subdivided by sutures, but seldom or never by the ordinary method of articulation which is by Ginglymus.

The animals we are now considering are qualified for executing different kinds of motion. The best walkers are

---

* Accurate representations of these remarkable organs are given by Sir EVERARD HOME, in the *Philosophical Transactions for the Year* 1816, Plates 18—21, from the beautiful drawings of Mr BAUER.

those which have long, slender limbs, as the crane-flies. Those which walk upon walls and trees are enabled to overcome the resistance of gravitation by their claws and the cushions, or suckers of their feet. By means of the last sort of organs, the common house fly securely walks upon the vertical glass of the window, or alights upon the ceiling of the room. In walking, they advance the fore and hind legs of one side, and the middle leg of the other side at the same time, alternately. Those that leap much, have usually the femoral joints of the limbs enlarged, and the hind legs are the longest. Among insects, the skip-jack (Elater) is able to leap to a considerable distance by the elasticity of the process of the thoracic sternum suddenly thrust into the cavity of the pectoral sternum, and again withdrawn. The elastic process of the Podura is attached to the tail, and, when at rest, is bent under the body. By suddenly unbending it, the body is thrown to a considerable distance. The flying insects do not possess rapidity of flight proportional to the number or size of their wings. In the coleoptera, the body hangs down during flight, while in the other classes it preserves nearly a horizontal position. In the tribes which swim, the legs are either flattened like the blade of an oar, or produced and ciliated on the edges. Some swim upon their back, others upon their belly. Some keep always floating upon the surface, others dive and perform their movements at various depths, regulated by the condition of the organs of respiration.

The *nervous system*, in the class of animals now under consideration, exhibits a greater uniformity of structure and disposition than any of the other great systems of organs which they possess. The brain is situate in the head, on the dorsal surface of the gullet. It consists of two lobes, which exhibit slight variations of form, are frequently so intimately united, that they appear as one, marked, however, in the middle by a groove. These lobes furnish the optic nerves, and frequently send filaments to the mouth. Near the posterior edge, two cords arise, which, after proceeding backwards, and, in many cases, embracing the gullet, unite to form a ganglion, from which nervous filaments proceed to the neighbouring parts. From this ganglion two cords again proceed and form a second ganglion, and the same process is repeated until the cords reach the anal extremity, where they terminate. The number of ganglia which are formed, differ in different genera, according to the number of articulations of the body. The cords, in some species, appear to unite and exhibit only the appearance of one.

Of the organs of the senses, the eye alone is the most perfect; and next to it may be classed the organ of touch. The existence of the other senses common to the vertebral animals, is rather inferred, in many cases, from the actions performed than from the structure of their parts.

The organs of vision are of two kinds, simple and compound eyes. The *simple eyes* are usually in the form of small black circular dots, three in number, situate on the crown of the head towards the neck. They are denominated *ocelli* or *stemmata*. They are not present in many tribes. The *compound eyes* are present in all insects and many crustacea. They are in general two in number, and situate one on each side of the head. In the Gyrinus, however, each eye appears to be divided by the marginal band of the head, so that there is one eye above and another below. Similar appearances may be observed in the Geotrupes stercorarius. In some crustacea they are united. In some cases the compound eyes are imbedded in a cavity of the head; in others they are continuous with its surface, or elevated on a fixed or articulated process above it. Their aspect varies with the habits of the species *.

---

*. The following observations of the celebrated Cuvier on the anatomical structure of the eye, here deserve a place :

" The structure of the eye of insects is so very different from that of other animals, even the mollusca, that it would be difficult to believe it an organ of sight, had not experiments, purposely made, demonstrated its use. If we cut out, or cover with opake matter, the eyes of the *dragon-fly*, it will strike against walls in its flight. If we cover the compound eyes of the *wasp*, it ascends perpendicularly in the air, until it completely disappears ; if we cover its simple eyes also, it will not attempt to fly, but will remain perfectly immoveable.

" The surface of a compound eye, when viewed by the microscope, exhibits an innumerable multitude of hexagonal facets, slightly convex, and separated from one another by small furrows, which frequently contain fine hairs, more or less long.

" These facets form altogether a hard and elastic membrane, which, when freed of the substances that adhere to it posteriorly, is very transparent.

" Each of these small surfaces may be considered either as a cornea, or a crystalline, for it is convex externally, and concave internally, but thicker in the middle than at the edges; it is also the only transparent part in this singular eye.

" Immediately behind this transparent membrane there is an opaque substance, which varies greatly as to colour in the different species, and which sometimes forms, even in the same eye, spots or bands of different colours. Its consistence is the same as that of the pigment of the choroides ; it entirely covers the posterior part of the transparent facets, without leaving any aperture for the passage of light.

" Behind this pigment we find some very short, white filaments, in the form of hexagonal prisms, situate close to each other, like the stones of a pavement, and precisely equal in number to the facets of the cornea ; each penetrates into the hollow part of one of these facets, and is only separated from it by the pigment mentioned above. If these filaments are nervous, as in my opinion they appear to be, we may consider each as the retina of the surface, behind which it is placed ; but it will always remain to be explained, how the light can act on this retina, through a coat of opaque pigment.

" This

The organ of touch is not generally distributed over the body, for there are few parts capable of receiving an impression, or, at least, of being applied to the surface of bodies. The *antennæ*, organs peculiar to annulose animals, are generally considered as appropriated to this sense. These organs are two or more in number, and are present in all the crustacea and insects, but wanting in the arachnida. They are situate on the head, usually between the eyes and the mouth. They consist of a number of joints, determinate in the individuals of a species, and, in general, capable, by their flexibility, of examining the condition of the surface of a body. Those which have long setaceous antennæ, upon approaching a body, move them along its surface with considerable rapidity, thrust them into its cavities, and in this manner appear to become acquainted with its form. Individuals of the same species meeting together, examine each other by means of their antennæ.

" This multitude of filaments, perpendicular to the cornea, have behind them a membrane which serves them all as a base, and which is consequently nearly parallel to the cornea. This membrane is very fine, and of a blackish colour, which is not caused by a pigment, but extends to its most intimate texture. We observe in it very fine, whitish lines, which are tracheæ, and which produce still finer branches, that penetrate between the hexagonal filaments as far as the cornea. By analogy, we may name this membrane the *choroides.*

" A thin expansion of the optic nerve is applied to the posterior part of the choroides. This is a real nervous membrane, perfectly similar to the retina of red blooded animals. It appears that the white filaments, which form the particular retina of the different ocular surfaces, are productions of this general retina, which perforates the membrane I have named choroides, by a multitude of small and almost imperceptible holes,

" To obtain a distinct view of all these parts, it is necessary to cut off the head of an insect that has the eyes large, and dissect it posteriorly ; each part will then be removed in an order the reverse of that in which I have described them."

Traces of the organs of hearing may be detected in some of the larger crustacea. In the cray-fish, there is a purse-shaped labyrinth at the base of each antenna, covered externally by an elastic membrane or tympanum, and open at its central extremity for the passage of the auditory nerves. No traces of the organs of hearing have been detected in insects. Yet insects emit a variety of sounds by the friction produced by their mandibles, their wings, and their legs, which are communicated to others, and understood by them. The proofs of the existence of taste and smell in the different tribes, rest on the same foundation, the evidence of the function being performed. These senses are chiefly used in the animal economy in subserviency to the digestive system. The organs in which they reside are probably the palpi, or the other more flexible parts of the mouth. But these parts are so different in their form from the organs employed for the same purpose in the higher classes of animals, and so diminutive in size, that neither analogy nor dissection can be called in to illustrate the subject.

In treating of the important object of *nutrition*, it is necessary to begin with a description of the parts employed in procuring and preparing food for the gullet. In many cases, as in the higher orders of animals, the legs, particularly the first pair, execute the first movement of seizing the food and conveying it to the mouth; in general, however, the parts of the mouth, unassisted, seize, cut, bruise, and prepare the food for the gullet.

The *masticating* organs, in their simplest form, include the four following parts (which we shall designate by their Latin appellations, for want of appropriate English terms), Labrum, Mandibulæ, Maxillæ, and Labium. These form the mouth, and are denominated *Instrumenta cibaria*.

The *labrum* is analogous to the upper lip of the higher classes of animals. It is articulated to the fore part of the head (*frons*, or *clypeus*, *chaperon* of the French), either directly by a simple suture, or by the intervention of a plate, to which Mr KIRBY, from its situation, has given the name *nasus*. The labrum itself, at its free edge, exhibits great variety of character in the condition of its margin, of which naturalists have availed themselves in the discrimination of species. This organ may be regarded as the cover to the other parts of the mouth above.

The *mandibulæ* are two in number. They take their rise immediately below the labrum on each side. They exhibit very remarkable differences in size, shape, and armature. They move horizontally, and serve to cut objects by their edges crossing like the blades of scissars.

The *maxillæ* are likewise two in number. They are united to the cheeks immediately underneath the mandibulæ, and between these organs and the labium. They are more complicated than the mandibulæ. On their inner margin, they are usually covered with stiff hairs. Externally they support the *palpi*, which are articulated appendages, consisting of two or more joints. Each maxilla is furnished with one of these, rarely with two. The use of the palpi has not been determined, although it is probable that they serve as organs of smell, and perhaps also of touch.

The *labium* is analogous to the under lip, and closes the under side of the mouth, pressing the maxillæ. Its free edge is variously marked, and there is usually a line in the middle indicating its tendency to be double. On each side it supports a palpus, consisting of two or three joints. Its base is connected, laterally, by ligament, with the base of the maxillæ, and behind, with a fixed plate jointed to the head, and termed *mentum*, corresponding, in position, with the nasus of the labrum.

When all these parts bear such a proportion to one another, as to be able to cut the substances used as food, and convey them to the pharynx, they are considered as in the most perfect condition. But all these parts in the different classes, exhibit very remarkable modifications. In many of the crustacea, the labium has been converted into a second pair of maxillæ, and the three pair of legs which, in insects, constitute the organs of motion, are here changed into three pair of auxiliary maxillæ, and close the mouth below, in place of a labium. The locomotive organs are, in reference to their position, abdominal legs. Modifications of this arrangement prevail in the arachnida. There are many species of insects which are destined to live chiefly or exclusively on fluid substances, to whom such masticating organs would be unsuitable. Those which hold a middle station between the gnawers (*mandibulata*), and suckers, have the labium transformed into a soft, fleshy plate, like the tongue of quadrupeds (and hence frequently termed *lingua*), capable of licking, and of being rolled inwards when at rest,—as in the Hymenoptera represented by the bee. In the genuine suckers (*Haustellata*), the organs of the mouth present two important modifications. In the first, the labrum is nearly obliterated. The labium is produced and either crustaceous (*rostrum*), or membranaceous (*proboscis*), with a groove on its upper side, for the reception of the four plates or hairs into which the two mandibulæ and maxillæ have been converted. These appearances are exhibited in the Hymenoptera and Diptera. In the second, both the labrum and mandibles are nearly obliterated, and the labium is short and fixed; and the two maxillæ are produced and so applied to each other, as to form a sucker. This suker is capable of being rolled up spirally, and as it appears in the Lepidoptera, the butterfly for example, is usually termed the proboscis or tongue.

All these different parts of the mouth, however much they may be modified in size and shape, may still be readily detected, either by their position or palpial appendages. We owe this discovery of the true nature of the parts of the mouth of insects, and the other pedate annulosa to M. SAVIGNY, who has accompanied his judicious observations with accurate representations of each organ, and its various changes *.

The orifice of the *gullet* is, in general, a simple aperture, into which the food is conveyed chiefly by the agency of the maxillæ. In some cases, however, among the Hymenopterous insects, there is an organ on the base of the mouth, more or less distinct, to which the term *lingua* ought to be restricted. M. SAVIGNY calls it *glossa* or hypopharynx. On the dorsal margin of the opening there is likewise, in some cases, a particular process, which is denominated by the above naturalist *epiglossa*, or epipharynx.

The gullet itself is usually membranaceous, and is either simple or furnished with an enlargement, denominated the first stomach. The walls of the true stomach, in some of the crustacea, present the appearance of muscular fibres, and in some of the orthoptera, these bands cover the whole surface, forming a true gizzard, fortified with teeth or scales pointing backwards. The intestines are variously convoluted, and frequently near the anus, exhibit an enlargement which has been denominated a rectum. In some cases, the canal is furnished with one or more side pouches, or cœcæ, near the stomach.

In the crustacea, the liver appears, in some, like a con-

---

* Memoires sur les Animaux sans Vertebres, premier partie, Paris, 1816.

glomerate gland attached to the intestines, in others, as a collection of intestinula cœca. The hepatic vessels of insects (for there is no secreting organ which can be compared to a liver), have their origin in the fluid contents of the abdomen, and even send their capillary extremities to the remotest part. They vary greatly in number and size. The walls are dense and cellular, and the bile which they secret is yellow, brown, or white, according to the species. These vessels, in some, terminate separately, in others, they unite into a common duct. The terminations vary greatly, being in different species at the pyloric extremity of the stomach, the middle of the intestine, or at the rectum.

The circulating system appears in this group, in its most perfect form, among the larger crustacea. There is a pulmonic ventricle situate dorsally, and a systemic ventrical situate ventrally, thus executing a complete circulation in which all the blood in its course is aërated. In the arachnida with pulmonary sacs, the heart seems to be lengthened, placed along the back, and to give out vessels laterally.

In insects, neither absorbing nor circulating vessels have been detected, although anxiously looked for by many celebrated anatomists and microscopical observers. The nutritious portion of the food appears to be absorbed by the walls of the intestines, and discharged into the cavity of the body, where there are neither veins, arteries, nor heart. Towards the back, indeed, there is an obvious vessel, placed longitudinally, to which some have given the name of heart, but which is more generally denominated the *Dorsal Vessel*. This vessel is widest in the middle, and diminishes in size at each extremity. Its walls consist of two membranes, the internal one muscular, and the external one cellular. This last is so much crowded with tracheal vessels, as to appear to be entirely composed of them. All the coats are liberally provided with nervous filaments. This vessel is kept in its position by the tracheal tubes, and by muscular fibres, which, in general, are disposed in triangles, and increase in breadth from the superior part of the body to its inferior extremity.

The contents of the dorsal vessel are fluid, but of such consistence, that, when its coats are punctured, no liquid flows out. The colour is usually similar to the adipose matter which is collected on or near its surface, and differs according to the species. When placed under a microscope, this humour appears to consist of grains or globules, containing other globules. It mixes readily with water, and when dried, resembles gum.

The humour of the dorsal vessel is subject to some degree of motion, arising from the contractions which it experiences. These are irregular as to time, and proceed from the one extremity to the other, by stages usually corresponding with the rings of the animal, and strongest in the abdomen. The nerves do not appear to exercise any influence on these contractions. The muscular fibres and tracheal filaments appear to exercise the greatest control. These contractions, which, in some species, are little more than thirty, in others amount to one hundred and forty in a minute, according to the species, have been denominated pulsations, and the organ itself has been termed a heart. To that viscus, however, it bears no resemblance, except in its contractions, which, however, are irregular. It neither receives nor gives motion to any circulating fluid. Its use appears to be to imbibe and convert the fluid of the abdomen into fat, to serve as a supply in those numerous cases where much nourishment is suddenly required, as, during the metamorphoses of youth, and the production of eggs in maturity. The peritoneal membrane appears destined to execute the same functions *.

There are three remarkable modifications of the aërating organs. In the first, exemplified in the crustacea, they are gills, in the form of tufts of plumes, to which the water has direct access. In the second, exhibited in part of the Arachnida, the lungs are in the form of sacs or lateral pouches, having the bloodvessels spread upon the walls. In the third group, including Insects and some Arachnida, there are, on each side of the body, small orifices, termed *stigmata*, differing in number according to the species. These are formed by a cartilaginous ring, and, in some cases, furnished with one or more valves. Each orifice is the extremity of a short tube, which opens internally into a cavity, one on each side of the body. From these lateral cavities arise innumerable tubes, termed *tracheæ*, most numerous at the termination of the stigmata, which convey the air to every part of the body. To enable them to do this, their coats consist of an external and internal cellular membrane, with a middle layer, consisting of a cartilaginous string, spirally twisted, resembling the spiral tubes of plants. These tracheæ, by their number and subdivisions, convey air to every part of the body, and form, indeed, the great bulk of its contents. In what manner the vitiated air is expelled, has not been ascertained.

From this view of the nutritive system of insects, it appears, that the chyle is absorbed by the inner surface of the alimentary canal; that it exudes from its external surface into the common cavity; that the tracheæ

---

* See " Observations on the use of the Dorsal Vessel," by M. MARCEL DE SERRES; translated in Annals of Philosophy, vol. iv. p. 346; vol. v. p. 191, 369; and vol. vi. p. 34.

aërate this mass; and that, while the dorsal vessel and the peritoneal membrane prepare fat, the hepatic filaments separate from it bile, and probably urine. The existence of this last excrementitious fluid, is evinced by the presence of urea in the excrement. The peculiar secretions are nowise remarkable *.

The pedate annulosa have the sexual organs on different individuals; impregnation takes place internally by their union, and the females are all oviparous.

In the male, the sexual organs bear a closer resemblance to those of the mammalia than some of the higher classes. The *testes* may be regarded as two in number, situate one on each side of the abdomen. In some cases, however, they appear a simple mass, while in others, they are sub-

---

* All those insects which live in society, when exposed to cold, are observed to cluster together as if to keep each other warm. Some, indeed, when exposed to cold become torpid, and revive upon the restoration of a suitable temperature; but there are others, as the honey-bee, which resist any reduction of their temperature below their ordinary digestive heat, and preserve it in their dwellings even during the winter season. JOHN HUNTER found a hive in July 18. at 82°, when the temperature of the air was only 54°, and in December 30. at 73°, when the air was 35°. (Phil. Trans. 1792, 136.) When cooled until they become benumbed, they seldom recover, while the wasp, belonging to the same natural order, can be rendered torpid, and again revived with safety.

There is one circumstance recorded, which ought to excite inquiry, in hopes of finding in insects electrical organs similar in their effects to those of the torpedo. It refers to the Reduvius serratus, commonly known in the West Indies by the name of the Wheel-bug. " The late Major-general DAVIES, of the Royal Artillery, well known as a most accurate observer of nature, and an indefatigable collector of her treasures, as well as a most admirable painter of them, once informed me, that when abroad, having taken up this animal and placed it upon his hand, it gave him a considerable shock, as if from an electric jar, with its legs, which he felt as high as his shoulders, and, dropping the creature, he observed six marks upon his hand where the six feet had stood." (KIRBY and SPENCE's Introduction to Entomology, vol. 1. p. 110.)

divided into a number of lobes, which may be regarded as so many separate organs. The *spermatic* ducts are two in number, varying greatly in length and the number of tubes, which, like roots, combine in their formation. They unite, in general, into a common duct near the penis; but, previous to doing this, they are joined by the ducts of the vesiculæ seminales, and other accessory tubes, which, in different genera, exhibit a great diversity of character. In this case, the penis is usually a simple tube, protected in some insects at the base by two scales, which separate upon entering the vagina of the female, and thus prepare a passage for the penis, and serve likewise for retention. In many of the crabs and spiders, the seminal ducts continue single, each opening into a separate penis. The external opening is usually situate at the extremity of the abdomen, beside the anus. In the libellulæ, or dragon-flies, it is seated at the base of the abdomen, and in the spiders, at the extremity of the palpi.

The males are seldom of so large a size as the females, and frequently exhibit peculiar characters in their abdomen, eyes, or antennæ, by which they may be distinguished.

The female organs exhibit fewer varieties of structure in the different genera than those of the other sex. The ovaria consist of numerous tubes, in which the eggs are prepared. These open, in insects, into a common oviduct, terminating in the vulva. Previous to the termination of the oviduct, it receives the ducts of one or more vesicles. The crabs have two oviducts and two vulvæ *.

The female may, in general, be distinguished from the male, by the superior size of the abdomen. In some cases, both males and females survive the process of generation,

---

* See Vol. I. p. 418.

to repeat it again in another season. In other cases, the female only survives, while, in many, death ensues, upon the eggs being prepared and excluded.

The apparatus with which some species are furnished, to enable them to place their eggs in a proper situation, has been denominated the *ovipositor*. It is a continuation of the vulva, more or less strengthened by bony spiculæ, according to the nature of the substance it is destined to penetrate.

The period which elapses between the union of the sexes, and the laying of the eggs, extends to days in some, and even to months in others. The eggs themselves are either excluded at once, or at particular intervals. They are deposited under one or other of the following conditions. In the first, the egg, upon being deposited, is left to the influences of external circumstances, and the young, when hatched, to the resources of their own instinct. In the deposition of the egg, the wisest arrangements are made for the welfare of the young. The mother attaches them, in general, to those substances on which, upon being hatched, they are destined to feed. The butter-fly attaches her eggs to a leaf; the flesh-fly deposits her's upon carrion; while others insert them into the young of other insects. Not a few females prepare a particular hole, in which they place the egg, and lay up for the young a suitable provision when they burst the shell. In the second, where the species live in society, the eggs are deposited within the dwelling, and the young are reared and fed by the mother's care. In the females of the crustaceous animals, the eggs, after exclusion, are collected under the tail, or upon the feet, and retained until the young come forth. Some of the spiders carry the eggs, and even young, for a time, in bags upon their backs.

In the aphides, or *plant lice*, as they are called, the female retains the eggs internally at one time until they are hatched, and at another, lays them like other insects.

When the eggs are hatched, the young ones, among the crustacea and arachnida, are perfectly formed, and resemble the parent. It is otherwise with insects. When they are hatched, the young are termed grubs, maggots, caterpillars, or technically, *larvæ*. In this state, they are proverbially voracious, and their digestive organs are of much greater dimensions than when arrived at maturity. In the condition of larvæ, insects possess a variety of members, as legs, suckers, hairs, and even stigmata, which they do not possess in their maturity. They are all, however, destitute of wings. Some of them live constantly in the water, instead of the land, their future residence,—swimming in youth, and flying in maturity. The food of the larvæ is often solid, requiring powerful jaws to gnaw it, while the food of the perfect insect is fluid, and sucked up. When the larva has attained a certain size, and acquired the requisite quantity of fat, having been nourished either by the food which it has acquired by its own industry, as the caterpillars, or by that which has been brought to its cell, as in the grub of the bee, it prepares to assume the forms of maturity, by passing through the third stage of existence as a *pupa*. In this state, the parts, which were suitable only in the larva condition, either become obliterated, or are changed into organs fit for maturity. The following conditions of the pupa state are recognised by naturalists:

1. Some insects of the apterous tribes, merely by repeated castings of the skin, arrive at the perfect state, without undergoing any sudden or remarkable change of form or structure. These are termed *Pupæ completæ*. They move and eat like the perfect insect.

2. Many insects, which have wings in the perfect state, are observed to acquire, first, the rudiments of them, and afterwards all their parts, and to assume the form of the perfect insect, while passing through this period of youth, without any particular transformation. In this state they are called *Pupæ semi-completæ*. Like the pupæ completæ, they likewise move and eat like the perfect insect. In the different kinds of pupæ which remain to be considered, the animal neither eats nor moves. It derives its nourishment from its stores of fat.

3. After retiring to some suitable place, the larvæ of some insects cast their skin, and disclose the body of an ovate form, enveloped with a coriaceous covering, forming, within, separate sheaths for the different external organs. In this covering all the changes of form and structure take place, which prepare it for maturity. These are termed *Pupæ incompletæ*.

4. Other insects, likewise, upon changing their skin for the last time in their larvæ state, appear within a coriaceous covering, destitute, however, of any sheaths within for the external organs; these last being closely applied to the body. These are termed *Pupæ obtectæ*.

5. In the last form of the pupa, the skin, instead of being cast off, is changed into the coriaceous covering. Such are termed *Pupæ coarctatæ*.

All insects which, in the pupa state, do not eat, and are motionless, are careful to retire to situations, sufficiently remote from enemies, and of suitable temperature and moisture. In many cases, the larva forms an exterior covering, in which the pupa may be lodged with greater safety. This covering is in some composed of threads of the well known substance termed silk. Sometimes only one or two threads are required to keep the pupa in a proper position; in others, the silk is woven into cloth, or so matted together, as to resemble paper. These external cases are termed *cocoons*. The matter of which they are fabricated is *prepared* by two long tubes, which take their rise in the abdomen, enlarge as they approach the head, and terminate by a duct, which opens under the labium. By pressing the orifice of this duct to one place, and then to another, the larva draws out the tenaceous threads.

The larvæ that live in cells ready fabricated for their reception, as the wasp and the bee, are not contented with these as a covering during the pupa state, but they line their sides and bottom, and cover their mouth with silk, thus making a complete cocoon. These, after the insect has been perfected, are left in the cell, and when it contains another larva, a second lining is likewise prepared. Each lining at the bottom, in the case of the bee, covers the excrement, which the animal had produced in its larva state. Hence, the walls of the bee-combs appear double or treble; nay, JOHN HUNTER, by whom the appearance was observed, has counted twenty different linings in one cell *.

The external covering of the pupa, in some cases, consists of pieces of earth or dried leaves, curiously joined together, and cemented by an adhesive secretion.

After the insect has remained in a pupa state for a certain period, exceedingly different in the various tribes, it bursts forth from its confinement in its state of maturity. In this perfect condition it is termed the *Imago*. The organs of reproduction now speedily enlarge, and preparation is made to increase and multiply.

The duration of the life of the pedate annulosa, in their mature state, is confined in some to a few days, while, in others, it extends to many years. Those which, like the annual flowers, are destined to breed only once during life,

---

* Phil. Trans. 1792, p. 193.

expire in a few days after the important end has been accomplished,—the males usually before the females. Such species, in general, require but little food in the mature state. Others outlive several impregnations and several years, and, during this period, require a constant supply of food.

There are many circumstances, however, which occasion a premature death to this tribe of animals, in all the stages of their existence. They are the food of many quadrupeds, birds, reptiles, and fishes. These animals seek them in their places of retreat; and but for the circumstance that a superabundant supply is provided, we may be surprised that the continuance of the species is preserved. They are not destitute, however, of means of escape or defence. Many of them so exactly resemble the substances on which they feed, as readily to escape the eye of the foe. Some of the crabs attach, by a glutinous matter, the leaves of sea-weeds to their body, so as to conceal completely their form, and secure them from detection. Some are protected by hairs and spines, the offensive attitudes they assume, and the disgusting odours they emit. The Brachinus crepitans, when pursued by its great enemy, the Calosoma sycophanta, a rare British beetle, emits from its anus repeated explosions of blue smoke, having a disagreeable smell, which stops the progress of the assailant. In another species, B. displossor, the scent of the smoke is pungent, and resembles that of nitric acid. Other insects, when irritated, emit a strong smelling fluid, from glands seated at the anus, which have been termed Osmateria. The jaws are employed by some species, and the sting by others, as means of defence in the hour of danger. This last organ is essentially connected with the business of reproduction, and is seated at the ovipositor of the female. In the bee, the apparatus of the sting consists of the coriaceous laminæ which cover it when at rest, the groove, in which it is contained when exerted, and the two spiculæ, with reversed serratures at the summit, which inflict the wound. The two tubular glands, which secrete the poison, take their rise in the abdomen, unite to form the bag, from which a duct issues, terminating at the base between the spiculæ: these, by uniting laterally, form a groove, by which the poison is conveyed to the bottom of the wound. But, besides these means of preserving life, many species are so vivacious, that they will live long without food, and even without air, and suffer sad mutilation before the vital spark be extinguished. Among the crustacea and arachnida, amputated limbs are even readily reproduced.

The annulose animals with articulated feet are conveniently divided by means of the characters furnished by the organs of respiration.

Order I.—Branchial aerating organs.

This order includes the class Crustacea of modern naturalists. Their external covering is usually crustaceous or coriaceous. The antennæ, in general, exceed two in number. The situation of the gills furnishes the characters for the next subdivision.

Section I.—Malacostraca.

Branchiæ at the base of the eight pair of legs, and concealed underneath the border of the shield.

The body appears to consist of a shield and tail above. The shield is destitute of sutures, is incorporated with the head, and supports anteriorly the two eyes, the antennæ, and the snout. Below, the thorax and abdomen are incorporated. The mouth is seated under the snout. The feet amount to eight pair, the three first of which are converted into auxiliary maxillæ, while the remaining five pair are subservient to locomotion. Each of these legs supports at the base, but concealed from view, a pyramidal plumose gill. This position of the gills induced LAMARK to denominate the section *Homobranches*. In many species, the anterior pair of locomotive feet are large, and the last joint of the tarsus furnished with a forceps *(cheliferous)*. In some of the species the posterior feet are broad, and fitted for swimming. The *tail* consists of several joints, is furnished beneath with several pair of fringed filaments for the retention of the eggs, and, in some, with terminal plates for swimming.

Tribe I.—PODOPHTHALMA.

Eyes pedunculated.

A. BRACHYURA. Tail short and simple at its extremity, and incurved upon the abdomen.

Dr LEACH, whose labours qualify him, in a peculiar manner, for arranging crustaceous animals, has given a methodical distribution of the genera of this group, which we shall here insert [*].

A. *Abdomen of the male five jointed, the middle joint longest; of the female seven jointed. Anterior pair of legs didactyle.*

Division I. Shell (shield) nearly rhomboidal. Two anterior legs very long, with deflexed fingers.

1. LAMBRUS.

Division II. Shell truncate behind. Two anterior legs of the male elongate, of the female, moderate.

Subdivision 1. Antennæ long, ciliated on each side.

---

[*] Article ANNULOSA in the Supplement to the Encyclopædia Britannica, likewise published in " The Entomologist's Useful Compendium," by GEORGE SAMUELLE, 8vo. London, 1819,—a work which gives many convenient indications of the British species.

2. Corystes. 3. Thia. 4. Atelecylus.

Subdivision 2. Antennæ moderate, simple. Hinder pair of legs with compressed claws.

5. Portumnus. 6. Carcinus. 7. Portunus. 8. Lupa.

Subdivision 3. Antennæ moderate, simple. Four hinder pair of legs compressed.

9. Mantua.

Subdivision 4. Antennæ simple, short. Four hinder pair of legs simple.

10. Cancer. 11. Xantho. 12. Calappa.

b. *Abdomen in both sexes seven jointed. Two anterior legs dedactyle.*

Division I. Eight hinder legs simple, and alike in form.

Subdivision 1. Shell anteriorly arcuated, the sides converging to an angle. Two anterior legs unequal.

13. Pilumnus. 14. Gegarcinus.

Subdivision 2. Shell quadrate, or subquadrate. Eyes inserted in the front.

* Shell quadrate. Eyes with a short peduncle.

15. Pinnoteres.

** Shell quadrate. Eyes with a long peduncle.

16. Ocypode. 17. Uca. 18. Gonoplax.

Subdivision 3. Shell quadrate. Eyes inserted at the anterior angles of the shell.

19. Grapsus.

Division II. Two hinder legs at least dorsal.

Subdivision 1. Two posterior legs dorsal. Eyes with the first joint of the peduncle elongated.

20. Homola.

Subdivision 2. Four hinder legs dorsal. Eyes with the first joint of the peduncle short.

21. Dorippe. 22. Dromia.

Division III. Shell rostrated in front. Eight hinder legs alike, and simple.

Subdivision 1. Fingers deflexed.

23. EURYNOME. 24. PARTHENOPE.

Subdivision 2. Fingers not deflexed. External antennæ with their first joint simple. Anterior pair of legs distinctly thicker than the rest.

25. PISA. 26. LISSA.

Subdivision 3. Fingers not deflexed. External antennæ with their first joint simple. Anterior pair of legs scarcely thicker than the others, which are moderately long.

27. MAJA.

Subdivision 4. Fingers not deflexed. External antennæ with the first joint simple. Anterior pair of legs about the thickness of the rest, which are very long and slender.

28. EGERIA. 29. DOCLEA.

Subdivision 5. Fingers not deflexed. External antennæ with the first joint externally dilated.

30. HYAS.

c. *Abdomen in both sexes six jointed. Two anterior legs didactyle.*

Division I. Fifth pair of legs minute, spurious.

31. LITHODES.

Division II. Second, third, fourth and fifth pairs of legs alike, and slender.

Subdivision 1. Eyes retractile.

32. INACHUS.

Subdivision 2. Eyes not retractile.

33. MACROPODIA.

D. *Abdomen of the male six jointed; of the female five jointed; the last joint very large. Eyes not retractile.*

34. LEPTOPODIA. 35. PACTOCLUS.

E. *Abdomen of both sexes four jointed. Two anterior legs didactyle.*

36. LEUCOSIA.   37. EBALIA.   38. IXA.

B. MACROURA. Tail about the length of the body, and not concealed under the shield.

The animals of this group differ from those of the preceding, in the produced form of their bodies. In the former, the breadth of the body usually equals or exceeds the length, while in these the length greatly exceeds the breadth. Dr LEACH distributes the genera in the following manner:

A. *Tail on each side with simple appendices. Legs ten; anterior pair largest and dactyle.*

1. PAGURUS.   2. BIRGUS.

B. *Tail on each side with foliaceous appendages, forming with the middle tail process a fan-like fin.*

a. *Interior antennæ with very long footstalks.*

Division I. External antennæ squamiform. Legs ten, alike and simple.

3. SCYLLARUS.   4. THENUS.

Division II. External antennæ setaceous, and very long. Legs ten, alike and simple.

5. PALINURUS.

Division III. External antennæ very long and setaceous. Legs ten, anterior pair didactyle, fifth pair spurious.

6. PORCELLARIA.   7. GALATEA.

b. *Interior antennæ with moderate peduncles.*

Division I. Exterior lamella of the tail simple. Antennæ inserted in the same horizontal line, the interior ones with two setæ, the exterior ones simple. Legs ten.

8. GEBEA.   9. CALLIANASSA.   10. AXIUS.

Division II. Exterior lamella of the tail bipartite. Antennæ inserted in the same horizontal line, the internal ones with two setæ, the external ones with a spine-shaped squama at the first joint of the peduncle. Legs ten, (anterior pair largest and didactyle).

11. Astacus.   12. Potamobius.   13. Nephrops.

Division III. External antennæ with a large broad squama, or scale, at their base. Abdomen with the second joint anteriorly and posteriorly produced below. Legs ten.

Subdivision 1. External antennæ inserted in the same horizontal line with the interior ones, which have two setæ. Tail with the external lamella composed of but one part.

14. Crangon.   15. Pontophilus.

Subdivision 2. External antennæ inserted below the internal ones; interior ones with two setæ inserted in the same horizontal line. Exterior lamella of the tail bipartite.

16. Atya.   17. Processa.

Subdivision 3. External antennæ inserted below the internal ones; interior ones with two setæ, one placed above the other. (External lamella of the tail composed of but one part).

* *Internal antennæ with the superior seta excavated below. Claws spinulose.*

18. Pandalus.   19. Hippolyte.   20. Alpheus.

** Internal antennæ with the superior seta not excavated. Claws simple.

21. Penæus.

Subdivision 4. External antennæ inserted below the internal; interior ones with three setæ. External lamella of the tail composed of but one part.

22. Palæmon.   23. Athanas.

Division IV. External antennæ inserted below the internal ones, with a large scale at their base. Legs sixteen.

24. Mysis.

c. *Tail with two setæ, one on each side.*
25. NEBALIA.   26. ZOEA *.

Tribe II. EDRIOPHTHALMA.

Eyes sessile.

Dr LEACH gives the following distribution of the genera;

SECTION I.

Body laterally compressed. Legs fourteen. Antennæ two, inserted one on each side of the front of the head. (Tail furnished with styles).

1. PHRONYMA.

SECTION II.

Body laterally compressed. Legs fourteen, with lamelliform coxæ. Antennæ four, inserted by pairs. (Tail furnished with styles).

Division I. Antennæ four-jointed, the last segment composed of many little joints; the upper ones very short.

2. TALITRUS.   3. ORCHESTIA.

Division II. Antennæ four-jointed, the last joint composed of several little joints; upper ones rather shortest.

4. ATYLUS.

Division III. Antennæ three-jointed, the last one composed of several little joints; upper ones longest.

5. DEXAMINE.   6. LEUCOTHOE.

Division IV. Antennæ four-jointed, the last segment composed of several little joints, upper ones longest.

Subdivision 1. Four anterior legs monodactyle; second pair with a much dilated compressed hand.

7. MELITA.   8. MÆRA.

Subdivision 2. Two anterior pair monodactyle and alike.

---

* See Dr LEACH's Description of the *Zoëa clavata*, TUCKEY's Nar. p. 414.

9. Gammarus. 10. Amphithoe. 11. Pherusa.

Division V. Antennæ four-jointed, under ones longest, leg-shaped. (Four anterior legs monodactyle.)

Subdivision 1. Second pair of legs with a large hand.

12. Podocerus. 13. Tassa.

Subdivision 2. Second pair of legs with a moderate sized hand.

14. Corophium.

Section III.

Body depressed. Antennæ four. Legs fourteen.

A. *Tail without appendices.*

Division 1. Body with all the segments bearing legs.

Subdivision 1. Body linear.

15. Proto. 16. Caprella.

Subdivision 2. Body broad.

17. Larunda.

Division II. Body with all the segments not bearing legs.

18. Idotea. 19. Stenosoma.

B. *Tail on each side with one or two appendices.*

Division I. Antennæ inserted in nearly the same horizontal line.

20. Anthura.

Division II. Antennæ inserted in pairs, one above the other.

Subdivision I. Tail with one lamella on each side.

21. Campecopea. 22. Næsa.

Subdivision II. Tail with two lamellæ on each side.

* *Superior antennæ with a very large peduncle. Claws bifid.*

23. Cymodice. 24. Dynamine. 25. Sphæroma.

** Superior antennæ with a very large peduncle. Claw single.

26. Æga.

*** *Superior antennæ with a moderate peduncle.*
27. Eurydice. 28. Limnoria. 29. Cymothoa.
c. *Tail terminated with two setæ.*
30. Apseudes.
d. *Tail furnished with styles.*
Division I. Interior antennæ distinct.
Subdivision 1. Styles of the tail exserted. Anterior legs monodactyle.
31. Janira. 32. Asellus.
Subdivision 2. Styles of the tail not exserted. Anterior legs simple.
33. Iæra.
Division II. Interior antennæ not distinct.
Subdivision I. Styles of the tail double, with a double footstalk.
34. Ligia.
Subdivision 2. Styles of the tail four, the latter ones biarticulate.

* *Body not capable of contracting into a ball.*
a. *External antennæ eight jointed.*
35. Philoscia. 36. Oniscus.
b. *External antennæ with seven joints.*
37. Porcellio.
** *Body contractile into a ball.*
38. Armadillo *.

Section 2.—Entomostraca.

Branchiæ irregularly placed with respect to the legs, and never concealed under the margin of the shield.

---

* The last five genera being terrestrial, and consequently having pulmonary aërating organs, ought not to be inserted among the Crustacea. They should probably form a group by themselves, in connection with the Myriapoda.

The following is the arrangement employed by Dr LEACH :

Division I. Body covered by a horizontal shield. Eyes sessile.

Subdivision I. Shield composed of two distinct parts.
1. LIMULUS.
Subdivision 2. Shield composed of but one part.
* With jaws.
2. APUS.
** With a rostrum, but no jaws.
a. Antennæ four.
3. ARGULUS.
b. Antennæ two.
4. CECROPS. 5. CALIGUS. 6. PANDARUS. 7. ANTHOSOMA.

Division II. Body covered by a bivalve shell. Eyes sessile.

Subdivision 1. Head porrected.
8. LYNCEUS. 9. CHYDORUS. 10. DAPHNIA.
Subdivision 2. Head concealed.
11. CYPRIS. 12. CYTHERE.

Division III. Body covered neither by a bivalve shell or shield. Eye one, sessile.

13. CYCLOPS. 14. CALANUS. 15. POLYPHEMUS (*Dichelestium*).

Division IV. Body covered by neither a bivalve shell or shield. Eyes pedunculated.

16. BRANCHIOPODA. (17. ARTEMIA. 18. EULIMENE).

There are two tribes of animals which have so many relations to the crustacea, in form at least, and partly in structure, that we have ventured to take some notice of them here.

The first, constituting the *Epizoariæ* of LAMARK *, contains the genera Chondracanthus, Lernæa, and Entomoda. The species reside in the gills, and different parts of the surface of fishes. Some species seem to attach themselves to the eyes, as appears to be the case in the one which adheres to the Greenland shark †.

The second, termed by Dr LEACH *Podosomata*, includes the genera Pycnogonum, Phoxichilus, Nymphum, and Amothea.

Order II.—PULMONARY AERATING ORGANS.

Section 1.—*ARACHNIDA*.

Head destitute of antennæ. Eyes simple.

1st Tribe.

Lungs in the form of lateral sacs.

The legs are always twelve in number, two pair of which serve as auxiliary maxillæ, and the remaining four pair execute locomotion. The eyes vary from six to eight, and have a dorsal aspect.

A. Palpi bearing the male organs in the last joint. The palpi themselves are of a moderate size, and do not terminate in a large moveable claw; neither are they cheliferous. The second pair of auxiliary maxillæ are pierced at the extremity of the termination of a poison duct. The abdomen is pedunculated, and without rings, and the pulmonary sacs, which are two in number, are seated near its base. The first stomach has numerous cœca, and the second is surrounded by the liver. The pores which give issue to the matter with which the threads for the construction of the web are formed, are seated at the anus. These threads they can eject to a considerable distance, and afterwards em-

---

* Hist. Nat. des An. sans Vert. vol. iii. p. 225.
† SCORESBY's Arctic Regions, vol. i. p. 538. tab. xv. fig. 5.

ploy them as a rope, on which to convey themselves [*]. The species feed on flies, and, in some instances, on mineral substances. Mr Holt found, that the *Aranea* (Salticus) *scenica* devoured the sulphat of zinc, and by the process of digestion deprived it in part of its acid [†]. The sexual organs of the male are double, complicated externally, and lodged in a cavity of each palpus. Those of the female are placed on the belly between the stigmata.

1. *Mandibles with their hooks folded transversely.*

This includes the genus Aranea of Linnæus, now formed by Walkenaer, Latreille, and others, into the following genera:

(1.) Segestria, Dysdera, Clotho, Aranea, Filistata, Drassus, Clubiona, Argyroneta. (2.) Scytodes, Theridium, Episinus, Pholcus. (3.) Linyphia, Uloborus, Tetragnatha, Epeira. (4.) Sparassus, Selenops, Thomisus. (5.) Ctenus, Sphasus, Dolomedis, Lycosa. (6.) Eresus, Salticus.

2. Mandibles with their hooks folded ventrally. This includes the following genera: Mygale, Avicularia, Atypus, Eriodon.

B. Palpi large, unconnected with the sexual organs, which are seated at the base of the abdomen. Anus destitute of spinners.

1. *Abdomen pedunculated.* Stigmata two. This includes the genera Thélyphonus and Phrynus.

2. *Abdomen sessile.* At the base, behind the legs where it joins the thorax, are two diverging pectinated scales, and it terminates in a jointed tail, with a poison sting. The stigmata are eight in number. This includes the genera Scorpio and Buthus.

---

[*] Annals of Phil. vol. ix. p. 306.　　[†] Ib. vol. xii. 454.

2d Tribe.

Lungs tracheal.   Eyes two or four.

A. Thorax and abdomen distinct. Eight feet. There are three genera. Solpuga, Chelifer, and Obisium.

B. Thorax and abdomen united.

1. *Mandibles distinct*, of two or three pieces, the last of which is cheliferous.

 a. Mandibles prominent.   Phalangium and Sirio.

 b. Mandibles not prominent.   Trogulus.

2. *Mandibles either in one piece, or in the form of an haustellum.*

 a. Legs eight.

 (1.) Legs formed for walking.

 (A.) With a haustellum, with or without palpi. Ixodes, Argas*, Uropoda, Smaridia, and Bdella.

 (BB.) Mandibles distinct, and palpigerous.

 (*a.*) Palpi with simple extremities. Mandibles cheliferous. Acarus, Cheyletus, Gamasus, and Oribata.

 (*b.*) Palpi subcheliferous. Mandibles with a claw. Erythræus and Trombidium.

 (2.) Legs fringed, and formed for swimming. Hydrachna, Elais, and Limnochares.

 b. Legs six.   Caris, Leptus, Atoma, and Ocypete.

Section II.

Head furnished with antennæ, and compound eyes.

This section includes the groups termed INSECTA and MYRIAPODA.

I.—INSECTA.   INSECTS.

Feet six, supported by the thorax.

The arrangement of insects has occupied the attention of many acute and accomplished naturalists. As yet, how-

---

* A species nearly related to this genus, found on Falco tinnunculus, is given in Plate IV. f. 7.

ever, a considerable difference of opinion prevails, as to the characters which should be employed in the formation of the orders and other subordinate divisions.

Since the days of SWAMMERDAM, there have not been wanting naturalists who have regarded the metamorphoses of insects, as furnishing the most suitable characters for primary divisions. But, instead of adopting the four forms of metamorphoses of that author, they have divided insects into such as do not undergo changes of form in the third or pupa state, and such as do undergo changes. Thus WILLOUGHBY (*in Raii Historia Insectorum*, p. 3.) divides insects into αμεταμορφωτα and μεταμορφυμενα, terms for which Dr LEACH has substituted Ametabolia and Metabolia. Strong objections may be urged against the classification of insects from such characters. All insects, even the αμεταμορφωτα, undergo changes, as they cast their skins repeatedly, and reach maturity by degrees. The extent of difference between the forms of the larva and imago state is not the same in the genera of the same family or order. The differences which insects exhibit in their changes are merely in degree, not in kind. But the strongest objection to which this method of classification is liable, arises from the circumstance of employing the characters furnished by the immature state of insects, in order to classify insects in their mature state when these characters do not appear.

LINNÆUS, to whom zoology in all its branches was greatly indebted, gave to insects a uniformity of nomenclature, and a methodical arrangement, greatly superior to all his predecessors. His primary divisions were taken from circumstances connected with the condition of the wings. The simplicity of this method, and the obviousness of the characters which have been employed, have secured for this system a decided preference among the entomologists of Britain. Perhaps the strongest objection which can be urged against this method is its limited nature, arising from the great increase of species, and the consequent influx of new characters calling for the formation of additional divisions.

FABRICIUS introduced a method of classifying insects, founded on the organs of the mouth, which has met with many admirers. These *instrumenta cibaria* furnish permanent and definite characters, and exhibit, in the modifications which they present, marks well calculated for fixing the limits of species and genera. They are, however, in many cases minute; careful dissection is requisite for their display; and not unfrequently the aid of the microscope is necessary.

These considerations have induced modern entomologists to combine the two systems of LINNÆUS and FABRICIUS, and to add that of SWAMMERDAM. The characters employed by LINNÆUS occupy the first rank, because they are most obvious to the senses. Those of FABRICIUS are resorted to in cases where the locomotive organs do not furnish marks sufficiently characteristic.

Insects may be divided into two classes, according as the mouth is fitted for cutting hard substances, or sucking fluid matter.

I.—INSECTA MANDIBULATA.

This class is usually subdivided into the seven following Orders:

Order I.—COLEOPTERA.

The insects of this order have their integuments of a coriaceous consistence, approaching, in some genera, to the hardness of shell. The wings are four in number. The upper part, denominated *elytra*, are of the same texture with the skin. They are, convex above, concave below, and, when at rest, their mesial margins join by a straight suture. In some genera the elytræ are inseparably united at their suture, and, in others, the elytræ at the base of the suture are in contact, or lap over each other, while at the extremity they recede from each other. The inferior pair of wings are membranaceous, strengthened by anastomosing ribs, and when at rest they are folded obliquely and transversely and concealed under the elytra. When the insect is about to fly, the elytra are raised, and remain fixed, while the under wings unfold and execute their motions. In some species, the under wings are imperfectly developed, or nearly obliterated, and in such cases the elytra are cemented at the suture. The abdomen consists of six or seven rings, more or less covered by the elytra, each having two sigmata, one on each side. The antennæ exhibit a great variety of character in their situation, length, form, and number of the joints. The eyes are compound, and two in number. Each eye, in some, is divided by the continuity of the marginal band of the head. There are no ocelli. The instrumenta cibaria are formed for cutting and masticating solid substances. The food, however, which is consumed by them, differs greatly, according to the species, in kind and consistence, so that all the parts of the digestive system exhibit extensive modifications of form. All the species are oviparous. The larva is lengthened, frequently destitute of antennæ and eyes, having twelve or thirteen rings, a scaly head, with the parts of the mouth similar to the perfect insect. It has usually six feet. After continuing months, or even years, in the larva state, the insect changes into a pupa obtecta, through the skin of which the different members may be distinctly perceived. The Coleoptera, in their perfect state, require a regular supply of food, and live to a greater age than other insects. They are very numerous in species and genera. LINNÆUS subdivided this extensive order into three groups, according as the antennæ were clavate, filiform, or setaceous. M. GEOFFROY employed the number of joints in the tarsus as the basis of his subdivisions. This method, though it separates a very few naturally connected genera, and is liable to some exceptions, is nevertheless so simple and easy of application, that it has been universally received throughout Europe.

I COLEOPTERA PENTAMERA.—Under this head are included those species which have five joints in each tarsus. The genera which they form are distributed by LAMARK into three sections, according as the antennæ are *filiform*, including those which are moniliform and setaceous, *clavate* or *lamellate*. (*Hist. Nat. des Animaux sans Vertebres*, tom. iv. p. 439.) LATREILLE, whose industry and acuteness have contributed greatly to extend the limits of entomology, has subdivided them in the following manner. (*Régne Animal*, par M. CUVIER, iii. p. 173.)

I. PENTAMERA CARNIVORA.—The distinguishing character of the insects here referred to, and one which is peculiar to them, is the possession of six palpi, in place of four, the ordinary number, two of these attached to the labium, and two to each maxilla. The maxillæ are hooked, and covered on the inside with stiff hairs or spines. There are two stomachs. The first is short and muscular; the second more produced, with villous walls. The hepatic vessels are four in number, and terminate at the pylorus. The antennæ are filiform or setaceous. The thighs of the middle and hind legs furnished with large trochanters. They are carnivorous both in their larva and perfect state. In the former, the body consists of twelve rings; the head and the first ring scaly, with two short antennæ.

1. *Carnivora Terrestria.*—The feet are formed for walking. The mandibulæ are apparent, the body is ob-

long, and the eyes prominent. The intestine terminates in a large cloaca, with two vesicles secreting an acrid humour.

1. *Cicindeladæ.*—The maxillæ are furnished with an articulated claw. Mandibulæ prominent, straight, and denticulated; eyes large and full; labium short; the tibiæ of the fore legs destitute of a notch on the inner side. In the genera Manticora, Cicindela, Megacephala, and Therates, the breadth and length of the thorax are nearly equal, and all the joints of the tarsus are entire. In the genus Collyris of FABRICIUS, changed by LATREILLE into Colliuris, the thorax is narrow, and produced, and the penult joint of the tarsus is bilobated.

1. *Carabidæ.*—The maxillæ destitute of the articulated claw. In general, the head is narrower than the thorax, and the mandibulæ are destitute of teeth. This very numerous family has been divided by LATREILLE into seven sections.

1. The external maxillary and labial palpi, with the last joint equal or larger than the preceding. The tibiæ of the fore legs have a deep notch on the inner side; elytra truncated or obtuse; labium entire and oval, or nearly square; the head contracted behind, and, with the thorax, is narrower than the abdomen; the thorax is heart-shaped and truncated behind, and its length never exceeds, but frequently falls short of, its breadth. The section includes the following genera: Anthia, Graphipterus, Brachinus, and Lebia. This last genus has been subdivided into the following genera by BONELLI, viz. Hellus, Cymindus, Lamprias, Dromius, and Demetrias.

2. The genera of this section exhibit the same form of palpi and elytra. The head is deeply divided from the thorax, to which it is joined by a socket; the labium is furnished on each side with a lobe; the thorax is lengthened; the penultimate joint of the tarsi frequently bilobate.

The genera are, Zephium, Galerita, Drypta, Agra, and Odacantha.

3. The palpi and tibiæ present the same characters as the preceding. The elytra are not truncated; the suture of the mentum obsolete. This includes the genus Siagona.

4. The genera of this section differ from the preceding ones, in the tibiæ being denticulated externally. The second and third joints of the antennæ are nearly equal and moniliform; the elytra are entire, and the mentum articulated. This includes the genera Scarites and Clivina.

5. The elytra are entire; the mentum articulated; anterior tibiæ entire externally, with short terminal spines; labium pointed in the middle, with lateral lobes. This section contains the genera Ozena, Morio, Aristus, Harpalus, Feronia, Licinus, Badister, and Panagæus.

6. The elytra are entire, and the anterior tibiæ very slightly, or not at all notched; labium pointed. This includes Cychrus, Pamborus, Calosoma, Carabus, Nebria, Omophron, Pogonophorus, Loricera, and Elaphrus.

7. In this last section, two at least of the exterior palpi are pointed at the extremity. The anterior tibiæ are notched. Bembidion, Trechus, and Apolomus.

2. *Carnivora Aquatica.*—The feet of the insects here referred to are formed for swimming. The mid and hind legs are compressed or ciliated; the mandibules are concealed; the terminal hook of the maxillæ bent from the base; the thorax is broad. They live in the water both in the larva and imago state. The larvæ are long and narrow, with twelve rings; the head large, with strong, hooked mandibulæ, pierced at the apex. The body has six feet.

1. *Dyticidæ.* The antennæ are filiform, and longer than the head. In the males of many species, the three first joints of the tarsi of the mid and fore legs are dilated, and

furnished with complicated suckers. In a few genera, the tarsi of the mid and fore legs have only four joints, as Hyphydrus and Hydroporus; in the others, the tarsi are entire, as Haliplus, Pelobius, Noterus, Laccophilas, Colymbetes, Hydaticus, Acilius, and Dytiscus.

2. *Gyrinedæ.*—The antennæ are here clavate, with a subsidiary ciliated one at the base of each; each eye divided into two by the marginal band of the head. There is only one genus belonging to this family, Gyrinus.

II. PENTAMERA MICROPTERA.—The insects of this division constituted the genus Staphylinus in the LINNEAN system. They are characterized by their filiform or moniliform antennæ, sometimes thickening a little towards the end. The body is narrow, and the elytra scarcely reach to half the length of the abdomen. The coxæ of the fore and mid legs are remarkably large. Two bags are protruded at pleasure from the anus. The species run and fly readily. When pursued, they elevate their head and abdomen, and assume a very threatening attitude. They frequent moist places, in the neighbourhood of putrid animal or vegetable substances. A few are found in flowers, in pursuit of minute insects. Their first stomach is very short, and without folds; the second is long and villous, with a short intestine. The species are very numerous, and have been divided into many genera, which admit of the following distribution:

1. Head exposed, and separated from the thorax by an obvious mark. Among the insects of this group there are some which have the labrum deeply divided into two lobes. The *Staphylinidæ* are distinguished by their filiform palpi, and consist of the following genera: Staphylinus, Pinophilus, and Lathrobium. The *Oxyporidæ* have the four palpi, or at least the labial ones, terminated by an enlarged joint, as in Oxyporus and Astraphæus. In other genera, the labrum

is entire. In the Pœderidæ, the maxillary palpi are nearly the length of the head, as in Pœderus, Evœsthetus, and Stenus. In the Oxytelidæ, the maxillary palpi are greatly shorter than the head, and the antennæ are inserted in front of the eye, as in Oxytelus, Siagonium, Omalium, Piestes, Proteinus and Lesteva. The Aleocharidæ differ from the preceding family, in the antennæ being inserted between the eyes, as in Aleochara.

2. Head sunk in the thorax. as far as the eyes. In the Lomechusidæ, the tibiæ are entire, as in Lomechusa. In Tachinidæ, the tibiæ are spinous, as in Tachinus and Tachyporus.

III. PENTAMERA SERRICORNUA.—The elytra cover the abdomen; the antennæ are usually filiform, or slightly clavate; and, in the males particularly, serrated, pectinated or plumose. In some of the genera, the thoracic sternum is advanced in front, under the head, and likewise produced behind. This character is exhibited in those which have the mandibles notched or bifid at their extremity, as the Elateridæ, a numerous family, in which the natural genera have not yet been established; and, in those which have entire mandibles, as, Buprestidæ, having filiform palpi, and containing the genera Buprestis, Tracys, and Aphanisticus: and the Melasidæ, in which the palpi have an enlarged terminal joint, as Melasis and Cerophytum.

In those having the thoracic sternum destitute of the singular character exhibited by the preceding families, there are several genera in which the mandibles are forked at the apex, or furnished with a tooth beneath. In some, the body and elytra are soft, and the head furnished with a neck, as the Lymoxylonidæ, in which the elytra do not embrace the abdomen, as Lymoxylon, Hylecœtus, Atractocerus, and Cupes; and, in Mastigoidæ, in which the abdomen is embraced by the elytra, as in Mastigus and Scydmanus. In others, the neck is concealed, as in Malachiusidæ, which exhibit, as a peculiar character, four vesicles divided into lobes, under the thorax and the base of the abdomen, which can be withdrawn or exerted and inflated at pleasure. The Melyridæ have the palpi filiform, as Melyris, Dasytes, and Drilus; while in Clerusidæ, the palpi are securiform as Clerus, Tillus, and Enoplium.

In those families where the body and elytra are firm and crustaceous, the Ptinusidæ have the head and thorax narrower than the abdomen, with the antennæ about the length of the body as Ptinus and Gibbium ; and the Anobiumedæ, having the thorax of the size of the abdomen, and the antennæ much shorter than the body, as Anobium, Ptilenus, and Dorcatoma.

There are several genera which agree with some of the preceding in the softness of their bodies, but having mandibles entire at the apex. In some of these, the palpi are filiform, as in the Cebrionidæ, which have tarsal joints entire, as Cebrio and Hammonia ; and in Scirtesidæ, which have the penultimate joint of the tarsus bifid, as Scirtes, Elodes, Rhipicera, and Dascillus. In others, the palpi, especially the maxillary ones, become thicker towards the extremities. In the Lampyridæ the antennæ are approximate at the base, and the maxillary are longer than the labial palpi, as in Lampyrus, Lycus, and Omalisus. In the Telephoridæ, the antennæ at the base are remote, and the labial and maxillary palpi are nearly of equal length, as in Telephorus and Malthinus.

IV. PENTAMERA CLAVICORNUA.—The antennæ are here obviously club-shaped, perfoliated or solid, generally exposed at the base, and longer than the maxillary palpi. In a few genera, forming the family Dryopsidæ, the first and second joints of the antennæ are enlarged, and the remainder form a club nearly solid, so that they appear three jointed; as Dryops, Hydera, and Heterocerus. In the remaining families, the antennæ increase more gradually from the base, and the club consists of several joints. In some of these, the pectoral sternum is produced under the head towards the mouth. In the Histeridæ, the mandibles are prominent, and the antennæ geniculated, as Hister, Abræus, Onthophilus, Dendrophilus, and Platysoma. In the Byrrhidæ, the antennæ are straight, as Byrrhus, Throscus, Anthrenus, Chelonarium, Nosodendron, Elmis, Macronychus, Georessus, and Megatoma. In other genera, the pectoral sternum is abbreviated in the usual form, as the Dermestidæ, which have the mandibles short, thick, and straight at the extremity, as Dermestes and Attagenus. In the remaining families, the mandibles are lengthened, compressed, and hooked at their extremity. The Nitidulidæ, have the mandibles notched, bifid, or furnished with a tooth at their extremity, as Nitidula, Biturus, Cateretes, Micropaplus, Thymalus, Colobicus, Engis, Ips, Scaphidium, Scaphisoma, and Choleva. The Silphidæ have the extremities of the mandibles entire, as Silpha and Necrophorus. The insects of this family are reputed to feed on carrion, and to dig under dead mice and moles, and bury them, in order to feast upon them more securely. These statements are without foundation. These insects feed on maggots, and their pupæ; and, in penetrating the ground in search of the last of these, they loosen the soil so much, that the dead animal sinks under the surface, by his own weight, or, if light, is elevated on a hillock.

V. PENTAMERA PALPICORNUA.—The maxillary palpi nearly equal to, or surpassing the clavated antennæ in length. These last are inserted in a pit beneath a singular production of the anterior margin of the head. In the Hydrophilidæ, the first joint of the tarsus is abbreviated, and the legs are flattened, and formed for swimming, as Hydrophilus, Spercheus, Elophorus, and Hydræna. In the Spheridiadæ, the five joints of the tarsus are distinct, as in Spheridium and Cercyon.

VI. PENTAMERA LAMELLICORNUA.—The insects of this division are readily recognised by their club-shaped antennæ, the extremity of which is divided into laminæ, capable of receding or approaching at pleasure. The Lucanidæ differ from all the rest of this tribe in the laminæ of the club of the antennæ, being placed (not as in the other families, approximating at the base, and opening and shutting like the leaves of a book, but) like the teeth of a comb, perpendicular to the axis, as Lucanus, Sinodendron, Esalus, Lamprina, Platyceres, and Passalus. In the Copridæ, the membranaceous termination of the maxillæ is large and transverse; the antennæ have eight or nine joints; the labrum is concealed by the semicircular margin of the head; the mandibles are soft; and the last joint of the labial palpi comparatively small, as Copris, Aleuchus, and Aphodius. The Geotrupidæ have the terminal joint of the labial palpi as large as the preceding one; the antennæ have eleven joints, and the mandibles are horny, as Geotrupes, Lethrus, and Typhæus. In the Scarabeidæ the antennæ have nine joints. The labium is concealed by the mentum, as Scarabeus, Trox, Egialia, Oryctes, Hexodon, and Rutella. The Melolonthadæ have the mandibles greatly concealed by the head and the maxillæ, as Melolontha and Anoplognathus. In the Glaphyridæ the labium is advanced and divided into three lobes, as Glaphyrus, Amphicoma, and Anisonix. The Trichiadæ have membranaceous mandibles, as Trichius, Goliathus, Cetonia, and Crematoschalus.

II. COLEOPTERA HETEROMERA.—The insects belonging to this great subdivision have the tarsi of the fore and mid legs furnished with five joints, as in the preceding, but the tarsi of the hind legs have only four joints. In a few genera, as Rhinomacer, Rhinosimus, and Stenostoma, constituting the family Rhinomaceridæ, the front of the head is advanced into a snout, on which are seated the antennæ. In the remaining genera, the front exhibits the usual characters. In some, the head is triangular, or heart-shaped, and is furnished with a neck, and the maxillæ have no corneous tooth on their inner edge. Among these there are some which have the claws simple, as the Pyrochroidæ, with bilobate tarsi, including the genera Pyrochroa, Notoxus, Scraptia, and Dendrocera; in the Mordelladæ, in which the tarsi of the hind legs at least are simple, are included the Mordella, Rhipiphorus, Anapsis, Horea, and Apalus. Among others, the claws are double, or deeply divided. In the genus Tetraonix, the penultimate joint of the tarsi is bilobate, in the others entire. In the Mylabridæ, the antennæ are thickened at the extremity, as in Mylabris, Hycleus, Cerocoma, while in the Cantharidæ, the antennæ are of equal thickness throughout, or rather taper towards the point, as Cantharis, Meloe, Zonites, and Onas.

Among those genera which have the head oval and destitute of a neck, there are some which have the maxillæ furnished with a corneous tooth on the inner side. The elytra, in some, are free, and cover membranaceous wings, as the Tenebroiondæ, including Tenebrio, Opatrum, Crypticus, Sarrotrium, and Toxicum. In many other genera the elytra are united, and the membranaceous wings are nearly obliterated. Some of these have the maxillary palpi filiform, with the last joint nearly cylindrical. The Erodiusidæ have the maxillæ covered with the mentum, as Erodius and Pimelia. In the Scaurusidæ, the base of the jaws are exposed, as Scaurus, Tagenia, Sepidium, Moluris, Tentyria, Hegester, Eurychora, and Akis. Others have the extremity of the maxillary palpi enlarged, or securi-form. In the ASIDADÆ, the base of the maxillæ is concealed by a large mentum, as Asida and Chiroscelis. In the Blapsidæ, the base of the maxillæ is exposed, as in Blaps, Misolampus, and Pedinus.

The genera, in which the maxillæ are destitute of a corneous tooth on the inner side, are likewise numerous. Many of these have the antennæ cylindrical, or slightly tapering. The Melandriadæ have the penultimate joint of all the tarsi bilobate, as Melandria, Lagria, Calopus, Nothus, Odemera, Stenostoma, and Rynomacer. In the Helopsidæ, the joints of the tarsi, at least those of the hind legs, are entire, as Helops, Serropalpus, Hallomenus, Pytho, Nilio, and Cistela. Others have the antennæ more or less club-shaped, and generally perfoliated. The Heleadæ, including the genera Helea and Cossyphus, have the head concealed, or received into a notch in the front of the thorax. The remaining families have the head exposed and projecting. In the Diaperidæ, the insertion of the antennæ is concealed by the lateral margin of the head, as in Diaperis, Hypophleus, Trachyscelis, Eledona, Cnodalon, and Epitargus. In the Leiodesidæ, the insertion of the antennæ is exposed, as in Leiodes, Tetratoma, Eustrophus, and Orchesia.

III. COLEOPTERA TETRAMERA.—The tarsal joints of all the feet are four in number. They are phytivorous, and live chiefly in wood or on flowers. In one extensive group, the head is produced in front, in the form of a snout. Among these there are two genera, Bruchus and Anthribus, in which the snout is short, and the labrum and palpi distinct. In the remaining genera the snout is long, and the labrum and palpi obscure. In the Curculionidæ, including the genera Curculio, Rhynchænus, Cionus, Calandra, and Rhina, the antennæ are distinctly geniculated. The Brentusidæ have antennæ destitute of the knee, as Brentus, Orchestes, Rhamphus, Brachycerus, Cylas, Apoderus, Attelabus, and Apion. In another group, equally numerous, the forehead is of the usual size. Among these, there are some which have eleven joints in the antennæ, and the third joint of the tarsi bilobate. The antennæ, in some, terminate in a perfoliate club, as the Erotylusidæ, including Erotylus, Triplax, Lanugria, and Phalacrus. In others, the antennæ are filiform. In the Cerambicidæ, the labium is dilated and heart-shaped at the extremity, including the genera Cerambix, Prionus, Callidium; Necydalis, Saperda, Lamia, Stencorus, Leptura, together with Spondylis and Parandra. In the Chrysomelinidæ, the antennæ are shorter than in the preceding family, and the labium is plain. It includes the following genera, Chrysomela, Cassida, Cryptocephalus, Clythra, Galeruca, Altica, Hispa, Crioceris, Donacia, and Sagra.

In those which have not eleven joints in the antennæ, and the third tarsal joints bilobate at the same time, there are some which have the third tarsal joints entire. The Mycetophagidæ have eleven joints in the antennæ, as Mycetophagus, Uleiota, Cucujus, Agathidium, Zylophila, Meryx, and Trogossita. In the Bostrichidæ, the joints of the antennæ do not exceed ten, as Bostrichus, Cerylon, Nemosoma, Cis, Cerapterus, and Pausus. The Scolytusidæ have the penultimate tarsal joint bilobate, as Scolytus and Phloiotribus.

IV. COLEOPTERA TRIMERA.—The tarsi in this division are all three-jointed. The antennæ are clavate. In the Coccinellidæ, the antennæ are shorter than the thorax, as Coccinella and Chilocorus. In Endomychidæ, the antennæ are longer than the thorax, as in Endomychus, Lycoperdina, Dascarus, and Eumorphus.

V. COLEOPTERA DIMERA.—The tarsal joints are only three in number. The genera hitherto determined amount only to two, as Claviger and Pselaphus. The latter, however, has been recently constituted into a family, Pselaphidæ, including Pselaphus, Euplectus, Bythenus, Arcopagus, Tychus, and Bryaxis.

Order II.—DERMAPTERA.

In this order, the elytra are short and coriaceous, with a straight suture. The wings are membranaceous, with longitudinal ribs, connected in the margin by a transverse one; they are folded when at rest longitudinally and transversely. The mandibles are bidentate. The maxillæ have a scaly cylindrical appendix or galea. There are no pyloric cæca. The tarsal joints are three in number. The metamorphosis is semicomplete. This class comprehends the following genera: Forficula, Labia, and Labidura.

Order III.—ORTHOPTERA.

The elytra, in the insects of this order, are coriaceous, and at their inner margin overlap each other. The under wings are membranaceous, and have numerous longitudinal ribs crossed alternately at right angles by many transverse ones, so that their reticulations, or little squares, are usually arranged like bricks in a wall; when at rest these are folded longitudinally, and unfold like a fan. The parts of the mouth are similar to the coleoptera, with the addition of the galea protecting the maxillæ at the sides. The alimentary canal is furnished with a membranaceous crop, and a muscular stomach, armed with corneous scales. The pyloric cœca receive the biliary vessels, a few of which likewise terminate in the intestine. The larvæ exhibit a pupa semicompleta. The insects in all the stages of life live on the land.

Among the Orthoptera, there are several genera with their wings, when at rest, roof-like. These have either the tarsi with four joints, as the Locustadæ; or the tarsi have only three joints, as the Achetadæ, including the genera Acheta, Truxalis, Ziphicera, Acrydium, and Pneumora.

In the remaining genera, the wings are horizontal. Among these the Gryllidæ, including Gryllus, Tridactylus, and Gryllotalpa, have the body not flattened, nor the sides truncated, but the abdomen is furnished with appendages. In the genus Blatta, the abdomen is depressed, and the sides truncated, with abdominal appendages. The remaining genera, with horizontal wings are destitute of the abdominal appendages. These are the Mantidæ, including Mantis, Empusa, Phasma, and Spectrum.

Order IV.—NEUROPTERA.

The wings in this order are generally four in number, wholly membranaceous, transparent, and greatly reticulated by the anastomosing ribs. The under wings are either larger or longer than the upper ones. The labrum, mandibulæ, maxillæ, and labium, are of the ordinary size, and formed for cutting. There are two large eyes, and two or three ocelli. The segments of the thorax are united, support the six legs, and are distinct from the abdomen.

In some of the families, the antennæ are about the length of the head, subulate, and consist of from three to seven joints, the last of which is setaceous. Among these the Libelluladæ, including Libellula, Aeshna, and Agrion, have three tarsal joints, the mandibulæ and maxillæ corneous, and the terminal ring of the abdomen, furnished with hooks or scales. In the Ephemeradæ, the mandibles are obscure, the tarsal joints four in number, and the terminal ring of the abdomen furnished with setæ, as in Ephemera, Baetis, and Cloeon.

In other families, the antennæ are much longer than the head, and consist of sixteen joints and upwards. Among these the Panorpadæ, including Panorpa, Nemoptera, Bittaces, and Boreus, have the front produced into a snout. The remaining families have the front short. In the Myrmelionedæ the antennæ are clavate, and the palpi six in number, as Myrmelion and Ascalapus. The two remaining families have filiform antennæ. The Termesidæ have from two to three tarsal joints, as Termes and Psocus. The Hemerobiadæ have four or five tarsal joints, as Hemerobius, Raphidia, Mantispa, Scalis, Corydalis, Chauliodes, and Osmylus. The Perladæ, including Perla and Nemoura, have the inferior wings folded longitudinally.

Order V.—Trichoptera.

The wings of the trichopterous insects are four in number, and membranaceous. The upper ones are usually of a darker colour and firmer consistence than the lower ones The ribs, which are usually hairy, are disposed longitudinally, and when they do anastomose, the intervening spaces are lengthened. The lower wings fold longitudinally. The mouth has a distinct labrum; the maxillary palpi have five joints. The labial palpi have only three joints, the last of which is a little enlarged. The maxillæ and labium are united, but do not form a sucker. There are two large compound eyes, and two ocelli. The legs are spinous, and the tarsal joints five in number. The larvæ live in the water in tubular dwellings, which they construct, and move about with, open at both ends, and consisting of bits of stick, sand, or shells. Hence they are usually termed *case-worms*. They change into a pupa incompleta in the tube, which they inhabited when larvæ, and, when ready for exclusion, by means of the sheathed antennæ and fore and mid legs, they crawl out of the water, throw off the covering, and become inhabitants of the land.

Dr Leach has subdivided this class into two families. The Leptoceridæ have the antennæ much longer than the whole body, as Leptocerus and Odontocerus. The Phryganidæ have the antennæ only the length of the body, as Phryganea and Limnephilus.

Order VI.—Hymenoptera.

The wings are four in number, membranaceous, and divided into large unequal meshes by the anastomosing ribs. The under wings are the smallest. The organs of the mouth are adapted both for cutting and sucking. For the former operation the labrum and mandibulæ are sufficiently strong; while the maxillæ are, together with the labium, more or less produced, and by their union, form a sucker. They have two labial and two maxillary palpi. The eyes are large, and the ocelli three in number. The females are armed with a sting or piercer. Many of the species live in society, and exhibit, in the magnitude and regularity of their operations, the most striking displays of the attributes of the social instinct. The insects of this order admit of a division into two orders.

I. Hymenoptera Terebrantia.—The females of this order are furnished with a produced ovipositor, frequently of sufficient strength to pierce solid bodies that the eggs may be deposited. Among these, there are two families in which the piercer is tubular, and does not consist of separate valves. The Chrysidæ have the piercer formed of the last rings of the body, retractile, and furnished with a small sting, as Chrysis, Parnopes, and Cleptes. In the

Oxyuridæ, the piercer is protuberant, without a sting, as Oxyurus and Drynus. In the remaining families the piercer consists of several valves. In some of these the abdomen is united to the thorax by a small portion of its transverse diameter. Among these there are some which have all the wings with ribs. The Ichneumonidæ have upwards of twenty joints in the antennæ, as in the following genera, Ichneumon, Zorides, Crypturus, Agathis, Sigalphus, and Alysia. In the Evaniadæ, including Evania and Fœnus, the joints of the antennæ do not exceed fifteen in number. In others, the under-wings are destitute of ribs. The Cynipsidæ have the antennæ broken, with from six to twelve joints, as Cynips, Leucopses, Chalcis, and Cynipsillum. The Diplolepidæ have the antennæ straight, with from eleven to sixteen joints, as Diplolepis and Eucharis.

The remaining families have the abdomen united to the thorax by the whole of its transverse diameter. In the Sirexidæ the piercer consists only of three valves, the lateral ones serving as sheaths, as Sirex and Oryssus. In the Tenthredadæ, including Tenthredo, Cimbex, Hylotoma, Xiphedria, and Pamphillius, the piercer consists of four valves, the internal pair serrated.

II. HYMENOPTERA ACULEATA.—In this order are included such as have no ovipositor or piercer. The abdomen in the females, however, is usually furnished with a sting, and poison-bags. The antennæ have thirteen joints in the male, and twelve in the female. These may be again reduced into two divisions. In the first, the feet are not formed for collecting pollen, and the first tarsal joints are cylindrical. Among these there are two families, in which the ocelli are indistinct, and the neuters or females are apterous. The Formicadæ, including Formica, Polyergus, Ponera, Myrmica, Atta, and Cryptocerus, have males, females, and neuters, the last of which are apterous They live in societies. The Mutilladæ have no neuters, and the females are apterous as Mutilla, Dorylus, and Labidus. They are solitary. The others have the ocelli distinct, and are all furnished with wings. Among these there are some families in which the wings are always expanded. In the Scoliadæ, the first segment of the thorax is large, and extends above to the base of the upper wings, as in Scolia, Tiphia, Sapyga, Thymus, and Pampilus. In the Sphexidæ, the first segment of the thorax is narrow and distant above from the base of the upper wings, as in Sphex, Bembex, Larra, Crabro, and Philanthus. There are other families in which the upper wings fold longitudinally. In the Masaridæ, the mandibles are narrow. There are only males and females, which are solitary, as Masaris, Synagris, Eumenes, and Zethus. In the Vespadæ, which are social, as Vespa and Polistes, there are males, females, and neuters; and the mandibles are large.

In nearly all the remaining genera, the hind legs are made for carrying pollen, having the first tarsal joint enlarged and compressed. Among these are some in which the tongue, or intermediate process of the labium, is as long or longer than its sheath, and deflected when at rest. In the Apidæ, which are social, there are males, females, and neuters, as Apis, Melepona, Bombus, Euglossa, Eucera, and Anthophora. While agreeing with the preceding in many particulars, the following genera are destitute of the expanded tarsal joint, for carrying pollen, Systropha, Panurgus, Zylocopa, Ceratina, Megachile, Phileremus, and Nomada. Others have the tongue shorter than the sheath, as the Andrenadæ, including Andrena, Halectus, and Colletus.

Strepsitera.

This class was instituted by Mr Kirby, one of the most acute and intelligent observers among the English entomologists. The elytra (if such they can be called) are coriaceous, and arise, not from the upper side of the breast, but from the base of the coxæ of the anterior pair of legs, consequently they are remote from each other. They first recede from the body, then approach, and lastly recede again, exhibiting a tortuous course. They do not cover the wings. These last are firmly membranaceous, and their ribs are simple, diverging from the base, and folding longitudinally like a fan. The parts of the mouth are obscure, apparently consisting of two minute, two-jointed palpi and two maxillæ, thus intimating that the imago consumes but little food, and is short lived. Each of the antennæ arises from a common jointed base, and afterwards divides. The eyes are pedunculated, two in number, and compound, with elevated septa, dividing the hexagonal lenses; the terminal segment of the abdomen ends in a reflected process. The larva inhabits hymenopterous insects, in which it changes into a pupa coarctata with the head exserted.

This order contains only two genera, Stylops and Zenos. In the former, the upper branches of the antennæ are jointed, in the latter they are simple. The Stylops melitta and tenuicornis are natives of Britain.

Since the institution of this class, doubts have been entertained as to the propriety of denominating the twisted processes which arise from the sides of the thorax, elytra. Latreille and Lamark, without indicating much reluctance to increase the synonimes of the science, or delicacy towards the naturalist who first instituted the class, have suppressed the term Strepsiptera, the former substituting that of Rhipiptera, the latter, with more classical propriety, Rhipidoptera (from ῥιπίς), in reference to the fanshaped wings.

II. INSECTA HAUSTELLATA.

This class includes three Orders.

Order I.—Hemiptera.

The insects of this order exhibit considerable differences with regard to their wings. In some, the upper wings are true elytra, crustaceous or coriaceous, with membranaceous extremities overlapping each other; while, in others, the upper wings are wholly membranaceous. In some of the genera the males only are winged, or they are all apterous. They agree, however, in the characters exhibited by the parts of the mouth. These are formed for sucking. The labium is produced, with a canal on its upper surface, and consists of several joints. The mandibulæ and maxillæ appear like four hairs, which, by their union, form the haustellum. The labrum is more or less produced as a covering to the base of these organs. The palpi are nearly obliterated. The metamorphosis is here semicomplete. This class is divided into two orders, which, in the opinion of some naturalists, ought to be elevated to the rank of classes.

I. Hemiptera Heteroptera.—These have the elytra crustaceous at the base, and the extremities folding over each other, and membranaceous. The rostrum is attached to the front of the head. The first segment of the thorax larger than the second; ocelli, two. The Heteroptera admit of subdivision into the terrestrial and aquatic.

The terrestrial heteroptera have two ocelli, the antennæ exposed, longer than the head, and inserted between the eyes near the inner margin. Some of these have the labium of four joints, and the labrum long, subulate, and striated above. Among these the Pentatomadæ have five-jointed antennæ, as Pentatoma, Cydnus, Tetyra, and Elia. The Coreidæ have only four joints in the antennæ, as Coreus, Berytus, Lygæus, Capsus, Miris, and Mydocha. In another group, the labium consists of only three joints, sheathing the labrum which is short. In the Reduviadæ, the rostrum is curved, as Reduvius and Plocaria; while in Cimicidæ it is straight, as in Cimex, Tingis, Aradus, and Phymata. In a third group the labium consists of two or three joints, and does not embrace the labrum, as the Acanthidæ, including the Acanthia and Galgulus.

The aquatic heteroptera live in or upon the water. They are destitute of ocelli. In the Hydrometridæ, including Hydrometra, Velia, and Gerris, the antennæ are long, and inserted between the eyes. In the remaining families, the antennæ are short, and inserted under the eyes, and are shorter than the head. In the Nepadæ, the tarsi of the fore-legs are indistinct, as Naucoris, Nepa, and Ranatra. In the Notonectadæ, the anterior tarsi are distinct, as in Notonecta, Plea, Sigara, and Corixa.

II. HEMIPTERA HOMOPTERA.—In the insects of this division, the rostrum seems to originate from the chin. The second segment of the thorax is as long as the first. There are two or three ocelli. The Cicadiadæ have three joints in the tarsi, the antennæ consist of six joints, and there are three ocelli, as Cicada. The Fulgoradæ have three tarsal joints, and only three joints to the antennæ, and two ocelli, as Fulgora, Flata, Issus, Tettigometra, and Delphax. The Cercopidæ differ from the preceding in the antennæ being inserted between the eyes, as Cercopis, Etalion, Ledra, Membracis, and Tettigonia. The Aphisidæ have two joints in the tarsi, as Aphis, Psylla, Thrips, and Aleyrodes. The Coccidæ have only one tarsal joint, as Coccus.

Order II.—Lepidoptera.

This extensive and beautiful order includes the butterflies and moths. The wings are four in number, membranaceous in texture, irregularly ribbed, and covered with coloured scales, in the form of a farinaceous powder. The parts of the mouth are formed for suction. There are only vestiges of the labrum and mandibulæ. The maxillæ are produced, with a groove on the inner edge. When united, as they are naturally, they form a tubular proboscis, through which the animal obtains its food. The maxillary palpi are inserted upon the base externally, and are minute. The labium is short and without joints, and supports two obvious palpi with three joints. There are two compound eyes, and in some species two ocelli. The antennæ consist of many joints, and are usually much longer than the head. The segments of the thorax are united. The tarsal joints are six in number. The larvæ have six feet with claws, and from four to ten others on the posterior portion of the body, which they use as suckers. They are changed into a pupa obtecta. The genera are now divided into three orders, corresponding with the Linnean genera Papilio, Sphinx, and Phalæna.

I. Lepidoptera Diurna.—The upper wings at least, in all the species, are vertical in a state of rest, and the lower ones are destitute of hooks. The antennæ are clavated, or filiform, with hooked extremities. Among these, the Papilionidæ have the hinder tibiæ furnished with two spines, situate at the tarsal end, as in Papilio, Parnassius, Thais, Pieris, Polyomatus, Heliconius, Danaus, Cethosia, Libythea, and Nymphalis. The Hesperiadæ, including Hesperia and Urania, have four spines to the hinder tibiæ, two in the middle and two at the tarsal extremity, as in the remaining lepidoptera.

II. LEPIDOPTERA CREPUSCULARIA.—The wings are horizontal in repose, and the under ones are furnished with a spine under the base at the external margin, which enters a hook on the lower side of the upper wings, as in the following division. The antennæ are prismatic and fusiform. The Glaucopidæ have the antennæ bipectinated in the male at least, as Glaucopis, Stygia, and Procris. The antennæ, in the remaining genera, are simple in both sexes. The Zygenadæ have the palpi slender and hairy, as Zygena, Sesia, and Macroglossum. The Sphingidæ have large scaly palpi, as Sphinx, Smerinthus, and Castinea.

III. LEPIDOPTERA NOCTURNA.—The insects included under this division have setaceous antennæ, diminishing in thickness from the base to the point. Among these, there are several families, in which the wings, when at rest, fold round the body. The Pterophoridæ have the margins of two of the wings, at least, divided into processes at each rib, as Pterophorus and Orneodes. The other families of this division have the four wings entire. The Tinneadæ have only two palpi apparent, as Tinea, Yponomeuta, Oecophora, Lithosia, and Adela. In the Alucitadæ, there are four palpi apparent, as Alucita, Crambus, and Galleria. Other families have the wings at rest, lying upon the body without inclosing it, and by their union form a lengthened triangle. The Aglossadæ have four apparent palpi, as Aglossa and Botys. The Pyralidæ have only two apparent palpi, as Pyralis, Hermenia, and Platyperix. In the third division, the wings do not rest upon the body so as to form a triangle. This extensive family includes the following genera, Phalæna, Campæa, Noctua, Collimorpha, Bombyx, Furcula, Hepialus, and Cossus.

M. SAVIGNY has observed, that, in those cases where the maxillary palpi have only *two* joints, the proboscis is naked or pubescent; and when they consist of three joints, the proboscis is always scaly.

Order III.—Diptera.

The diptera have only two wings; and, in a few instances, none. Many species are furnished with *halteres*, and squamulæ. The mouth is formed for suction. For this purpose, the labrum is more or less produced as a cover. The mandibulæ are obliterated, or in the form of threads. The maxillæ are produced into threads, and by their union in company with the mandibulæ, form the syphon. The labium is either double or single, and forms a sheath destitute of joints, for the reception of the syphon. In some cases, there are two maxillary and two labial palpi. The larvæ are destitute of feet, and pass into *pupæ obectæ*, or *coarctatæ*. In this numerous division, the labium, or sheath, is univalve, in others it is bivalve. We shall now advert to the first of these.

Among those with a univalvular sheath to the proboscis, there are some in which this sucking organ is entirely withdrawn, when not in use. Some of these have the sucker consisting of only the produced maxillæ. The Muscadæ have the eyes sessile, as Musca, Tephritis, Myoda, Macrocera, Scenopinus. The Achiasidæ have the eyes pedunculated, as Achias, Diopsis. The genus Oestrus, or Gadfly, agrees with the muscadæ in habit, but the parts of the mouth are imperfect. Others have the sucker of four filaments, formed from the mandibulæ and maxillæ. In the Syrphadæ, the front is produced like a beak, as Syrphus, Psarus, Chrysotoxum, Cerea, and Rhingia. In the Aphritidæ, the front is abbreviated, as Aphritis and Milesia. In the remaining genera of this group, the last joint of the antennæ is not simple, as in the preceding families, but annulated, and destitute of the lateral hair which they possess. These are termed Stratiomydæ, and include the genera Stratiomys, Oxycera, and Nemotelus.

In the remaining families, with a univalvular sheath, the proboscis is always more or less protuberant. Among these, there are some which resemble the preceding families, in having only three joints in the antennæ. The Conopsidæ have the sheath bent, and the sucker, with two filaments, as Conops, Zodion, Homoxis, Bucentes, and Myopa. The Bombylidæ have the sucker composed of from four to six filaments, and the sheath is destitute of large lips, as Bombylus, Ploas, Mithrax, Nemestrina, Panops, Cyrtus, Acrocera, Astomella; together with Empis, Asilus, and Dioctria. The Tabanidæ have the sheath furnished with large lips, and the third joint of the antennæ distinctly annulated; as in Tabanus, Pangonia, and Cænomya; together with Pachystoma, Rhagis, Dolechopus, and Mydas. Others have six joints or more in the antennæ. Among these the Bibionidæ have moniliform or perfoliated joints, about the length of the head, as Bibio, Scathopsis, and Simulium. The Tipuladæ have filiform or setaceous antennæ, as Tipula, Cetenophora, Trichocera, Psychoda, Tanypus, Limonia, Hexatoma, and Culex, all of which are destitute of ocelli; together with Asindulum, Ceroplatus, Mycetophila, and Rhyphus, which are furnished with ocelli.

In the remaining genera of this class, the sheath of the sucker is bivalve. The Hypoboscidæ have the head distinctly divided from the thorax, as in Hippobosca, Feronia, Ornithomyia, Craterina, Oxypterum, and Melophagus. In the genus Nycteribia, the head is united with the thorax.

Order IV.—Aptera.

Into this class we have placed tribes of insects which differ greatly from each other in the organs of digestion;

but which do not agree with any of the preceding classes. They possess one common character in wanting wings, in all the stages of their existence. They may be divided into three orders, which by some are elevated to a primary rank.

I. APTERA SUCTORIA.—The head, thorax, and abdomen intimately united. The mouth consists of two simple processes, the lowest of which is longest, and receives the superior one in a cavity in its upper side. On each side is a process of four joints covering the others. These unite to form a proboscis, which rests upon the sternum. Are the simple processes to be considered as the labrum and labium, and the articulated lateral ones as palpi, the mandibulæ and maxillæ being absent? The antennæ consist of four joints. This order contains at present only one genus, Pulex or Flea. There are several species. The P. irritans is every where common, and the P. fasciatus of Bosc may be met with on moles and mice.

II. APTERA THYSANURA.—The head in this and the following order is obviously separated from the thorax by a contraction or neck. The last segment of the body is furnished with long filaments. In the Lepismadæ the setæ of the tail are continually extended in the direction of the body as Lepisma and Forbicina. In the Poduradæ, the setæ, when at rest, are folded under the body, as Podura and Smynthurus.

III. APTERA PARASITA.—The tail in this order is simple, or destitute of the filaments which distinguish the preceding. The Nirmidæ, including numerous species of lice which infest birds, and included in the genera Nirmus *

---

* See figures of three species found on the Anser erythropus, 9th January 1819, Plate iv. f. 6. a. b, c.

and Pediculus of authors, have the mouth furnished with two teeth. The Pediculidæ have a tubular proboscis, and include the genera Pediculus, Phthirius, and Hæmatopinus.

II.—MYRIAPODA.

Feet numerous, supported by the abdomen.

The animals of this group are lengthened. The head and thorax are united, and the six thoracic legs corresponding with those in the insecta, are converted into auxiliary maxillæ. The abdomen is divided into numerous segments, each bearing a pair of legs. They have compound eyes, and no simple ones.

Order I.—CHILOGNATHA.

Antennæ seven jointed. Legs short, body crustaceous.

The following genera belong to this order: Glomeris, Julus, Craspedosoma, Polydesmus, and Pollyxenus.

Order II.—SYNGNATHA.

Antennæ of fourteen joints at least, legs long. In some of the genera the tarsus of the second pair of auxiliary maxillæ are pierced by the aperture of a poison-duct.

This order includes the following genera: Scolopendra, Scutigera, Lithobius, Cryptops, and Geophilus.

The *capture* of the articulated annulosa is accomplished with the hand, with forceps, or gauze-nets, according to the nature of the species. Care ought to be taken to preserve as entire as possible all the limbs, antennæ, and down upon the body. They should then be placed in separate boxes, or transfixed with a pin, through the thorax or side, according to circumstances, and fixed in a box, with the bottom lined with cork. Butterflies, when fixed in this state, by the motion of their wings greatly injure their beauty by rubbing off the fine coloured scales with which they are covered. It is convenient, therefore, to kill them by compressing their sides, or fixing them with the pin through the thorax laterally. Some, after killing them by compression, carry them home in the leaves of a book. In many cases the killing of the animal is a more difficult task than its capture. Some suffer them to writhe on the pin until they die from pain or hunger. Others shorten their sufferings by suffocating them with the fumes of burning brimstone, or by passing a red hot needle, or one dipped in aquafortis, through their bodies, while a few attempt to kill them by putting oil of turpentine or tobacco in their mouths. Fumigation, however, is the most expeditious method. When this is inconvenient, they may be put into a small tin box, which must be immersed half its depth in boiling water: the heat communicated to the box will speedily kill them.

When the animal is dead, it is then to be *set* in a natural position, in reference to its wings, legs, and antennæ, these organs being kept in their proper place by pins stuck in the cork below, or by slips of card fixed down with pins. When dry it is fit to be added to the collection.

The marine crustacea must be steeped in water before being dried. The larger kinds must be embowelled, and a little preserving powder introduced.

In order to exhibit the history of an insect, it is necessary to preserve it in the egg, larva, and pupa state, as well as the imago. The eggs and pupa are easily preserved by drying, but many of the fleshy larvæ require, previous to being dried, to be embowelled, and the cavity distended with air, cotton, or sand. When perfect insects are obtained in a dry state, without having been set, their different members may be readily relaxed for that purpose, by placing them on a piece of cork, floating on water in a basin, the mouth of which is covered lightly with a damp cloth. In a few hours the joints will become sufficiently supple.

The entomological collection is kept in drawers of hard wood, with moveable glass covers. The bottom of each drawer is covered with cork or wax, for the reception of the pins supporting the animal. It is washed over with arsenic, or corrosive sublimate, and covered with white paper glued to it. The insects are distributed in rows, with their names marked on the paper below, or with a number referring to a catalogue. The collection must be frequently inspected, to see if any insect depredators have got admission. These must be carefully removed, and their eggs destroyed, by baking the specimens to which they have been attached, in the sun, or before the fire. If they need washing, it may be done with a hair pencil, dipped in rectified spirit of wine.

Subdivision II.—ANNULOSE ANIMALS DESTITUTE OF ARTICULATED LIMBS FOR LOCOMOTION.

The methodical distribution of this great branch of annulose animals, presents a considerable degree of difficulty. The species, in general, are small in size, and the dissections of the organs must be conducted under the guidance of the microscope. The vessels are often too thin in the coats to suffer the injection of coloured fluids, or even inflation with air. The characters of many of the genera, therefore, still remain in obscurity, and the arrangements which have been proposed are necessarily defective. In the plan which we propose to follow, the *external* tribes, or those which live immediately exposed to the air or water, will occupy the first division, and the *internal* tribes, termed Entozoa, or those which dwell in the inside of the larger animals, will occupy the second. Objections to the formation of this division may be advanced, from the consideration, that the characters employed in arrangement ought not to be taken from *station*, but from *structure*, so that unless the internal animals exhibit some peculiarity in their organization, distinguishing them from the external ones, they should not occupy a division apart, but be introduced according to their analogies among the groups of external genera. To this it may be answered, that while the animals of the division Entozoa cannot be characterised by any peculiarity of structure, nay, while they differ remarkably from one another, we draw the conclusion, that they must necessarily differ from the external groups in the structure and function of their organs of respiration, since they can neither possess gills nor lungs of the ordinary kinds. But, after all, the *systematical convenience* of the division (a principle frequently acted upon, but seldom acknowledged), is the best apology for its employment.

Order I.

Habitation external.

Section I.—CIRRHIPIDES.

Body protected by a multivalvular shell.

This group includes those animals which constitute the Linnean genus Lepas. They are all sessile animals, and the base by which they are attached corresponds with the coronal or anterior aspect of the body. The tunic is closed laterally, dorsally, and anteriorly, but is more or less separated at the ventral aspect. The tunic is strengthened by testaceous plates, to which the body adheres by one or more muscles. The head appears as a slight eminence attached to the thorax. The thorax is followed by an abdomen supporting six feet on both sides, each of which consists of a short stem, which divides into two tapering, jointed, fringed filaments; these, by their motion towards the mouth, bring the water and its contents within the sphere of that organ. The abdomen terminates in a conical tubular body, which has improperly been termed a proboscis.

The nervous system consists of a cord encircling the gullet, and giving out filaments to the mouth; its two ends running along the thorax and abdomen, and uniting at the base of each pair of feet to form a ganglion, and give off filaments.

The mouth is furnished with an obvious upper lip, a pair of maxillæ on each side, with the rudiments of palpi. ELLIS says, " The mouth appears like that of a contracted purse, and is placed in front between the fore claws. In the folds of this membranous substance are six or eight horny laminæ, or teeth, standing erect, each having a tendon proper to direct its own motion. Some of these teeth are serrated, others have tufts of sharp hairs, instead of indentations on the convex side, that point down into the mouth, so that no animalcula that become their prey can escape back." The gullet is very short, and enters into a stomach, having two cœca and glandular walls, which serve as a liver. The intestine is short and simple, and terminates behind at the base of the tubular appendage. There are two salivary glands attached to the stomach. The gills are conical bodies, situate at the base of the feet. The organs of circulation are imperfectly understood. POLI observed the motion of the heart, but the vessels which are connected with it are unknown.

The organs of reproduction appear, according to CUVIER, to consist of an ovarium, giving out an oviduct, which traverses a body considered as a testicle, and both the canals unite in a common tube, which traverses the tubular appendage, and opens by a small pore at its extremity. This aperture CUVIER regards as simple, but ELLIS observes of the whole, " it is of a tubular figure, transparent, composed of rings lessening gradually to the extremity, where it is surrounded with a circle of small bristles, which likewise are moveable at the will of the animal. These, with other small hairs on the trunk, disappear when it dies." There is probably no union of individuals, each being a perfect hermaphrodite.

The spawn of the Lepas fascicularis is a spongy body, and adheres to fuci, feathers, or other bodies, with which it accidentally comes in contact. The rapidity of growth, after hatching, is truly astonishing. The spawn of full sized individuals, springing from spawn which has been deposited on a feather, (as I have stated in the Memoirs of the Wernerian Society, vol. ii. p. 243.), will become unfolded, and attain maturity, before the feather itself begins to exhibit any symptoms of decay. A ship's bottom comes covered with other species in a few months.

The animals of this class are all inhabitants of the sea. Many of them are attached to floating wood, and others to the skin of marine animals, so that they enjoy all the advantages of locomotion, without the exercise of the exertion requisite for its production. Their remains are seldom found in a fossil state. Some of the species have been used as food. In taking a view of the genera we shall distribute them, after the manner of ELLIS, *Phil. Trans.* vol. i. p. 848, into pedunculated and sessile.

Tribe I.—PEDUNCULATED.

The essential character of this tribe consists in the body being supported by a peduncle, the lower part of which is permanently fixed to other bodies.

The cloak consists of three membranes. The external one is the cuticle, and invests the whole external surface of the animal. Underneath this is the true skin, in which are formed the testaceous plates that protect the body. These plates or valves are evidently formed in the same manner as common shell, the layers of growth being indicated by the striæ on the surface. The inner membrane forms a sac for the body itself. This bag is closed on all sides, except opposite to the abdomen, where there is a slit, through which the feet are protruded.

The peduncle consists of the two external membranes of the integuments of the body. The cuticle covers its surface, and even the base by which it adheres. The true skin is covered on its central aspect with numerous muscular threads. The summit of the peduncle next the head is covered with the inner membrane of the cloak, through which, however, there is a perforation, corresponding to a large vessel which descends along one of the sides of its central cavity. This cavity, in the *Lepas anatifera*, CUVIER found filled with a white cellular substance soaked with mucus. ELLIS, on the other hand, found the peduncle of what has usually been regarded as the *L. aurita* of LIN., "full of a soft, spongy, yellow substance, which appeared, when magnified, to consist of regular oval figures, connected together by many small fibres, and no doubt are the spawn of this animal." This view of the subject entertained by ELLIS, may, upon investigation, lead to the conclusion, that the cavity of the peduncle, and its lateral vessel, are connected with the reproductive organs. It would be desirable to have the branched peduncles dissected with care, as a knowledge of their structure might throw some light on the mode of growth of these animals.

Family 1.—Tunic protected by five testaceous plates; the peduncle naked. This includes the following genera

LEPAS. The two lateral plates at the summit of the shell are very large, nearly covering the whole of the compressed body, and having attached to it the large adductor muscle. The two valves which protect the sides of the abdomen are much smaller, and somewhat triangular, while the dorsal one is narrow and convex externally. The branchiæ are four in number, two on each side of the thorax, near the origin of the first pair of feet. The British species are four in number, *L. anatifera, anserifera, sulcata,* and *fascicularis.*

OTION. This genus was instituted by our zealous and intelligent friend Dr LEACH, whose labours have greatly contributed to improve the classification of the genera of this class. The body is but slightly compressed, and the valves are very small and distant from one another, the whole being chiefly covered by its membranaceous cloak. At the extremity of the abdomen, the cloak terminates in two tubular appendages. Through these the water escapes which has been taken in at the ventral aperture, and has passed along the surface of the gills. The gills are sixteen in number, eight on each side, the first pair on each side resembling those of the Lepas, the remaining six are attached to the base of the feet. There are two British species. 1. *O. aurita,* CUVIER, *Mem. des Anatifes,* Fig. 12, 13. A specimen of this was found on the Dawlish coast, Devon, by Mr COMYNS. 2. *O. cornuta,* taken alive from the bottom of a transport stranded on the coast of Devon, by MONTAGU, and described and figured by him, *Lin. Trans.* vol. xi. p. 179, tab. xii. f. 1.

CINERAS. This genus was likewise instituted by Dr LEACH. The valves are equally minute and remote as in the preceding genus, but there is here no appearance of tubular appendages to the cloak. The *C. membranacea,* first described and figured by MONTAGU, *Lin. Trans.* vol. ix.

p. 182. tab. xii. f. 2. is the only species known to inhabit the British seas.

Family 2. The testaceous valves are numerous, greatly exceeding five. It consists of two genera.

SCALPELLUM. The testaceous valves are thirteen in number, and invest the body. The peduncle is covered with corneous wrinkles, having hairy interstices. The *S. vulgare*, the *Lepas scalpellum* of British authors, is the type of the genus.

POLLICIPES. The testaceous valves are ten in number, with numerous scales investing the base of the peduncle near the body. The *P. vulgaris*, or *Lepas polliceps*, is the type of the genus.

Tribe II.—SESSILE.

In this tribe the body adheres directly to foreign substances, without the intervention of a tubular stalk. The adhesion is effected in some by the coriaceous cloak, in others by a layer of testaceous matter. The testaceous covering usually assumes a conical form. The base is attached to rocks or other substances, and the apex is truncated and open as an entrance for the water. This cone consists of six valves, closely connected together, but capable of being disjoined by maceration, especially when young. In old shells, where the valves have attained their full growth, they appear to become cemented together, so that it is very difficult to effect their separation. The valves are so arranged, that one protects the belly, another the back, and two on each side the lateral parts. In some genera, all these valves are so united that the lines of separation are not perceptible, while in others the double lateral valves only are incorporated. Each valve consists of an elevated and depressed portion. The elevated portion is conical, with its base at the adhering part of the shell, while the depressed part, of the same form, has the base at the mouth. The former consists of conical vertical tubuli, while the latter appears solid. When the base is testaceous, it is either solid, or consists of horizontal tubuli, radiating from a centre united by a simple layer, exhibiting concentric circles.

The structure of the valves gives sufficient indications of the manner in which they have been formed. M. DUFRESNE, in a paper published in the *Annales du Museum*, vol. i. p. 465, advanced the singular opinion, that the animals quitted their old shells when they became too small, and formed new ones suited to their size. The arguments by which it is supported, indicate inattention to the structure of the shell, and the relation of the parts of which it consists. To us it appears obvious, that each valve is increased in two directions, the elevated part by an extension of the tubuli at the base, and the depressed part, by the application of fresh matter to the side. The striæ, which are the indications of successive depositions of matter, and the structure of the valves themselves, point out this mode of enlargement as the only one which can take place, even on the supposition that the shell is frequently renewed. By the growth of the elevated parts, the shell increases in height and diameter at the base, while the growth of the depressed parts, preserves to the mouth suitable dimensions for the corresponding increase of size in the parts of the operculum. It is obvious that this increase of diameter at the base must be accompanied by a corresponding enlargement of its covering. This takes place by the extension of the horizontal tubuli, and each enlargement is marked by a concentric ridge.

This opinion here advanced, and which we find indistinctly hinted at by LAMARK, in his " *Histoire Naturelle des Animaux sans Vertebres,*" vol. v. p. 389, is founded on the structure of the different valves, the indications of the layers of growth, and the manner in which the valves are separable from each other, and from the base; and by the morbid appearances of the shell, the restraints imposed on its growth by the situation in which it lives, but especially the manner in which fractures are healed, and abstracted parts restored by the secretion of new matter. It may be added, however, that in the case of the inversely conical shells, the increase probably takes place at the mouth.

The aperture of the shell is inclosed by the cloak of the animal, leaving in the centre a tubular or lineal opening for the protrusion of the feet and entrance of the water. This part of the cloak is protected by testaceous plates, which, by their union, form a lid to the mouth of the shell, for the protection of the contained inhabitant. The valves of the lid are four in number, two on each side of the mesial line, or orifice. In some genera the lateral valves are united. The operculum of this order may, with propriety, be compared to the shelly plates of the body of the preceding tribe,—and the shelly body of this tribe considered as corresponding to their peduncle, circumstances indicated by the muscular attachments of the animal. The continued action of the valves of the lid, obviously assist in wearing down and enlarging the aperture of the shell.

There is little known, either with regard to the organs of digestion or respiration, in the animals of this tribe. In their manner of reproduction, they appear to resemble those of the preceding order. ELLIS found the lower part of the shell containing a cavity equal to two thirds of the whole, full of spawn. The genera of this order divide themselves into three families, from circumstances connected with the shell.

Family 1.—The shell in this family consists of six valves, and the lateral valves of the lid are divided. It contains five genera.

1. TUBICINELLA. The form of the shell in this genus is inversely conical, and the apex which constitutes the base is truncated. It consists of a series of horizontal rings, which mark the successive periods of growth, and there are six vertical grooves, which indicate the divisions of the valves. The increase of the shell, with age, in this genus, probably takes place by the addition of a new ring to the mouth. The testaceous plates of the lid are all of equal size. The inferior aperture of the shell is open, or simply closed by the integuments of the cloak. The animal resides in the skin of the whale, the lower rings being inserted in the fat, while one or more of the upper ones appear above the cuticle. The *T. balænarum* is the only known species.

In the remaining genera of this family, and the others which follow, the shell is conical, its truncated apex being the mouth ; and its mode of growth such as is detailed in the general remarks on the order.

2. CORONULA. The base of the shell is open, but the valves of the lid are unequal in size, the dorsal ones being small. The animals included under this genus likewise inhabit the skin of the whale. Several species of this genus are known. The *C. diadema* holds a place in the British Fauna.

3. CHELONOBIA. In this genus the base of the shell is likewise open, but it differs from the preceding in the plates of the lid being all of equal size. The *C. testudinaria*, a species which resides on turtles, is the type of the genus.

4. BALANUS. The shell is closed below by a layer of shelly matter, which adheres to foreign bodies, and conforms to the inequalities of their surface. Nine species are described as natives of Britain.

5. ACASTA. The base of the shell is cup-shaped. The species reside in sponges, in the substance of which the base

and sides are imbedded. One species, the *Balanus spongiosus* of MONTAGU, is a native of the British seas.

Family 2.—The valves of the shell are only four in number.

1. CREUSIA. The base is funnel-shaped. The lateral valves of the lid are united. The *C. spinulosa* of Dr LEACH is the type of the genus. It is imbedded in the substance of Madrepores.

2. CONIA. The shelly base conforms to the substance to which it is attached. The lateral valves of the lid are separate. The *C. porosa* is the type of the genus.

3. CLISIA. The base of the shell is spread on the surface of the bodies to which it is attached. The lateral valves of the lid are united. The Balanus striatus of British writers is the type of the genus. An imperfect representation of the animal is given by CORDINER, in his " *Remarkable Ruins*," plate inscribed " Aggregate of Corals."

Family 2.—The shell is undivided in the only genus of this family which is known, termed *Pyrgoma*.

Section II.—ANNELIDES.

Shell not multivalvular, or wanting.

The body of the Annelides is of a lengthened form, bearing, more or less, a resemblance to the common Earthworm, and, like it, divided into numerous rings. The skin is furnished in many genera with different kinds of appendices. In some, it is protected by a shell secreted by the animal. In others, particles of sand or mud are cemented by mucus, in the form of a convenient dwelling, which the inhabitant never quits. The head, in a few genera, cannot be distinguished from the opposite extremity. In general, however, it is marked by the antennæ, the tentacula, and eyes. The sides of the body are in some species naked, and in others, at each ring they are furnished with complicated organs. These, in their most perfect form, may be denominated *sheaths*. They consist of a conical stalk, capable of being withdrawn into the body by a kind of inclusion. This sheath is furnished, at its base or extremity, with *filaments*, which are either simple, or jointed like the antennæ; with *spines*, which are solitary; and with *bristles*, which are numerous, and placed in tufts or rows. In some cases the filaments only occur, in others the bristles or spines, and the stalk is wanting; still, however, they are retractile. These lateral organs serve the purposes of feet, in burrowing in mud, in crawling on its surface; or in swimming. The eyes are simple, and, in general, are numerous.

The mouth is in the form of a sucker or proboscis, or a simple aperture. In some it is unarmed, in others it is furnished with corneous jaws. The intestine is usually straight, sometimes contracted into rings, and the anus terminal. The organs of circulation have, in some of the species, been successfully investigated, and seem to consist of lateral, dorsal, and central vessels, extending the length of the body, and executing the offices of veins and arteries. The blood is of a purplish colour. The aërating organs consist of a row of pouches on each side, of a row of plumose branchiæ at the base of the sheaths or seated on the neck, or of stellular filaments near the mouth,

The sexes are united in the same individual, requiring, however, mutual impregnation. The greater number of the species appear to be oviparous. A few are ovoviviparous.

1st Tribe.
Body furnished with a shell.

A. Free, not permanently attached to other bodies. Shell tubulo-conical.

1. *Shell with a longitudinal, lateral, subarticulated slit.* This includes the genus Siliquaria (*Serpula anguina*), the characters of which remain to be determined.

2. *Shell destitute of a lateral slit.* In the genus Dentalium, the tube is open at both ends. The sides of the body are furnished with tufts of bristles, and the posterior extremity terminates in a radiated disk, where, probably, the branchiæ will be detected. In the genus Cæcalium (which, when instituted, Edin. Encyc. vol. vii. p. 67, I inadvertently termed Cæcum), the smaller end of the shell is closed. Three British species, formerly included in the genus Dentalium are known, viz. *C. imperforatum, trachea* and *glabrum.* The animal is unknown.

B. Fixed or permanently attached to other bodies.

1. *Shell closed at both ends.* The cover of the extremity corresponding with the mouth, is covered with tubular elongations. There is only one genus, Penicillus, formed from the *Serpula penis,* LIN.

2. *Shell open at the mouth.*

a. Shell irregular.

(A.) Operculum multivalve. This includes the genus Galeolaria of LAMARK.

(B.) Operculum simple. There are two genera.

1. Vermilia (LAM). Shell attached throughout, with the aperture round, and the margin toothed. Operculum shelly. *V. vermicularis, triquetra** and *serrulata* are native species.

2. Serpula. Shell not attached throughout. Mouth round, with a simple margin. *S. tubularia.*

---

* The observations of MONTAGU, indicate several species to be included under the name of Triquetra.—Testacea Britannica, Supp. p. 156.

b. Shell discoid and spiral.

In this group there are at least four genera, with well marked characters exhibited by the shell. 1. Magilus, (MONTFORT). The shell posteriorly is fixed, with régular spires, the last of which is produced into a tube. The animal is unknown. *M. antiquus.* The *Serpula recta anfractibus tribus contiguis regulariter involutis,* and *Serpula recta umbilico pervio anfractu apicis unico involuto* of WALKER *, either belong to this genus, or deserve a separate place. 2. Spirorbis. Spires discoid, not produced into a straight tube, and dextral. *S. communis, spirilium, granulatus, carinatus, corrugatus* and *corneus,* are natives. 3. Heterodisca. Like the preceding, but the spires are sinistral. *H. heterostrophus, sinistrorsus, minutus, conicus, lucidus* and *reversus,* are natives. 4. Lobatula. Discoid, with the tube divided into numerous unequal cells. *L. farcta* and *concamerata,* described by MONTAGU as Serpulæ, are native examples. In the genera Serpula, Vermilia and Spirorbis, the branchiæ constitute a tuft of fan-shaped filaments, on each side the mouth.

2d Tribe.

Body destitute of a shell.

A. Branchiæ external.

1. *Furnished with antennæ.* This division includes the Linnean genera Aphrodita and Nereis, now formed into numerous genera, which LAMARK distributes into the following families. 1. *Aphroditæ,* including Aphrodita †, Halithea, Palmyia and Polynoë. 2. *Nereides,* including Lycoris, Nephtys, Glycera, Hesione, Phyllodoce, Syllis, Spio and Diplotis. 3. *Eunicæ,* including Leodice, Lysi-

---

* Testacea Minuta Rariora, p. 3. Tab. f. 11, 12.

† The structuré of the aerating organs in this genus are exhibited by Sir E. Home, Phil. Trans. 1815, tab. xiii.

dice, Aglaura and Oenone. 4. *Amphinomæ*, including Amphinome, Pleione and Euphrocine.

11. *Destitute of antennæ.* No eyes.

a. Body protected by an artificial tube.

With the exception of the genus Clymene, in which the branchiæ are unknown, but supposed to reside in the funnel-shaped termination at the anus, all the others belong to the family *Amphitritées* of LAMARK, consisting of Pectinaria, Sabellaria, Terebella, and Amphitrite.

b. Body destitute of an artificial tube. This includes the genus Arenicola of LAMARK, formed from the *Lumbricus marinus* of LINNÆUS *.

B. Aerating organs internal, in the form of lateral sacs, or unknown.

1. *Body furnished with filaments, or bristles, or spines.* In the genus Thalassina, distinguished by the subglobular body and funnel-form mouth, the intestine is longer than the body, and folded. In Lumbricus, the intestine is straight †. The characters of the remaining genera of this group, as Cirratulus, Nais, Stylaria and Tubifex, have not been determined with any degree of precision. The history of the genus Derris of ADAMS is equally obscure ‡.

2. *Body naked.*

a. One or both extremities furnished with suckers. This includes the species of the Linnean genus Hirudo, now

---

* Its structure is developed by Sir E. HOME, Phil. Trans. 1817, p. 2. Tab. iii. p. 1, 2, 3.

† The structure of the common earthworm is delineated by SIR E. HOME, Phil. Trans. ib. The *L. vermicularis* and *variegatus* of MULLER are common natives.

‡ LIN. Trans. III. p. 67., Tab. xiii. fig. 1, 2.

subivided as follows: 1. Mouth armed with teeth, Hirudo (*H. medicinalis* and *sanguisorba*, are British examples *). 2. Mouth with a proboscis, Glossopora †, (*G. complanata*, LIN., Trans. vol. ii. tab. xxix. and *bioculata*, URES Hist. of Rutherglen, p. 234. are British species). 3. Mouth simple. This includes three genera. Erpobdella, (Br. sp. *E. vulgaris, tessulata* and *lineata*). Pisciola (Br. sp. *P. geometra*). Pontobdella, (Br. sp. *P. muricata*.) To these may be added Phyllene (of M. OCKEN), and Trochætia (of M. DUROCHET.) The last of these, however, breathes air, and should probably be placed near the Lumbricus.

The leeches are oviparous; but as they greedily devour different species, (when confined), and even the young ones of their own species, and readily vomit them again, when writhing in agony on a table, it is probable that such circumstances, which have frequently presented themselves to us, have given rise to the belief that some of the species are ovoviparous.

b. Destitute of terminal suckers. This includes two very distinct families.

1. PLANARIADÆ.—Ventral surface flattened, and capable of acting as a foot. There are two genera: 1. *Planaria*, (MULLER) mouth towards the middle, or posterior portion of the ventral surface, in the form of a retractile funnel-shaped proboscis. There is a contiguous pore, which is

---

* These species were formerly considered as destitute of eyes; but the presence of these organs has been recently ascertained by Professor CARENA. See his valuable Monograph of the genus Hirudo, published in the Memoirs of the R. Acad. of Sc. of Turin, vol. xxv. p. 281. The aerating organs are delineated by Sir E. HOME, Phil. Trans. 1815, Tab. xiii.

† This genus was instituted by Dr JOHNSTON of Bristol, Phil. 1817, p. 339. In the same vol. p. 13., there are some curious observations on the Erpobdella vulgaris.

probably the orifice of the oviduct. The intestines are in the form of diverging cæca, and the observations of MULLER leave little room to doubt that the mouth serves also as an anus. Some of the species spin threads, like the slugs. They are oviparous, or propagate by spontaneous division. *P. nigra* is the type of this genus. 2. *Dalyellia.* Mouth a simple slit, placed anteriorly. *D. graminea.* This genus I have named in honour of an acute observer of nature, JOHN GRAHAM DALYELL, Esq. whose work on the Planariæ * exhibits much patient research, and should be perused with care by all who devote themselves to the study of the habits of minute aquatic animals.

2. GORDIUSIDÆ. Ventral surface not fitted to act as a foot in crawling. Locomotion is executed by twistings of the body, which is narrow, and very long. There are two genera: 1. *Gordius.* Mouth terminal, tail divided. The *G. aquaticus* is the type (of which *G. argillaceus* is only a variety.) It is probable that some species of *Filariæ*, which have been found in soil, detached from the animals in which they ordinarily reside, may have been referred to this genus. 2. *Lineus.* Mouth a longitudinal slit under the snout. This genus was instituted in SOWERBY's British Miscellany, Tab. 8. (1806,) for the reception of the sea long-worm of BORLASE, (" *Cornwall*," p. 255, Tab. xxiv, 113.) CUVIER instituted, in 1817, his genus Nemertes for the reception of the same animal. There are several species natives of this country.

* " Observations on some interesting phenomena in Animal Physiology exhibited by several species of Planariæ."—Edin. 1814. In this work, the conjectures of MULLER (Hist. Ver. i. p. 62.), regarding the use of the proboscis, have been confirmed, the anterior mouth in Dalyellia ascertained, and the manners of eight native species determined.

Order II.

Habitation internal.

The animals of this order were included by LINNÆUS, in his Vermes Intestina, and have more recently been denominated *Entozoa* by RUDOLPHI, who has successfully illustrated their physiological and systematical characters *. They live and propagate within the bodies of other animals, inhabiting not only the intestines, but the vessels, glands, and cellular substance.

1st Tribe.

Mouth and anus at the opposite extremities, with an obvious intervening simple alimentary canal.

This tribe forms the *Nematoidea* of RUDOLPHI, and the *Cavitaires* of CUVIER. The form of the body is lengthened, nearly linear, round, transversely striated, and elastic. In the integuments, transverse and longitudinal muscular fibres may be perceived. In some species, the nervous collar and double longitudinal filaments may be detected. The mouth exhibits, in its lips, and surrounding papillæ, useful characters for classification. There is a distinct abdomen, containing the alimentary canal. There is no trace in this or the following tribe, of aërating organs. The sexes are distinct, and the organs placed posteriorly. The following genera belong to this group: Filaria, Hamularia, Trichocephalus, Oxyuris, Cucullanus, Ophiostoma, Ascaris, Strongylus, Liorhynchus, Prionoderma, and Shistura.

2d Tribe.

Mouth and anus not at the opposite extremities, nor is there a distinct simple alimentary canal.

The animals of this tribe constitute the Parenchymateux of CUVIER. The body consists of cellular substance, without a distinct abdomen, in which the organs of nutrition and reproduction are imbedded. There is seldom an

---

* " Entozoorum sive Vermium Instinalium Historia Naturalis, auctore C. A. RUDOLPHI," 2 vols. 8vo. Amst. 1808.

obvious anus, and the intestines usually appear as cæca arising from one orifice. This tribe has been subdivided into the four following groups.

A. Head furnished with a produced snout, armed with reflected bristles. This includes the following genera, (forming the Acanthocephala of Rudolphi,) Echinorhynchus, and Hæruca.

B. Head furnished with two or four pores or processes. *a*. Tail simple. Tenia, Scolex, Tricuspidaria, Bothryocephalus, Horiceps, and Tetrarynchus. *b*. Tail vesicular. Cysticercus and Cænurus.

C. Body with one or more suckers for adhesion. Monostoma, Amphistoma, Caryophyllæus, Distoma, Polystonia, and Tristoma.

D. No external organs. Ligula.

In the investigation of this singular group of animals, considerable assistance, in ascertaining the species, is derived from a knowledge of the particular animals in which they reside. Several lists of animals, with their intestinal inmates, have been published, as the one by RUDOLPHI, vol. ii. part ii. p. 295; and a still more extensive one, as an Index to the collection of ENTOZOA, at Vienna.

II.
RADIATA.

Nervous system disseminated through the body, and not appearing in the form of a collar round the gullet, nor a longitudinal cord.

The animals of this extensive division of the Invertebrata invariably reside in the water. In form, they are in general more or less stellular; neither organs of hearing nor sight have been detected. In the following brief view of the different tribes, we shall consider them under the four established classes, Echinodermata, Acalepha, Zoophyta, and Infusoria.

Class I.—Echinodermata.

Skin coriaceous or crustaceous. Intestinal canal distinct, and contained in an abdomen. Numerous vessels likewise appear connected with circulation and reproduction.

Order I.—Intestine open at both extremities.

Section 1. Furnished with suckers for locomotion.

The coriaceous skin is pierced by numerous pores, through which issue tubular tentacula, connected centrally with a vascular system, and furnished with terminal suckers. There are three tribes, two of which are formed from the Linnean genera Echinus and Holothuria.

1st Tribe Echinida.

The integument consists of an immoveable testaceous perforated plate. The body is orbicular or depressed, and not divided into arms. The mouth is seated in the under surface. The intestine terminates in an anus. LAMARK distributes the genera in the following order:

A. Anus in the margin, or inferior disc.

1. *Mouth central.*

a. Avenues of pores circumscribed, as Scutella, Clypeaster and Fibularia.

b. Avenues of pores complete, as Echinoneus and Galerites.

2. *Mouth lateral*, as Ananchytes and Spatangus.

B. Anus in the upper disc.

(1.) Anus lateral, as Cassidulus and Nucleolites.

(2.) Anus vertical, as Echinus * and Cidarites.

---

* The species of the genus Pedicellaria, instituted by MULLER, and adopted by many naturalists, are, in fact, the appendages of the Echinus. The late Dr MONRO stated, that, " in the interstices of the thorns, there are three kinds of bodies, soft at the ends, supported on calcareous stalks, en-

2d Tribe, Fistulida.

This tribe resembles the preceding in the intestine being open at both extremities, but the skin is coriaceous and moveable. The body is lengthened, and the head is surrounded with tentacula, and furnished with a circle of osseous pieces, as teeth. Gills arborescent near the anus. Ovarium, consisting of multifid vesicles opening by a common oviduct near the mouth. The genus Holothuria, of which this tribe consists, may, with propriety, be subdivided into several genera.

Section 2. Destitute of suckers for locomotion.

This section includes the following genera, the relations of which with one another are not very intimate: Molpadia, Menyas, and Sipunculus.

Order II.

Intestine open at the mouth only. No anus.

1st Tribe, Stellerida.

Suckers for locomotion. The integuments are moveable, and the depressed body is divided into angles or arms. The mouth is inferior. The stomach is a central bag. Traces of a nervous system have been discovered. There are four genera, Asterias, Ophiura, Euryale, and Comatula.

2d Tribe, Crinoidea.

No suckers for locomotion. Body, opposite the mouth, produced into an articulated stalk, the extremity of which is probably fixed to other bodies. Only one recent species, Pentacrinus caput Medusæ *, is known; but the characters

---

closed in a membrane, and articulated with the shell by means of muscular membranes."—Struct. and Phys. of Fishes, p. 66. I lately had an opportunity of satisfying myself that these organs in the E. esculentus are in organical connection, and consequently that the genus Pedicellaria is a spurious one.

* Phil. Trans. 1761, T. xiii. xiv.

of many fossil species, and the genera under which they may be distributed, have been successfully elucidated by Mr J. S. MILLER, in his valuable work on the animals of this tribe *.

Class II.—ACALEPHA.

This class was instituted by CUVIER, for the reception of the animals of the genera Actinia and Medusa of LINNÆUS. The integuments are soft, and frequently gelatinous. The stomach and intestines never float distinct in a particular cavity. The traces of circulating vessels are obscure. CUVIER divides the genera into two tribes, such as are fixed, and such as are free.

Order I.

Base opposite the mouth, adhering to other bodies. In this order there are two kinds of adhesion, temporary and permanent. The first kind is exhibited in the genera Actinia and Lucernaria; the latter in Zoanthus. The first and last of these genera, however, agree in so many other respects, as to forbid their separation, by the intervention of the second. Indeed, in this case, (or as is still better illustrated in the case of the Oyster and Scallop) the circumstance of being fixed or free, exercises but little influence on the organization.

1st Tribe.

Tentacula uninterruptedly surrounding the oval disc. Of the genus Actinia we have many species in our seas, and the Mammaria mammilla, likewise a native, appears to belong to Zoanthus.

2d Tribe.

Tentacula on the margin of the oral disc disposed in tufts. This includes only the genus Lucernaria, five spe-

---

† " A Natural History of the Crinoidea," 4to, Bristol, 1821.

cies of which have been described by LAMOUROUX, in his
" *Memoire sur la Lucernaire campanulée* *."

Order II.

Base opposite the mouth incapable of adhering to other bodies. This includes the numerous kinds of Medusæ which float about in the water.

1st Tribe.

Body floating in the water, without air-bags.

A. Body closed opposite the mouth.

1. Body strengthened internally by a cartilaginous plate. This includes the genera Velella and Porpita †.

2. Body destitute of an internal cartilaginous plate. The numerous genera of this group have been arranged with considerable care by PERON and LESEUR, in their valuable paper in Annales du Museum, vol. xiv. p. 325. Many characters are employed, which it may be proper here to notice. Some of the species have neither mouth nor cavity, and are termed Agastrique; while others possess both. Some have a mouth and stomach, without lateral openings; while, in others, the lateral openings exist, and were considered as mouths by the authors just cited; while CUVIER now regards them as the orifices of the oviducts. In some, the mouth is produced into proboscis, either simple at the extremity, or divided into fringed lips. In others the mouth is sessile. In some, the margin of the body is simple. In others, fringed with tentacula. The body, in some, is smooth; in others, covered with ridges of fine hairs, which are used as fins.

B. Body open at both extremities. This includes the animals arranged under the genus Beroe. This genus ap-

---

* The Hydra glomerata, Lin. Am. Acad. iv. T. iii. f. 1. may probably be included in this genus.

† Two species of Velella, and one of Porpita, are described by Dr LEACH, in TUCKEY's Narrative, p. 418.

pears capable of subdivision. Those with ciliated ribs, without tentacula, may remain under the old genus, as *B. ovata* * and *fulgens*, while I include under Pleurobrachia, the *Beroe pileus*, which is furnished with two long ciliated tentacula.

2d Tribe.

Buoyant air-vessels in the interior of the body. There is no apparent mouth, but there are many appendages which have been considered as supplying the place of mouths, suckers, tentacula, and ovaria. The following genera have been established: Physalia, Physsophora, Rhizophyza, and Stephanomia.

Class III.—Zoophyta.

It is not practicable, in the present state of our knowledge, to draw a line of distinction between the animals of this and of the preceding class. In general, the mouth is surrounded with a circle of tentacula, seldom with a double circle, by which they are distinguished from the first order; while they may be recognized as not belonging to the second, not only by the oral tentacula, but by the tendency to become compound. It is, however, with the first order of the preceding class that the resemblance is most complete. The divisions of CUVIER and LAMARK, by appearing to be neither natural, definite, nor convenient, intimate the extreme difficulty attending the methodical distribution of the species. Some of the genera, as Corallina and Spongia, exhibit no signs of sensibility, and but indistinct traces of irritability. It has been proposed to form these into a separate class, to be placed after the Infusoria; but, in consequence of the numerous analogies of structure

---

* I have given the result of some observations on this Beroe, in Mem. Soc. vol. iii. p. 400, Tab. xviii. p. 3, 4.

RADIATA. 613

and composition which may be traced in them, with the well known genera of zoophytes, the change would not be of advantage.

The portion of the animal exhibiting the tentacula, head and body, continuously connected, is termed the *Polypus;* and the substance, whether it be fleshy, membranaceous, crustaceous, or testaceous, to which the body adheres, is denominated the *Coral.*

Order I.—CARNOSA.

Polypi, connected with a fleshy substance, and furnished with eight rays.

The animals of this order seem much more complicated than those which follow, and bear a close resemblance to Actinia, and especially Zoanthus They are furnished with a stomach, intestines, and ovaria.

1st Tribe.

Aggregated animals, free and floating in the water. The fleshy coral is supported by a loose bony axis. In one genus (Veretillum), CUVIER found the stomach separating into fine tubes, which, after entering the fleshy substance of the coral, reunited with other tubes from the stomachs of the neighbouring polypi[*]. The following genera belong to the group: Pennatula, Veretillum, Funiculina, Renilla, Virgularia, and Umbellularia. In the remaining tribes, the compound mass is always permanently attached to other bodies.

2. Tribe.

Aggregated animal fixed. The coral is fleshy, convex, or lobed. The polypi have the central mouth surrounded with eight pectinated tentacula. The viscera of each polypus is enclosed in a double tunic, the inner one of which is

---

[*] Leçons, vol. iv. p. 146.

covered with eight longitudinal folds, dividing the cavity into eight cells, the walls incomplete at the centrel. The stomach is large, and opens at the bottom into eight intestines, which attach themselves to the partitions. Two of these intestines proceed to the base of the abdomen, and penetrate the fleshy coral without anastomosing. The other six terminate in as many pedunculated glands at the bottom of the lateral cells *. The following genera belong to this tribe. Anthelia, Zenia, Ammothea, Lobularia and Cornularia. As nearly connected with this tribe in form and the condition of the coral, we may notice the curious natural family of Sponges, the polypi of which are unknown. The fleshy matter is similar in its nature, and supported by intermixed fibres of a denser substance. In the Spongia and Ephydatia †, the fibres are soft, irregularly dispersed, and appear to consist of a substance like coagulated albumen. In the genus Tethya, the fibres are stony, and diverge from the centre to the circumference of the mass. The form of the first is irregular or branched, that of the second globular.

In the remaining tribes, the fleshy matter covers a stony or corneous support, or axis.

3d Tribe.

The axis is branched, without stellular pores, and the fleshy matter spread over it, with the polypi irregularly distributed over the surface.

A. Axis corneous, as in Antipathes and Gorgonia.
B. Axis stony, as Corallium, Melitea and Isis.

---

* Lamark, Hist. Nat. 11. 403.

† I have given a delineation of the *Ephydatia canalium*, from an Irish specimen, Plate v. f. 4.

4th Tribe.

Axis stony, covered with stellular discs, to which the fleshy substance bearing the polypi is confined. This includes the extensive group included by LINNÆUS under his genus Madrepora, which LAMARK has subdivided into eighteen genera. The fleshy matter appears to be thin and deciduous, like the Antipathes, easily separating from the stellular discs. The nature of the connection of the fleshy with the stony matter, its form, and the characters of the polypi, have not been distinctly explained.

Order II.

Polypi not connected with a fleshy covering, and having a greater number of tentacula than eight.

In this order, the tentacula appear to be irregular in number, even on individuals of the same species. The stomach seems to be a simple bag destitute of intestines, and the ovaria to be external.

1st Tribe.

Coral, with the polypi in cells which are closed at the bottom. Each polypus is, in this manner, separated from its neighbour, and lodged in its own cell. This tribe is capable of distribution into groups; but their characters of distinction are far from being precise. LAMARK has formed them into two divisions, *Polypiers a reseau,* and *Polypiers foramines.* There is not any one character common to all the genera of the first group, except the porosity of the coral; and this character, which is stated as wanting in the second, may be readily observed in several species of Millepora, for example, which he quotes.

A. Cells imbedded in a stony coral. Here the cells may be considered as excavated out of the surface of the coral, and are usually arranged perpendicular to the axis. The orifices are simple and minute. This group is represented by the Millopora (LAMARK), Retipora, &c. and includes the curious family Corallinadæ.

B. Cells originating from a stony base, and forming tubes which are prominent, as Tubulipora of LAMARK.

C. Tubes parallel, and joined by transverse plates, as *Tubipora musica.*

D. Cells vesicular, and somewhat external, as Cellipora and Eschara.

E. In the preceding groups, the coral is brittle and stony; in this it is flexible, as Flustra, Cellularia, and Alcyonium, (represented by *A. gelatinosum* [*].)

2d Tribe.

Polypi surrounded by a membranaceous tube, covering all the subdivisions of their compound body, and allowing the head and tentacula to protrude at the simple or cellular summit. This includes, as the types of as many subdivisions, the genera Sertularia [†], Tubularia and Alcyonella, (LAMARK).

Order III.

Polypi naked, or destitute of a coralloid covering. In this order there is one genus, which seems to be fixed, with a tough subcoriaceous integument, termed Coryna [‡], and probably more complicated in its organization than the other, named Hydra, which is capable of displacing the base by which it adheres.

---

[*] It is probably at the conclusion of this tribe that we should place the fresh-water genera Cristatella (LAM.) and Difflugia, which are free, and the Plumatella (Bosc.), which is fixed.

[†] The *Sertularia gelatinosa* is delineated, Plate v. f. 3. a nat. size, b. mag. The numerous subdivisions of this LINNEAN genus are unfolded by M. LAMOUROUX in his " Histoire des Polypiers Coralligenes," Paris, 1816.

[‡] Two British species are delineated, Plate v. *C. squamata,* f. 1. a. nat. size, b. mag. *C. glandulosa,* f. 2. a. nat. size, b. mag.

INFUSORIA.

Order IV.

Polypi, with the mouth furnished with hairs. These vibrate rapidly, and produce currents in the water. Very minute, and require the aid of the microscope in their examination.

1st Tribe.
Oral hairs in tufts. This includes the genera Folliculina, Brachionus, Furcularia, Urceolaria, Vorticella and Tubicolaria.

2d Tribe.
Vibrating hairs not collected in tufts, as Rattullus, Trichocerca and Vaginicola.

Class IV.—INFUSORIA.

The animals of this last class exhibit the greatest simplicity of parts. They have no visible mouth, stomach, or internal vessels. They possess, however, the power of swimming about in the water, by the action of parts, the structure of which the microscope cannot help us to unfold. They appear to propagate by buds, eggs, or spontaneous division. They are usually met with in infusions of vegetable and animal substances, both in fresh and salt water. They are too minute for the eye to perceive their forms, and require the aid of the microscope. The work of MULLER on the " Animalcula Infusoria," is the storehouse from which naturalists derive their descriptions of all the genera of this class, and of the last order of the preceding. LAMARK has contributed a good deal towards distributing the genera into convenient groups, but a great deal yet remains to be done, even with the materials which MULLER has furnished.

Order I.

The animals of this order have always some external organs, as cuticular processes or hairs. Two of the genera

Cercaria and Furocerca. A very common species of Furocerca, found in ditches in spring, is delineated in Plate V. f. 5. in Trichoda and Kerona are furnished with a tail, which is wanting.

Order II.

In this order there are no appendices of the skin. In one tribe the body is globular or rounded, and of sensible thickness, as Monas, Volvox, Proteus, Enchelis and Vibrio. In the other tribe, it is expanded or concave, and very thin, as Bursaria, Kolpoda, Paramecium, Cyclidium and Gonium.

" O LORD, how manifold are thy works? in wisdom hast thou made them all: the earth is full of thy riches. So is this great and wide sea, wherein are things creeping innumerable, both small and great beasts. There go the ships: there is that leviathan whom thou hast made to play therein. These wait all upon thee: that thou mayest give them their meat in due season. That thou givest them they gather: thou openest thine hand, they are filled with good. Thou hidest thy face, they are troubled; thou takest away their breath, they die, and return to their dust. Thou sendest forth thy spirit, they are created: and thou renewest the face of the earth."

FINIS.